NANOCOMPOSITES WITH UNIQUE PROPERTIES AND APPLICATIONS IN MEDICINE AND INDUSTRY

NANOCOMPOSITES WITH UNIQUE PROPERTIES AND APPLICATIONS IN MEDICINE AND INDUSTRY

Editor

Felix Ryler

scitus academics

Nanocomposites with Unique Properties and Applications in Medicine and Industry
Edited by **Felix Ryler**

ISBN: 978-1-68117-642-0
Library of Congress Control Number: 2016946503

SCITUS Academics LLC,
616, Corporate Way, Suite 2, 4766,
Valley Cottage, NY 10989

www.scitusacademics.com

Notice

Reasonable efforts have been made to publish reliable data and views articulated in the chapters are those of the individual contributors, and not necessarily those of the editors or publishers. Editors or publishers are not responsible for the accuracy of the information in the published chapters or consequences of their use. The publisher believes no responsibility for any damage or grievance to the persons or property arising out of the use of any materials, instructions, methods or thoughts in the book. The editors and the publisher have attempted to trace the copyright holders of all material reproduced in this publication and apologize to copyright holders if permission has not been obtained. If any copyright holder has not been acknowledged, please write to us so we may rectify.

Printed in United States of America on Acid Free Paper

PREFACE

Due to the combination of exclusive qualities, nanocomposites appeal to researchers from a theoretical and practical perspective. In the past two years, many efforts have been made to obtain nanomaterials with determined functionality using new nanotechnology and nanoscience knowledge. Many recent studies reveal the use of nano-materials in different medical fields, an application which has led to a brand new discipline called nanomedicine. The processes of diagnosis, treatment, cure, disease prevention and treatment of traumatic injury, pain relief and the preservation/improvement of human health through the use of nano-materials are included. A key place for the group of synthetic and natural polymer nanocomposites is nano materials. Nanocomposites offer modern medicine new opportunities to generate antibacterial treatment products, drug delivery, dental applications, medical imaging, tissue engineering, cancer therapy, etc. The possibilities of producing materials with tailor-made electronic and physical properties at low cost could lead to interesting applications ranging from drug delivery to corrosion prevention to electronic/automotive components and several other industrial equipment.

This book focuses on polymer nanocomposites and their possible divergent applications in medicine and industry. There was a great deal of interest in the marketing of nanocomposites for a variety of uses and a selection of these applications can be found in the industry. This book therefore serves as a good resource for these analytical methods valuable for researchers and practitioners.

TABLE OF CONTENT

Chapter 1 On the Prediction of the Residual Behaviour of Impacted Composite Curved Panels 1

Chapter 2 Applications of nanoparticles in biology and medicine 27

Chapter 3 Damage Assessment of Short Glass Fiber Reinforced Polyester Composites: A Comparative Study 43

Chapter 4 Composites of Polymer Hydrogels and Nanoparticulate Systems for Biomedical and Pharmaceutical Applications 69

Chapter 5 Micromechanisms Controlling the Structural Evolution of Tribosystems 177

Chapter 6 Recent Advances in the Synthesis and Biomedical Applications of Nanocomposite Hydrogels 221

Chapter 7 Nanoparticle-Integrated Hydrogels as Multifunctional Composite Materials for Biomedical Applications 259

Chapter 8 Synthesis and Properties of Discontinouosly Reinforced Aluminum Matrix Composites 283

Chapter 9 Development of β Type Ti23Mo-45S5 Bioglass Nanocomposites for Dental Applications 315

Index 343

1

ON THE PREDICTION OF THE RESIDUAL BEHAVIOUR OF IMPACTED COMPOSITE CURVED PANELS

Philippe Viot[1], Ludovic Ballere [1] and Jean-Luc Lataillade [1]

[1] Arts et Métiers -ParisTech, Institut de Mécanique et d'Ingénierie, UMR CNRS n° 5295, Esplanade des Arts et Métiers, F-33405 Talence, France

1. INTRODUCTION

Composite materials are very often used in the aeronautical industry, because of their high specific strength, they are more appropriate for such applications than metals are. However, one disadvantage of such materials is the problem of detecting damage initiated by impact (e.g., dropping tools, collisions with foreign objects and other accidents), particularly when the reinforcement used is carbon fibre because may not be visible. Therefore, since it is difficult to avoid accidents, it is necessary to evaluate the effects of such damage on the residual resistance of the structure. This approach is related to the concept of the damage tolerance of structures. The structures considered here are filament-wound vessels subjected to high internal pressure loading and damage can be initiated in the carbon-epoxy shell during all their life cycle (manufacturing, storage, etc). In order to qualify the behaviour of these impacted structures, preliminary validation tests have to be done. However, these specific tests are generally very expensive and difficult to perform, especially when the structures' dimensions are large. An alternative way must be developed and a first one is to employ small-scale models.

The use of these reduced scale structures requires the identification of similitude models allowing the extrapolation of the small-scale model behaviour to the real structure. Although largely used in the case of homogeneous materials, such similitude techniques are not significantly developed for composite materials, mainly because of the interactive character of the different and multiscale damage mechanisms. As a first attempt, two scaling rule methods were developed based on a dimensional analysis using Buckingham's Pi theorem (Buckingham, 1914) or defined from dynamic equation of the system (Qian & Swanson, 1990). From these similitude models, some authors (Morton, 1988, Nettles et al., 1999) studied scale effects on composite structures taking into account the damage evolution during an impact but the gap was important between experimental results and predictions issued from scale models. For our study, in a preliminary phase of this research, scale models were evaluated (Viot et al., 2008) : a first approach consisted to apply similitude laws currently used on two scales (A and B) of composite structures. The purpose of this preliminary study was to predict the behaviour of the composite structure (scale A) from the knowledge of the response of the second scale model (scale B). It has been shown that existing similitude laws can be used to evaluate the elastic response of the two scales of composite structure but these models do not allow simulating the behaviour of the different scales when one of them is damaged ; it is due to non linearities.

For composite structures of large dimensions, an alternative and new approach of small-scale models must be developed since the experimental cost of impact study can be too expensive. Then, the main objective of our work is to predict the residual behaviour of impacted structures when the residual behaviour of small-scale structures is known. And because classical similitude laws cannot be used for damaged composite structures, another approach can be the use of a numerical model coupled with experimental data to predict the residual behaviour of impacted structures.

The proposed method is in three steps and must be applied on small scale panels to predict the behaviour of damaged vessels loaded by internal pressure. Before any numerical simulation, the analysis of the critical damage initiated during an accident must be quantified (point 0, figure 1). It is not the accident, the impact which is really important to qualify, even if it is interesting to know the impact conditions (mass, velocity, impactor's geometry, boundaries conditions...), but mostly the different kinds of damages (matrix cracks, fibre breakages or delamination) initiated during this impact which have to be precisely determined. And from the observations and analysis of the impacted composite microstructure, these damages must be classified from their critical effects on the performances of damaged structure. For vessel structures investigated, damages initiated during the impact were mainly delamination between carbon plies and fibre breakage. However, if delamination is a critical phenomenon for composite structures under bending load, this damage has not a drastic effect on vessel residual behaviour because the gap appearing between two delaminated plies decreases when the vessel is under the pressure and the propagation of delamination is then not effective. On the contrary, the breakage of fibres under tensile loading can obviously have a significant effect on the residual performance of the vessels.

From this preliminary study "identification of damage on real structure", the critical kind of damage was quantified and its effect must be experimentally and numerically evaluated on small scale structure. The main objective of the step 1 is then the development and the calibration of a numerical model, able to estimate the response of impacted small-scale structure: first, the critical damage has to be experimentally reproduced on small-scale structure by impact (working package1, figure 1). Secondly, this damage must be controlled and precisely quantified (nature and size) at the scale of the composite microstructure (working package2, figure 1). Finally, the residual behaviour of small-scale structure is estimated (working package 3, figure 1) by imposing a state of loading close to the one imposed on the real structure (in order to initiate the propagation of the critical damage on small scale model in similar loading conditions than the ones imposed on real structure). This experimental approach is essential to calibrate the numerical model which has to be developed (working package 4, figure 1), in taking into account the damage at the scale of the microstructure, in order to estimate the residual response of the small-scale structure.

The second step of this methodology is the evaluation of the performance of the numerical model. The same experimental study "impact – analysis of the damage- identification of the residual performance" is carried out on a second small-scale structure to obtain experimentally the effect of the damage on the residual behaviour of a second composite structure. This effect is also numerically evaluated in using the model developed during step 1. The performance of this model to predict the residual response of damaged composite structure is obtained from the comparison between experimental and numerical results obtained on the second small-scale structure (working package 5, figure 1).

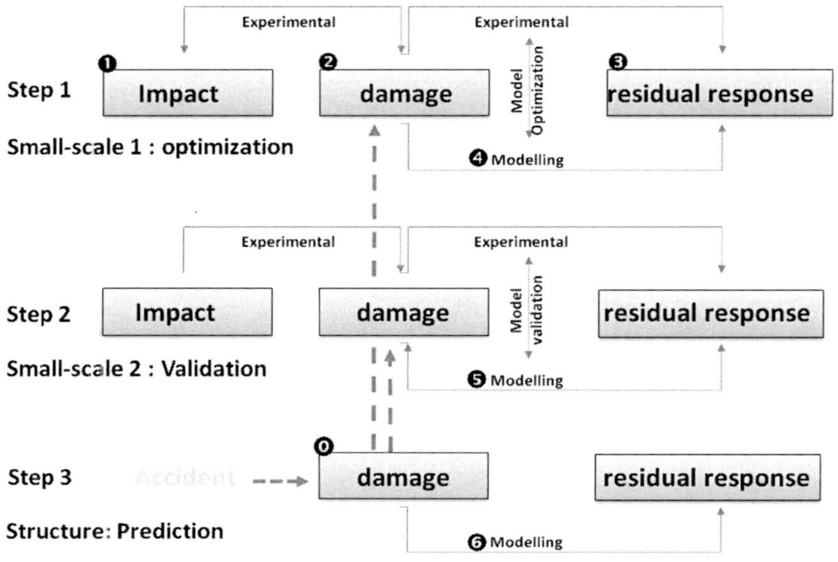

Figure 1. Scheme of the multi-scale methodology

Finally the third step of the method is the prediction of the residual behaviour of the real structure. The critical damage identified on the real structure at the beginning of this methodology is implemented on the numerical simulation of the real structure. The residual behaviour of this structure can be then numerically estimated (working package 6, figure 1).

This methodology was carried out for the study of the behaviour of impacted carbon-epoxy vessels under pressure. As an experimental study of damage tolerance using this type of structure is very expensive, the experiments were performed on curved panels extracted from tubes which had the same geometrical and mechanical properties as the vessels. The experimental procedure was carried out on these curved panels and the whole of the results were presented in a previous paper (Ballère et al., 2008): Firstly, the specimens were impacted to simulate an accident which can occur on such structures. Then, they were loaded in tension, according to their longitudinal axes, to reproduce the axial stresses caused by internal pressure being applied to the vessels' bottoms. The residual tensile strength was determined according to the initial damage states of the specimens.

The objective of this paper is to present a progressive failure analysis for the prediction of the residual properties of impacted curved specimens loaded in tension. First, the numerical results are compared with the experimental results obtained from undamaged specimens. Then, the damage observed experimentally is implemented numerically and the residual tensile strength is compared to the experimental results.

This methodology uses two scales of specimens. The first - close to the real scale - is used to validate the numerical modelling. The second- half the size of the first - is employed to highlight the mechanisms which have to be taken into account for the high-scale reduction of curved composite structures.

2. NUMERICAL SIMULATION

The modelling proposed in this study is based on a progressive failure analysis at the mesoscale (i.e., at the scale of the layer and the interface). Three steps are needed to build this model: i) choose a failure criterion; ii) choose damage kinetic; and iii) determine the consequences of the criterion activation on the elastic properties of the layer. Many approaches can be found in the literature to describe the progressive failure of a laminate, e.g., a state of the art approach was presented during the World Wide Failure Exercise (Kaddour et al., 2004). Different criteria are used in these approaches: for example, Ambur (Ambur et al., 2004) and Laurin(Laurin et al., 2007) use the Hashin-Rotem multi-criterion (Hashin and Rotem 1973); Bogetti (Bogetti, 2004) a 3-D maximum strain criterion and Zinoviev (Zinoviev, 2002) a maximum stress criterion. For this study, we have chosen the following maximum strain criterion.

2.1. Criterion: Maximum Strain

The numbering of the orthotropic axes of the layer is shown in Figure 2a. The failure criterion is based on a damage variable, d_{ij}, defined as

$$\frac{\varepsilon_{ij}^{n}}{\varepsilon_{ij}^{R}} = d_{ij}^{n} \qquad (1)$$

where i and j correspond to the orthotropic axes of the layer (i,j=1,...3), ε_{ij}^{n} is the component ij of the strain tensor at increment n and ε_{ij}^{R} its value to failure. The failure occurs whendij≥1dij≥1. In this formulation, it should be noted that, for $\varepsilon_{ij} < 0$ (i.e., in compression), d_{ij} is always negative so that there is no failure in compression. This assumption can be justified here since this modelling is applied to the prediction of residual strength in tension.

As soon as the failure occurs (i.e.,dij≥1), the elastic modulus, E_{ij}, is reduced instantaneously to a residual value close to 0 (Figure 2b). This value is maintained regardless of the post-failure loading. This property avoids the healing of the damaged material. This approach proposes a degradation of the elastic properties of the layer in an independent way and the interface is considered non-damaging.

This progressive failure analysis has been implemented in the Finite Element Code, ZéBuLoN.

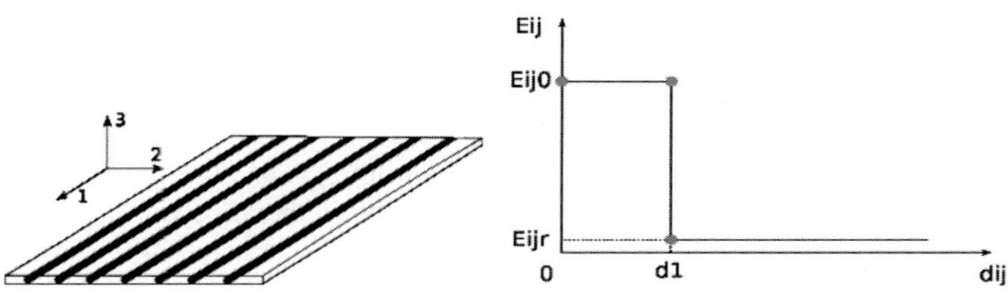

Figure 2. Asdsd(left) Numbering of orthotropic axis, b (right) variation of elastic properties

2.2. Numerical Simulation Of Curved Panel

In order to validate this numerical approach, experimental tests were performed on two scales of composite curved panels. The methodology used and the results obtained for one of these scales of specimens (called «specimens Ø600») is presented in (Ballère et al., 2008). The first step in the numerical modelling is to check that the behaviour of an undamaged specimen is well-predicted.

2.2.1. Undamaged Curved Panel

The stacking sequence of the laminate used here is:

Inner-$(90°)_2/[(\pm 20°)_2/(90°)_2]_3\}$-Outer

For this modelling, the stacking sequence has been simplified and has been chosen to model the layers oriented at +20° and -20° independently (see Figures 3 and 4). For this scale of specimens, n is equal to 1.

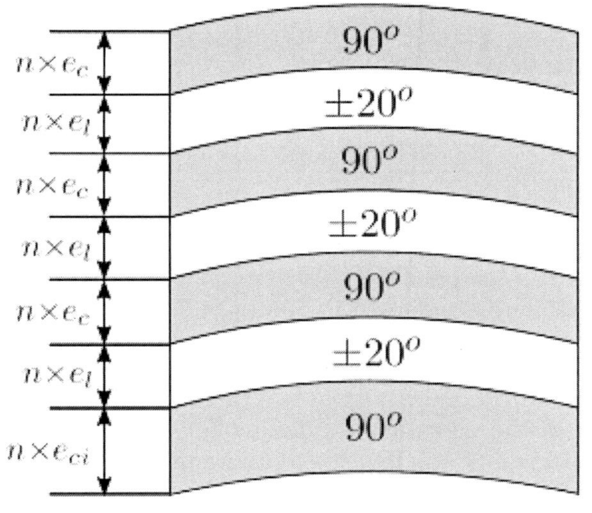

Figure 3. Stacking sequence of the real structure

The laminate is not symmetrical because the inner circumferential layer is thicker (e_{ci}}) than the others (e_c). Nevertheless, the choice of the stacking sequence for the numerical specimen was made in order to try to create a nearly symmetrical laminate according to the mid-plane so that the modelling would be easier. The elastic properties of the carbon/epoxy used are:

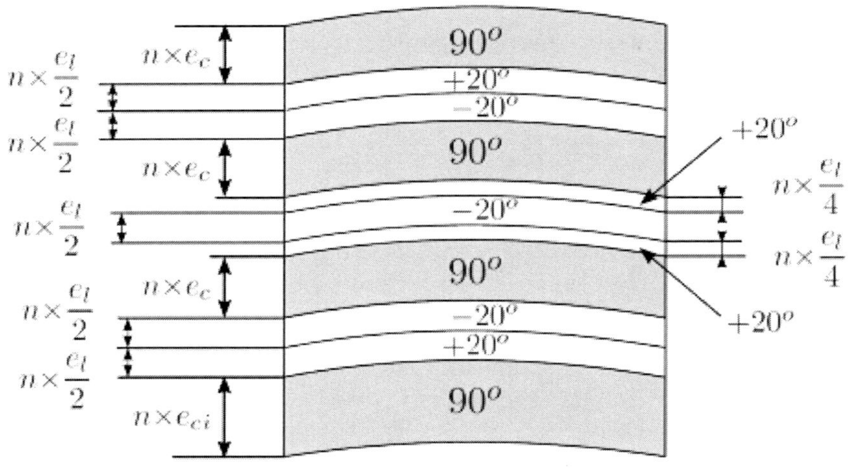

Figure 4. Stacking sequence of the numerical specimen

Table 1. Mechanical properties

$E_1\,(GPa)$	$E_2\,(GPa)$	$G_{12}\,(GPa)$	ν_{12}	ν_{21}	$\sigma_1^R\,(MPa)$	$\sigma_2^R\,(MPa)$
165	7.1	3.9	0.39	0.015	2610	38

The dimensions of the panels and the boundary conditions used in this numerical modelling are shown in Figure 5. They correspond to those used to perform the experimental tests. The radius of the curvature is 278 mm.

The influence of the element formulation for the failure prediction of these curve specimens was investigated in a previous study in which it was shown that the through-thickness displacement field is non-linear in tension. Therefore, in this research, an element denoted C3D20 (quadratic brick element) in ZeBuLoN (Carrère et al., 2009) has been chosen to model, through the thickness, each layer of a different orientation making it possible to detect the non-linearity of the displacement field.

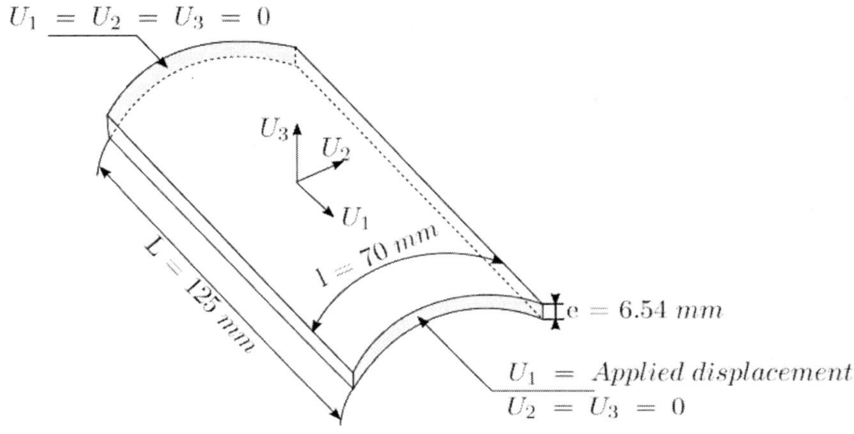

Figure 5. Geometry and boundary conditions

2.2.2. Damaged Curved Panel Φ 600

To implement the damage numerically, the typology of the damage mechanisms generated by impact had to be observed. This observation was undertaken during the experimental study (Ballère et al., 2008) and the results are summarized below.

For the impact tests, the specimen was clamped between two aluminium blocks and tightened with screws. It was fully supported on both surfaces except for a circular region of 30 mm in diameter in the centre corresponding to the impact zone. With this specimen-mounting device, classical damage mechanisms were observed. Delamination initiates and propagates during impact, even at low energy, but the delamination zone is always restricted in the centre because of the specimen-mounting device. Since impact energy is not fully dissipated by delamination, intra-laminar failures also occur (fibre failure, matrix cracking). Impacted specimens were loaded in quasi-static tension in order to evaluate their residual behaviour. It is well-known that the most prejudicial damage mechanism for laminates loaded in tension is fibre failure. For this reason, specific attention was paid to this phenomenon.

Table 2 presents some results of microscopic observations performed on specimens impacted with different impact energy levels. Each row is associated with a specific layer of the laminate. The columns of this table are ranked in order of increasing impact energy. The ≋ symbols indicate layers in which fibre breakages were observed. The number of layers damaged during the impact increased with the increase in the impact energy.

Table 2. Damage levels experimentally evaluated

Structure		Damage1 (E = 22 J)	Damage2 (E = 30 J)	Damage3 (E = 38 J)	Damage4 (E = 71 J)
90°	Circ. 1	≋	≋	≋	≋
±20°	Long. 1	≋	≋	≋	≋
90°	Circ. 2		≋		≋
±20°	Long. 2			≋	≋
90°	Circ. 3				≋
±20°	Long. 3				≋
90°	Circ. 4				≋

In order to model the damage observed experimentally in a cylinder of 30 mm in diameter, an equivalent zone of the numerical specimen was defined (Figure 18). One or many layers can be damaged independently in this zone by decreasing all the elastic modulusEijEij to their residual value. This assumption can be justified since, at each time a fibre failure was observed in a layer, all the primary damage mechanisms (i.e., matrix cracking, fibre-matrix shear failures) were also observed. It is possible to suspect that the elastic properties of the layer decrease along all directions in the impact zone. In a first approach, the Poisson's ratio is not degraded.

All these observations were used to validate this modelling in the case of impacted specimens.

3. MODEL OPTIMISATION ON A FIRST SMALL SCALE STRUCTURE Φ 600

3.1. Numerical Results On Non Impacted Panels: Effect Of The Mesh Size

Most progressive damage laws are very dependent on the mesh fineness. Therefore, two meshes of different element sizes were first considered in order to evaluate this effect (Figures 7 and 8). Mesh B consists of four times more elements in its surface than does mesh A. There is the same number of elements through the thickness in each mesh.

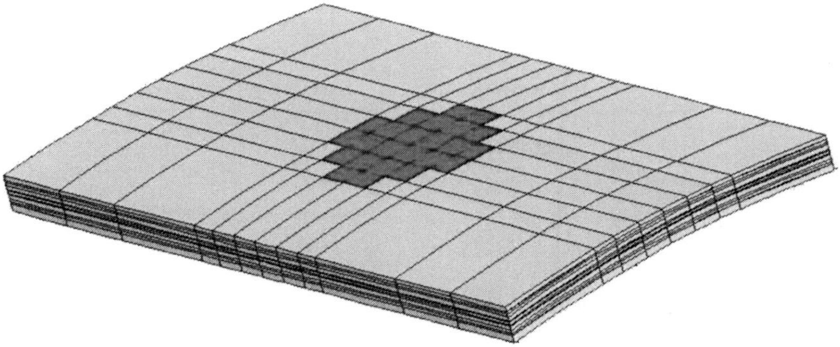

Figure 6. Damage implementation

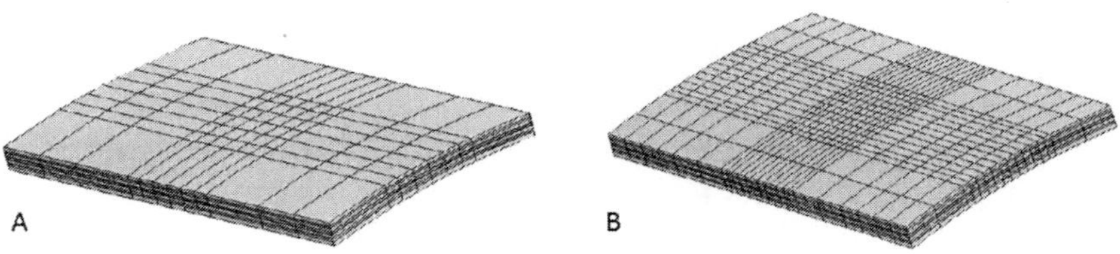

Figure 7. Mesh A and Mesh B

Figure 8 shows a comparison between the stress-strain curves obtained using these two meshes. The horizontal line corresponds to the mean ultimate stress determined during experimental tests. Obviously, the two curves are similar in the first part of the loading, but a plateau occurs for them close to the experimental failure value and then, after this plateau, the main difference appears. For mesh A, the loading increases to reach a final failure value which is very far from the experimental value. Using mesh B, the final failure occurs just after this plateau with a stress value close to the experimental failure (3%). The decrease in the element size allows the failure value reached to be close to the experimental results. This can be explained by focusing on a zone of the graph located around the plateau (Figure 8).

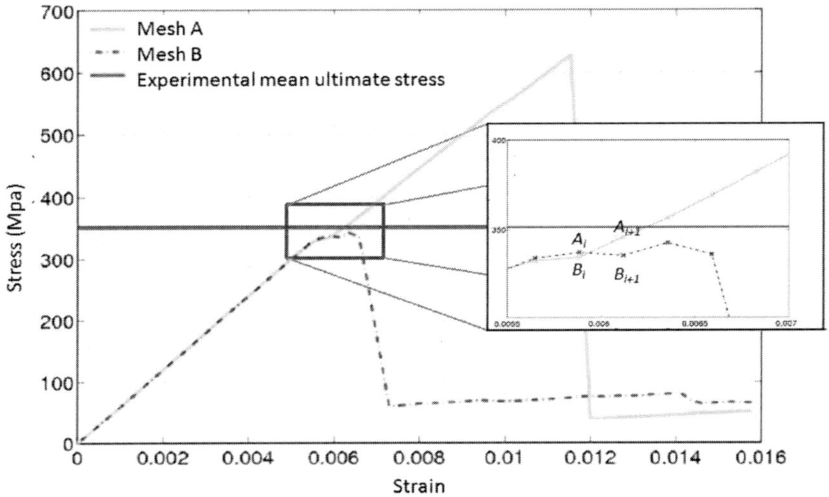

Figure 8. Influence of the element size on the numerical stress vs. strain response

In order to identify the mechanisms which change according to the mesh fineness, attention has been paid to the criterion activation at particular points: i) points A_i and B_i, located just before the change of behaviour; and ii) points A_{i+1} and B_{i+1}, located at the next increment for mesh A and mesh B respectively. At points A_i and B_i, the criterion is highly activated by the damage variable d_{22}, (related to the orthoradial strain) in all the circumferential layers. The stress plateau observed for the two meshes is mainly due to this failure mode. For this level of loading, the damage variable, d_{33}, is also equal to 1 in many elements of the circumferential layers. Failures due to transverse shear stresses (damage variables d_{13} and d_{23}) also appear in all the layers of the laminate. The main difference between the behaviour obtained using these two meshes is in the detection of in-plane shear failures. The criterion is activated (i.e., d_{12}=1) in many elements with mesh B (Figure 9, left) but not in any element of the mesh A. The increase in the mesh fineness allows the detection of in-plane shear failures to be earlier.

This difference between the predictions from these two meshes is amplified at higher loading levels (i.e., points A_{i+1} and B_{i+1}). The criterion is still not activated with mesh A for in-plane shear stresses whereas there are many in-plane shear failures detected for mesh B in all the layers of the laminate and they are close to the free-edges of the specimen (figure 9, right). Experimentally, these failures lead to the delaminations observed post-mortem.

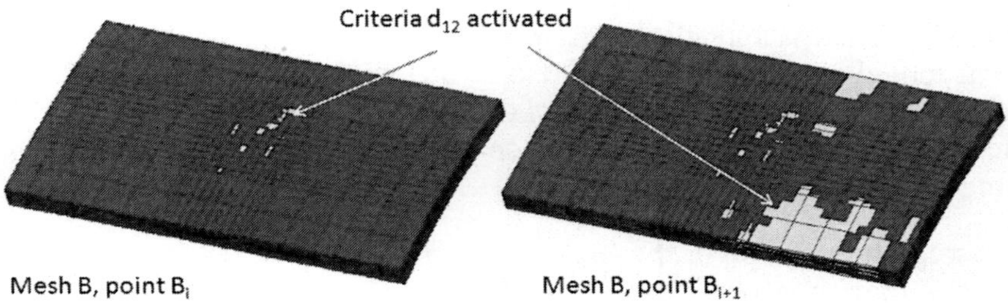

Figure 9. Damage variable d_{12} calculated with mesh B at the point B_i (left) and at the point B_{i+1} (right)

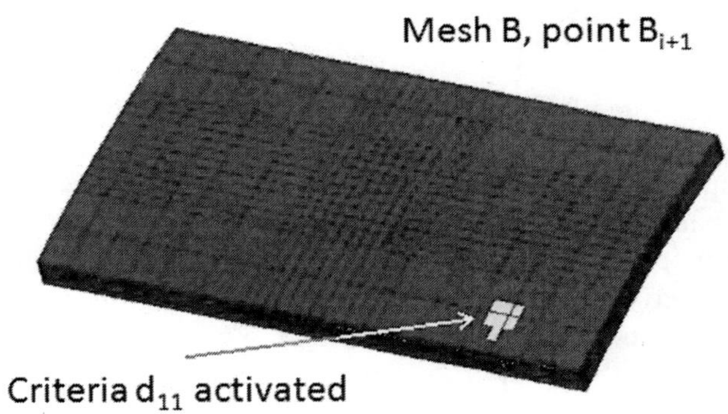

Figure 10. Damage variable d_{11} calculated with mesh B at the point B_{i+1}

The existence of in-plane shear failures exhibited when using mesh B leads to failures in the fibre mode (i.e. damage variable d_{11}) in the longitudinal layers, as shown in Figure 10. This phenomenon leads to the global failure of the specimen. By decreasing the element size, it was possible to detect earlier the initiation of two damage mechanisms strongly prejudicial to the integrity of the specimens: in-plane shear failures and fibre breakages.

3.2. Numerical Results On Impacted Panels

Since the proposed modelling was validated in the case of undamaged specimens, the next step was to use this model to predict the residual behaviour of impacted specimens. Each damage level presented in Table 2 was modelled and the residual tensile behaviour assessed. The stress-strain curves of the pre-damaged specimens are presented in Figure 11 and are compared with the response of the undamaged specimen (in using the mesh A).

Focusing on the slope of the first linear part of these curves, it appears that the more the pre-damage state is important, the more the slope decreases. This slope can be related to the homogenized elastic modulus of the specimens. Obviously, the damage generates decreasing stiffness. This phenomenon has been observed experimentally (Ballère et al., 2008) and could be analyzed in detail using this approach. Nevertheless, because the aim of this study is to predict the residual tensile strength of impacted specimens, particular attention was paid to the ultimate stress.

For each curve, the ultimate stress was extracted and then plotted versus the associated impact energy (figure 12a). Concerning an undamaged specimen, it was shown that an increase in mesh fineness leads to a global failure located at the stress plateau level. Thus, for this curve, the ultimate stress chosen was equal to this plateau value. These numerical predictions are in good agreement with the experimental results. Experimentally, there is a bi-linear evolution of the ultimate stress according to the impact energy. The impact energy yields (i.e., damage), when the ultimate stress starts to decrease. For this scale of specimens, this yield energy is equal to 40 J. This phenomenon is also observed in this numerical analysis since the ultimate stress obtained for the «damage 3» level is more or less the same as that for an undamaged specimen.

It is interesting to note that an increase in the damage state of a specimen does not involve, systematically, a decrease in its ultimate stress. For example, the ultimate stress of a specimen damaged according to the «damage 3» level is almost equal to that of a specimen associated with the «damage 2» level (figure 12a). For instance, the «damage 2» level corresponds to the failure of two first circumferential layers (oriented at 90°) and one intermediate longitudinal layer (long. 1, oriented at ±20°, table 2) while the «damage 4» level is associated with the failure of the first circumferential layer and the two longitudinal layers (long. 1 and long. 2, table 2). For flat laminates, degradation of the longitudinal fibres is very harmful to the strength of the specimen in tension compared with degradation of the fibres oriented at 90°. It seems that, for these curved specimens, fibres oriented at 90° play a very significant role in their tensile strengths. Also, the influence of the damage organization through the thickness of the laminate has to be studied.

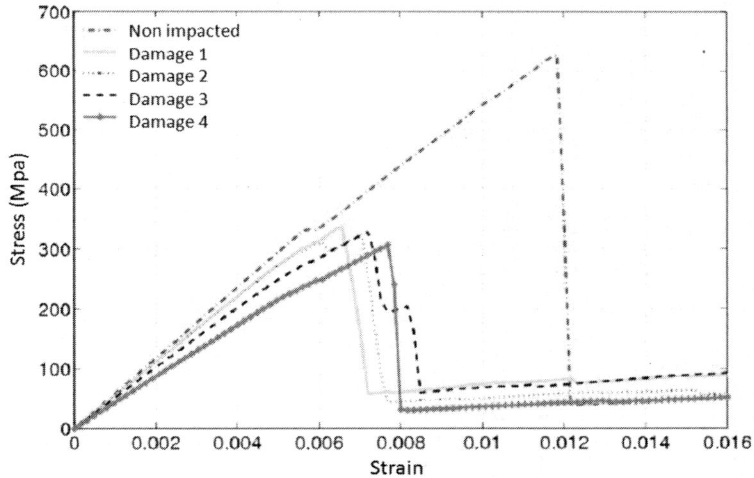

Figure 11. Numerical stress vs. strain response for impacted specimens

3.2.1. Influence Of The Damage On Residual Behaviour Prediction

Experimentally, damage assessment was conducted using optical microscopy with a limited number of specimens. From the results, it was possible to correlate a residual tensile strength and an initial damage state of the specimen. Nevertheless, after the dynamic test, a variability of the damage could be suspected due to: i) dispersion of the properties of the different components; ii) variability in the mechanical properties of the specimens introduced by the manufacturing process;

and iii) the boundary conditions used for the impact tests possibly being slightly different for each test. This damage variability was reflected in the experimentally observed strength dispersion. However, numerically, it cannot be taken into account without implementing a random damage variable.

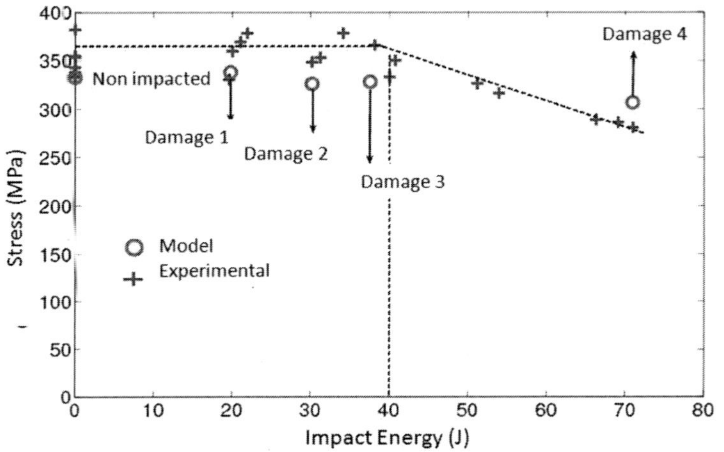

Figure 12a. Comparison between model predictions and experimental results

In this approach, it was decided to quantify this strength variability by studying different cases of damage. »). Two new damage cases (called « Virtual damage A» and «Virtual damage B, table 3) were modelled to evaluate the residual behaviour of composite specimens if these kinds of degradation are imposed during an impact. The « virtual damage A» level was established from the «Damage 3» by adding the fibre breakage in the second circumferential layer. The « virtual damage B» was a damage level located between the «Damage A» level and the «Damage 4» level where five layers are damaged.

An increase in the initial damage state (from «Damage 3» to «Damage «A») obviously leads to a decrease in the ultimate stress (7%). This decrease is equal to 16% when one circumferential layer more than «Damage A» is damaged («Damage B»). For the damage B, the residual tensile strength is lower than the ultimate stress calculated for the damage 4 and experimentally obtained in any case of impact energy.

Table 4. Add of two damage level cases (A and B) for specimens ϕ600 mm

Structure		Dam.1	Dam. 2	Dam. 3	Virtual Dam. A	Virtual Dam. B	Dam. 4
90°	Circ. 1	≋	≋	≋	≋	≋	≋
±20°	Long. 1	≋	≋	≋	≋	≋	≋
90°	Circ. 2		≋		≋	≋	≋
±20°	Long. 2			≋	≋	≋	≋
90°	Circ. 3					≋	≋
±20°	Long. 3						≋
90°	Circ. 4						≋

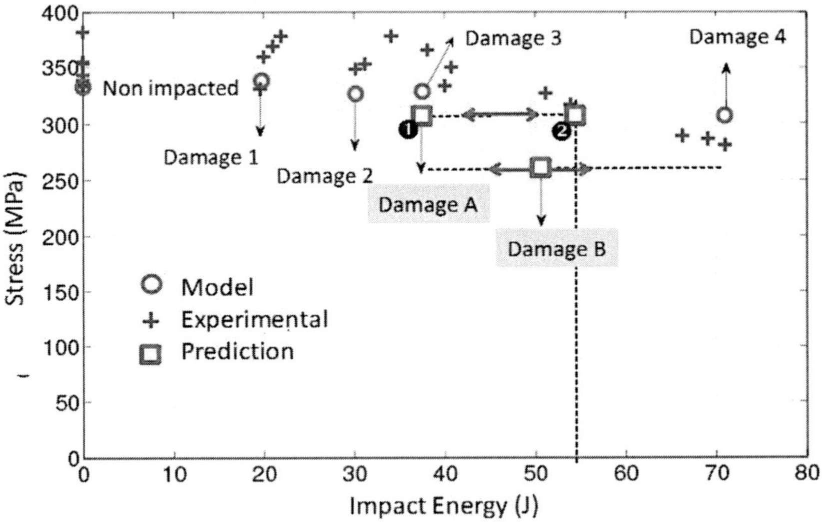

Figure 12b. Comparison between model predictions for two new damages (A and B)

These two new results were plotted on the ultimate stress vs. impact energy diagram. It is clear that we cannot determine the level of impact energy necessary to obtain these two damage states A and B, we can just consider that the energy levels to reach these damages are in the range of the impact energies of the damage 3 and 4 (if we do not consider any variability of the composite specimen damage to the impact energy). For the damage A, the ultimate stress of which is really close the value numerically computed for the damage 3, it can be firstly assumed that the effect of supplementary damaged circumferential layers is not significant for the damage A (location1, figure 12b). Or, in another way, the impact energy to obtain the damage A can be estimated to 55 J (location2, figure 12b) by comparing the ultimate stress obtained experimentally to those one derived numerically.

In a first conclusion the used damage model is able to represent the residual resistance of a curved panel under tensile loading. However, it requires an identification of the damage induced during the impact. In the case of specimens Ø600, an increase in the number of damaged layers leads, beyond a specific threshold, to a global decrease in the residual strength. This decrease appears regardless of the type of damaged layers (i.e., circumferential or longitudinal). The shape of this decrease depends slightly on the location of the damaged layers through the thickness of the laminate. With the knowledge of the morphology and the size of the damage in the microstructure, the behaviour of larger specimen or larger structure previously damaged by an impact could be predicted. The second part of the paper presents the limits of this method and the difficulties which can appear if the scale of the specimen induces supplementary phenomena.

4. PREDICTION OF THE BEHAVIOUR OF IMPACTED STRUCTURES

The aim of the approach presented here is to predict the residual behaviour of specimens through the knowledge of the mechanical responses of small-scale specimens. In order to reveal the limits of the scale reduction in the case of curved composite structures, this approach was used to predict the residual behaviour of specimens which are half the size of specimens "Ø600".

The dimensions of these small-scale specimens (denoted as specimens"Ø300") were chosen particularly to observe the influence of curvature and thickness on their mechanical responses. The radius of the curvature was divided by two (i.e., 139 mm). The thickness was determined by the stacking sequence of the laminate. The «ply-level scaling» technique has been chosen to design the stacking sequence of the small-scale specimens. Thus, the thickness of each layer of a different orientation was divided by two, thereby leading to a global thickness of the laminate equal to 3.27 mm. This technique allows the same stacking sequence to be maintained between each scale while only the layer thickness changes (i.e., in Figures 3 and 4, n=0.5). The width and length are the same for both scales.

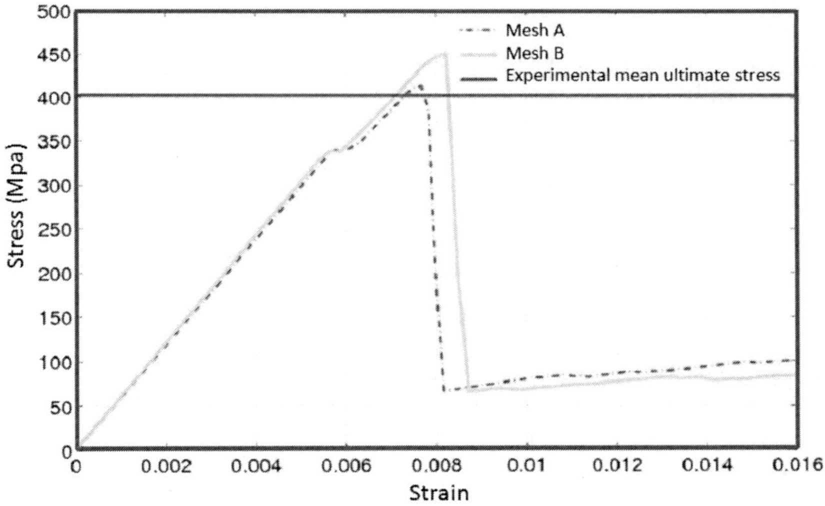

Figure 13. Influence of the element size on the numerical stress vs. strain response

The mesh used for this scale of specimens is equivalent to mesh A (Figure 6). Only the element thickness, which is related to the layer thickness, and the radius of the curvature are divided by two.

The influence of the mesh fineness was quantified using four more elements (equivalent to mesh B (Figure 7) for specimens "Ø600"). The stress-strain curves obtained using these two meshes are presented in Figure 13.For this scale of specimens, it seems that the numerical results are less dependent on the element size. A plateau appears in these two curves which, for the same reasons as mentioned in the case of specimens "Ø600", should be related to the global failure of the specimens. For specimens "Ø300", the damage mechanisms responsible for this plateau and, therefore, for the global failure, appear for the same stress value. In order to reduce computation times, mesh A was used for the modelling of the impacted specimens.

These small-scale specimens were tested experimentally and some of the results are presented in the next sub-section.

4.1. Numerical Results On Impacted Panels Φ 300

The different levels of damage modelled for this scale of specimens are presented in Table 5. They correspond to the damages observed experimentally.

Table 5. Damage levels experimentally evaluated

Structure		Damage1 (E = 7 J)	Damage2 (E = 12 J)	Damage3 (E = 14.5 J)	Damage4 (E = 24 J)
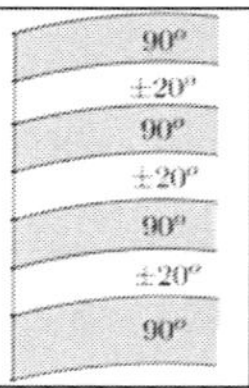	Circ. 1	≋	≋	≋	≋
	Long. 1		≋	≋	≋
	Circ. 2	≋	≋	≋	≋
	Long. 2		≋	≋	≋
	Circ. 3		≋		≋
	Long. 3			≋	≋
	Circ. 4				≋

Experimentally, a significant dispersion of the ultimate stress of the undamaged specimens is observed (figure 14). This dispersion decreases as soon as the specimens are pre-damaged by impact. Indeed, the defect initiated by the dynamic test restricts and governs the possible locations of failure initiation. Considering only one point, located at the mean ultimate stress for the undamaged specimens, it seems that there is a yield damage point from which the tensile strength starts to decrease.

The numerical predictions are in good agreement with the experimental results for the most-damaged specimens (i.e., «damage 4» level). They are located in the same area as the experimental dispersions for the undamaged specimens. Between these two damage levels, the numerical results do not match with those of the experiments. Moreover, the bi-linear evolution of the ultimate stress according to the impact energy cannot be predicted by this modelling.

It has been shown that this progressive failure analysis allows the residual tensile strength of one scale of curved specimens to be predicted. Unfortunately, it cannot be used for the prediction of the residual behaviour for smaller scale of specimens.

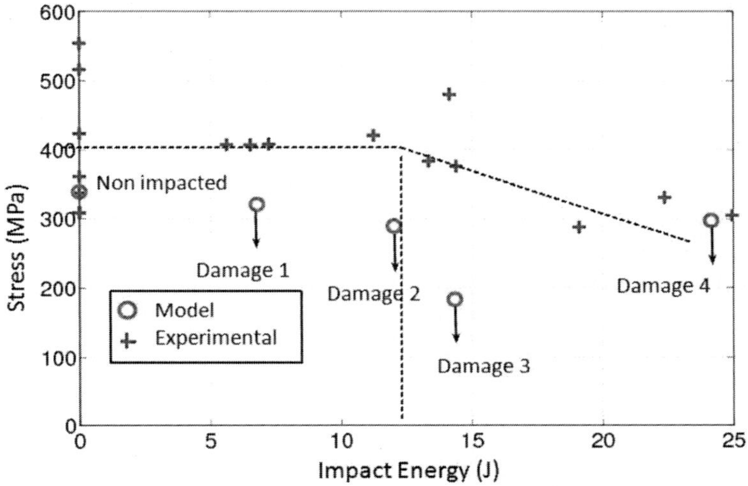

Figure 14. Comparison between model predictions and experimental results

Because the numerical results are very different from the experimental ones in the case of small-scale specimens, a complementary study on the influence of the implemented damage on the ultimate stress was carried out. The results of this study are presented in the next section.

4.1.1. Influence Of The Damage On Residual Behaviour Prediction

It was decided to quantify this strength variability by studying different cases of damage. This methodology was used for both scales of specimens starting with specimens "Ø300".

Table 6. Add of two damage level cases (A and B) for specimens ϕ300 mm

Structure		Dam.1	Virtual Dam. A	Dam. 2	Dam. 3	Virtual Dam. B	Dam. 4
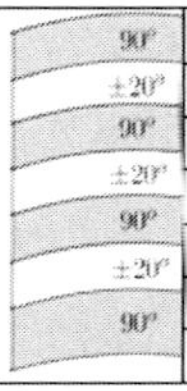	Circ. 1	≈	≈	≈	≈	≈	≈
	Long. 1		≈	≈	≈	≈	≈
	Circ. 2	≈	≈	≈	≈	≈	≈
	Long. 2			≈	≈	≈	≈
	Circ. 3			≈		≈	≈
	Long. 3				≈	≈	≈
	Circ. 4						≈

Thus, the «Damage A» level was created from the «Damage 1» level by adding the degradation of the first longitudinal layer (see Table 6). The «Damage B» level was established from the «Damage 3» level by adding the degradation of the third circumferential layer. These damage cases were chosen specifically to quantify the influence of one more failed layer on the ultimate stress of a specimen.

Figure 15 shows the ultimate stress evolution according to the impact energy (i.e., the initial damage state) from experimental and numerical analyses.

The ultimate stress obtained in the case of «Damage A» is 46% higher than in the case of «Damage 1» while the damage level is more important. In the former case, the ultimate stress is also higher than that of the undamaged specimens. It seems that the degradation of all the layers located in the convex part of the curved specimen (i.e., above the mid-plane) increases the residual tensile strength. This is not true in the case of a partial degradation where only the circumferential layers are failed.

Moreover, increasing strength appears in the case of «Damage B». The ultimate stress is 60% higher than that obtained in «Damage 3» while «Damage B» considers one more circumferential layer failed. All these observations lead to a conclusion regarding the high sensitivity of the ultimate stress according to the damage organization through the thickness of these small-scale specimens Ø300.

It seems that any changes in the radius of the curvature and/or the thickness of the damaged specimens modify the phenomena which occur when they are stressed in tension.

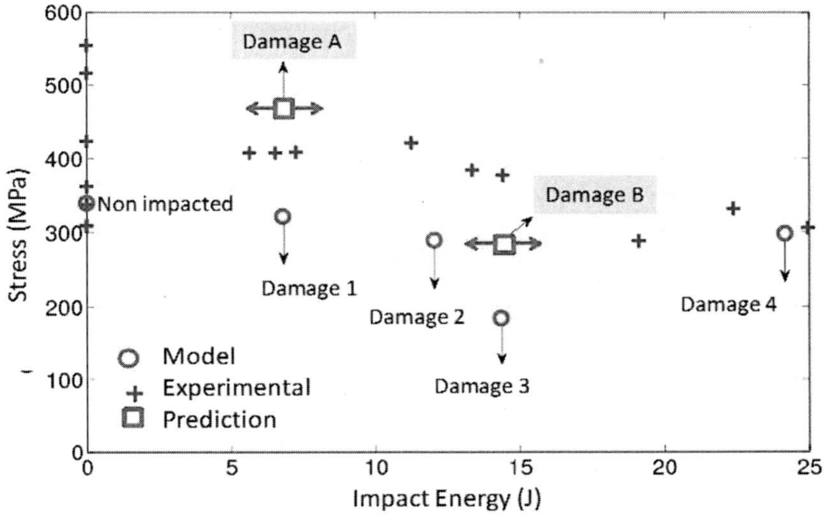

Figure 15. Comparison between model predictions for two new damages (A and B)

Experimentally, during the tensile test, due to the lay-up and the boundary conditions, the specimen curvature tends to increase between the jaws [Ballère et al., 2008]. Circumferential layers are then progressively damaged according to the inter-fibre mode, thereby causing a change in the orientation of the fibres oriented at ±20° towards the loading direction (0°). This phenomenon is accentuated by the initial curvature of the specimen. In the case of a partial pre-damage, the type of layers damaged (circumferential or longitudinal), as well as their locations through the thickness of the specimen's boundary conditions, cause this change of orientation.

This phenomenon only exists because the interface between two layers of different orientations can be damaged. Looking at the delaminations observed after the global failure of specimens, the interface integrity strongly influences the residual tensile strength of curved panels. Because damage of the interface is not considered in this model, there are high variations in the ultimate stress in the case of specimens Ø300.

5. CONCLUSION

A progressive failure analysis, based on a maximum strain criterion, is presented in this paper. The aim of this approach is to predict the residual tensile behaviour of impacted curved panels. Applied on a scale of specimens (Ø600), it governs the progressive damage which appears at the layer level and leads to a residual tensile strength close to the experimental results. Using this approach, a scenario of failure was proposed and it was shown that the integrity of the circumferential layers (oriented at 90°) is very important for ensuring the tensile strength of the curved panels. Indeed, their failure leads to a change of a specimen's curvature between the jaws. This phenomenon alters the orientation of fibres oriented at ±20° towards the loading direction. High shear stresses then appear and lead to the global failure of the specimen.

This approach has been used for the prediction of the residual tensile strength of a second scale of specimens (Ø300), thinner and more curved than the previous ones. The numerical results are in good agreement with the experimental results in the cases of undamaged specimens and specimens fully damaged in the impact zone (i.e., all the layers are failed). But, for partially damaged specimens, the results do not match those of the experiments. Because of the higher curvatures of the specimens, the residual tensile strength seems to be very dependent on the organization of the damage through the specimen thickness.

This study indicates the need to pay attention to phenomena which can appear during a scale reduction in the case of composite curved structures. A significant decrease in the scale of the specimens (i.e., a reduction in the thickness and the radius of the curvature) can change the nature of the main damage mechanisms responsible for the global failure of pre-impacted specimens. For slightly curved specimens, it has been shown that the damage which occurs at the interface has to be taken into account to accurately predict the residual tensile strength of pre-impacted specimens.

This possible degradation of interfaces could be conducted using cohesive elements (Elder et al., 2004, Pinho et al., 2006). These elements are implemented between two volumic elements representing two layers of different orientations. Their thickness is equal to 0 before loading. Choosing an appropriate behaviour law, they govern the displacement between two opposite nodes.

First results were obtained by taking into account the degradation of the interfaces. They show that the behaviour is different as soon as the interface is damaged. The change of the curvature, initiated by the failure of the circumferential layers, leads to high stresses which progressively damage the interface. The load transfer cannot take place anymore. This leads to criterion activation according to the fibre-mode in the longitudinal layers and the global failure of the specimen happens. These results confirm the fact that the interface damage has to be considered in order to reflect the physical mechanisms which lead to the failure of curved panels loaded in tension. Nevertheless, the computation time is very important when the cohesive elements are used with an intra-laminar progressive failure analysis. An alternative way has to be found to decrease this high computation time while maintaining the combined modelling of intra-laminar and inter-laminar damage mechanisms.

REFERENCES

1. D. R. Ambur, N. Jaunky, M. Hilburger, C. G. Dávila, Progressive failure analyses of compression-loaded composite curved panels with and without cutoutsComposite Structures, 2004 2004 65 143 155

2. L. Ballère, P. Viot, J. Lataillade, L. , L. Guillaumat, S. Cloutet, Damage tolerance of impacted curved panels International Journal of Impact Engineering, 2009 2009 36 243 253

3. T. A. Bogetti, C. P. Hoppel, V. M. Harik, J. F. Newill, B. P. Burns, M. Hinton, A. Kaddour, P. . Soden, Eds, the. Predicting, response. nonlinear, failure. progressive, composite. of, Failure. laminates, in. Criteria, Composites. Fibre-Reinforced-Polymer, Elsevier, 2004 2004 402 428

4. E. Buckingham, On physically similar systems; illustration of the use of dimensional equations. Phys.Review, 4 41914 1914

5. N. Carrere, Y. Rollet, F. Leroy, H. , J. Maire, F. , Efficient structural computations with parameters uncertainty for composite applications, Composites Science and Technology, 2009 2009 69 1328 1333

6. D. J. Elder, R. S. Thomson, M. Q. Nguyen, M. L. Scott, Review of delamination predictive methods for low speed impact of composite laminates Composite Structures, 2004 2004 66 677 683

7. Z. Hashin, A. A. Rotem, failure. fatigue, for. criterion, reinforced. fibre, materials, Journal of Composite Materials, 7 448 464 1973

8. S. Kaddour, M. J. Hinton, S. P. D. A. &a, of. comparison, predictive. the, of. capabilities, failure. current, for. theories, laminates. composite, contributions. additional, Science. Composites, Technology, 2004 2004 64 449 476

9. F. Laurin, N. Carrère, J. Maire, F. A. multiscale, failure. progressive, for. approach, laminates. composite, on. based, viscoelastic. thermodynamical, models. damage, Part. A. Composites, Science. Applied, Manufacturing, 2007 2007 38 198 209

10. S. Pinho, L. Iannucci, P. Robinson, Formulation and implementation of decohesion elements in an explicit finite element code Composites Part A: Applied Science and Manufacturing, 2006 2006 37 778 789

11. Y. Qian, S. R. Swanson, An experimental study of scaling rules for impact damage in fibre composites. J. of Composite Materials, 24 559 570 , May 1990

12. P. . Viot, L. . Ballère, L. Guillaumat, Lataillade, J. L. , Scale effects on the response of composite structures under impact loading. Engineering Fracture Mechanics Journal, 75 9 2725 2736 , 2008

13. P. A. Zinoviev, O. V. Lebedeva, L. P. A. Tairova, analysis. coupled, experimental. of, results. theoretical, the. on, deformation, of. failure, laminates. composite, a. under, of. state, stress. plane, Composites Science and Technology, 2002 2002 62 1711 1723

2

APPLICATIONS OF NANOPARTICLES IN BIOLOGY AND MEDICINE

OV Salata

Sir William Dunn School of Pathology, University of Oxford

ABSTRACT

Nanomaterials are at the leading edge of the rapidly developing field of nanotechnology. Their unique size-dependent properties make these materials superior and indispensable in many areas of human activity. This brief review tries to summarise the most recent developments in the field of applied nanomaterials, in particular their application in biology and medicine, and discusses their commercialisation prospects.

KEYWORDS

Nanotechnology, nanomaterials, nanoparticles, quantum dots nanotubes, medicine biology, applications

1. INTRODUCTION

Nanotechnology [1] is enabling technology that deals with nano-meter sized objects. It is expected that nanotechnology will be developed at several levels: materials, devices and systems. The nanomaterials level is the most advanced at present, both in scientific knowledge and in commercial applications. A decade ago, nanoparticles were studied because of their size-dependent physical and chemical properties [2]. Now they have entered a commercial exploration period [3, 4].

Living organisms are built of cells that are typically 10 μm across. However, the cell parts are much smaller and are in the sub-micron size domain. Even smaller are the proteins with a typical size of just 5 nm, which is comparable with the dimensions of smallest manmade nanoparticles. This simple size comparison gives an idea of using nanoparticles as very small probes that would allow us to spy at the cellular machinery without introducing too much interference [5]. Understanding of biological processes on the nanoscale level is a strong driving force behind development of nanotechnology [6].

Out of plethora of size-dependant physical properties available to someone who is interested in the practical side of nanomaterials, optical [7] and magnetic [8] effects are the most used for biological applications.

The aim of this review is firstly to give reader a historic prospective of nanomaterial application to biology and medicine, secondly to try to overview the most recent developments in this field, and finally to discuss the hard road to commercialisation. Hybrid bionanomaterials can also be applied to build novel electronic, optoelectronics and memory devices (see for example [9, 10]). Nevertheless, this will not be discussed here and will be a subject of a separate article.

2. APPLICATIONS

A list of some of the applications of nanomaterials to biology or medicine is given below:

- Fluorescent biological labels [11–13]
- Drug and gene delivery [14, 15]
- Bio detection of pathogens [16]
- Detection of proteins [17]
- Probing of DNA structure [18]

- Tissue engineering [19, 20]
- Tumour destruction via heating (hyperthermia)[21]
- Separation and purification of biological molecules and cells [22]
- MRI contrast enhancement [23]
- Phagokinetic studies [24]

As mentioned above, the fact that nanoparticles exist in the same size domain as proteins makes nanomaterials suitable for bio tagging or labelling. However, size is just one of many characteristics of nanoparticles that itself is rarely sufficient if one is to use nanoparticles as biological tags. In order to interact with biological target, a biological or molecular coating or layer acting as a bioinorganic interface should be attached to the nanoparticle. Examples of biological coatings may include antibodies, biopolymers like collagen [25], or monolayers of small molecules that make the nanoparticles biocompatible [26]. In addition, as optical detection techniques are wide spread in biological research, nanoparticles should either fluoresce or change their optical properties. The approaches used in constructing nano-biomaterials are schematically presented below (see Figure 1).

Nano-particle usually forms the core of nano-biomaterial. It can be used as a convenient surface for molecular assembly, and may be composed of inorganic or polymeric materials. It can also be in the form of nano-vesicle surrounded by a membrane or a layer. The shape is more often spherical but cylindrical, plate-like and other shapes are possible. The size and size distribution might be important in some cases, for example if penetration through a pore structure of a cellular membrane is required. The size and size distribution are becoming extremely critical when quantum-sized effects are used to control material properties. A tight control of the average particle size and a narrow distribution of sizes allow creating very efficient fluorescent probes that emit narrow light in a very wide range of wavelengths. This helps with creating biomarkers with many and well distinguished colours. The core itself might have several layers and be multifunctional. For example, combining magnetic and luminescent layers one can both detect and manipulate the particles.

The core particle is often protected by several monolayers of inert material, for example silica. Organic molecules that are adsorbed or chemisorbed on the surface of the particle are also used for this purpose. The same layer might act as a biocompatible material.

However, more often an additional layer of linker molecules is required to proceed with further functionalisation. This linear linker molecule has reactive groups at both ends. One group is aimed at attaching the linker to the nanoparticle surface and the other is used to bind various moieties like biocompatibles (dextran), antibodies, fluorophores etc., depending on the function required by the application.

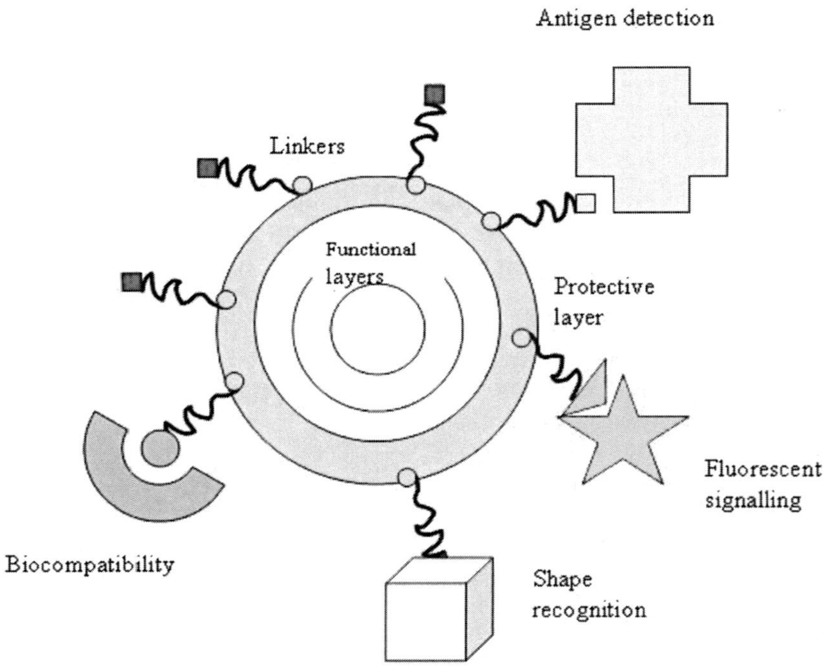

Figure 1 Typical configurations utilised in nano-bio materials applied to medical or biological problems.

3. RECENT DEVELOPMENTS

3.1 Tissue engineering

Natural bone surface is quite often contains features that are about 100 nm across. If the surface of an artificial bone implant were left smooth, the body would try to reject it. Because of that smooth surface is likely to cause production of a fibrous tissue covering the surface of the implant. This layer reduces the bone-implant contact, which may result in loosening of the implant and further inflammation. It was demonstrated that by creating nano-sized features on the surface of the hip or knee prosthesis one could reduce the chances of rejection as well as to stimulate the production of osteoblasts. The osteoblasts are the cells responsible for the growth of the bone matrix and are found on the advancing surface of the developing bone.

The effect was demonstrated with polymeric, ceramic and, more recently, metal materials. More than 90% of the human bone cells from suspension adhered to the nanostructured metal surface [27], but only 50% in the control sample. In the end this findings would allow to design a more durable and longer lasting hip or knee replacements and to reduce the chances of the implant getting loose.

Titanium is a well-known bone repairing material widely used in orthopaedics and dentistry. It has a high fracture resistance, ductility and weight to strength ratio. Unfortunately, it suffers from the lack of bioactivity, as it does not support sell adhesion and growth well. Apatite coatings are known to be bioactive and to bond to the bone. Hence, several techniques were used in the past to produce an apatite coating on titanium. Those coatings suffer from thickness non-uniformity, poor adhesion and low mechanical strength. In addition, a stable porous structure is required to support the nutrients transport through the cell growth.

It was shown that using a biomimetic approach – a slow growth of nanostructured apatite film from the simulated body fluid – resulted in the formation of a strongly adherent, uniform nanoporous layer [19]. The layer was found to be built of 60 nm crystallites, and possess a stable nanoporous structure and bioactivity.

A real bone is a nanocomposite material, composed of hydroxyapatite crystallites in the organic matrix, which is mainly composed of collagen. Thanks to that, the bone is mechanically tough and, at the same time, plastic, so it can recover from a mechanical damage. The actual nanoscale mechanism leading to this useful combination of properties is still debated.

An artificial hybrid material was prepared from 15–18 nm ceramic nanoparticles and poly (methyl methacrylate) copolymer [20]. Using tribology approach, a viscoelastic behaviour (healing) of the human teeth was demonstrated. An investigated hybrid material, deposited as a coating on the tooth surface, improved scratch resistance as well as possessed a healing behaviour similar to that of the tooth.

3.2 Cancer therapy

Photodynamic cancer therapy is based on the destruction of the cancer cells by laser generated atomic oxygen, which is cytotoxic. A greater quantity of a special dye that is used to generate the atomic oxygen is taken in by the cancer cells when compared with a healthy tissue. Hence, only the cancer cells are destroyed then exposed to a laser radiation. Unfortunately, the remaining dye molecules migrate to the skin and the eyes and make the patient very sensitive to the daylight exposure. This effect can last for up to six weeks.

To avoid this side effect, the hydrophobic version of the dye molecule was enclosed inside a porous nanoparticle [28]. The dye stayed trapped inside the Ormosil nanoparticle and did not spread to the other parts of the body. At the same time, its oxygen generating ability has not been affected and the pore size of about 1 nm freely allowed for the oxygen to diffuse out.

3.3 Multicolour optical coding for biological assays [29]

The ever increasing research in proteomics and genomic generates escalating number of sequence data and requires development of high throughput screening technologies. Realistically, various array technologies that are currently used in parallel analysis are likely to reach saturation when a number of array elements exceed several millions. A three-dimensional approach, based on optical "bar coding" of polymer particles in solution, is limited only by the number of unique tags one can reliably produce and detect.

Single quantum dots of compound semiconductors were successfully used as a replacement of organic dyes in various bio-tagging applications [7]. This idea has been taken one step further by combining differently sized and hence having different fluorescent colours quantum dots, and combining them in polymeric microbeads [29]. A precise control of quantum dot ratios has been achieved. The selection of nanoparticles used in those experiments had 6 different colours as well as 10 intensities. It is enough to encode over 1 million combinations. The uniformity and reproducibility of beads was high letting for the bead identification accuracies of 99.99%.

3.4 Manipulation of cells and biomolecules [30]

Functionalised magnetic nanoparticles have found many applications including cell separation and probing; these and other applications are discussed in a recent review [8]. Most of the magnetic particles studied so far are spherical, which somewhat limits the possibilities to make these nanoparticles multifunctional. Alternative cylindrically shaped nanoparticles can be created by employing metal electrodeposition into nanoporous alumina template [30]. Depending on the properties of the template, nanocylinder radius can be selected in the range of 5 to 500 nm while their length can be as big as 60 μm. By sequentially depositing various thicknesses of different metals, the structure and the magnetic properties of individual cylinders can be tuned widely.

As surface chemistry for functionalisation of metal surfaces is well developed, different ligands can be selectively attached to different segments. For example, porphyrins with thiol or carboxyl linkers were simultaneously attached to the gold or nickel segments respectively. Thus, it is possible to produce magnetic nanowires with spatially segregated fluorescent parts. In addition, because of the large aspect ratios, the residual magnetisation of these nanowires can be high. Hence, weaker magnetic field can be used to drive them. It has been shown that a self-assembly of magnetic nanowires in suspension can be controlled by weak external magnetic fields. This would potentially allow controlling cell assembly in different shapes and forms. Moreover, an external magnetic field can be combined with a lithographically defined magnetic pattern ("magnetic trapping").

3.5 Protein detection [31]

Proteins are the important part of the cell's language, machinery and structure, and understanding their functionalities is extremely important for further progress in human well being. Gold nanoparticles are widely used in immunohistochemistry to identify protein-protein interaction. However, the multiple simultaneous detection capabilities of this technique are fairly limited. Surface-enhanced Raman scattering spectroscopy is a well-established technique for detection and identification of single dye molecules. By combining both methods in a single nanoparticle probe one can drastically improve the multiplexing capabilities of protein probes. The group of Prof. Mirkin has designed a sophisticated multifunctional probe that is built around a 13 nm gold nanoparticle. The nanoparticles are coated with hydrophilic oligonucleotides containing a Raman dye at one end and terminally capped with a small molecule recognition element (e.g. biotin). Moreover, this molecule is catalytically active and will be coated with silver in the solution of Ag(I) and hydroquinone. After the probe is attached to a small molecule or an antigen it is designed to detect, the substrate is exposed to silver and hydroquinone solution. A silver-plating is happening close to the Raman dye, which allows for dye signature detection with a standard Raman microscope. Apart from being able to recognise small molecules this probe can be modified to contain antibodies on the surface to recognise proteins. When tested in the protein array format against both small molecules and proteins, the probe has shown no cross-reactivity.

3.6 Commercial exploration

Some of the companies that are involved in the development and commercialisation of nanomaterials in biological and medical applications are listed below (see Table 1). The majority of the companies are small recent spinouts of various research institutions. Although not exhausting, this is a representative selection reflecting current industrial trends. Most of the companies are developing pharmaceutical applications, mainly for drug delivery. Several companies exploit quantum size effects in semiconductor nanocrystals for tagging biomolecules, or use bio-conjugated gold nanoparticles for labelling various cellular parts. A number of companies are applying nano-ceramic materials to tissue engineering and orthopaedics.

Most major and established pharmaceutical companies have internal research programs on drug delivery that are on formulations or dispersions containing components down to nano sizes. Colloidal silver is widely used in anti-microbial formulations and dressings. The high reactivity of titania nanoparticles, either on their own or then illuminated with UV light, is also used for bactericidal purposes in filters. Enhanced catalytic properties of surfaces of nano-ceramics or those of noble metals like platinum are used to destruct dangerous toxins and other hazardous organic materials.

Table 1 Examples of Companies commercialising nanomaterials for bio- and medical applications.

Company	Major area of activity	Technology
Advectus Life Sciences Inc.	Drug delivery	Polymeric nanoparticles engineered to carry anti-tumour drug across the blood-brain barrier
Alnis Biosciences, Inc.	Bio-pharmaceutical	Biodegradable polymeric nanoparticles for drug delivery
Argonide	Membrane filtration	Nanoporous ceramic materials for endotoxin filtration, orthopaedic and dental implants, DNA and protein separation
BASF	Toothpaste	Hydroxyapatite nanoparticles seems to improve dental surface
Biophan Technologies, Inc.	MRI shielding	Nanomagnetic/carbon composite materials to shield medical devices from RF fields
Capsulution NanoScience AG	Pharmaceutical coatings to improve solubility of drugs	Layer-by-layer poly-electrolyte coatings, 8–50 nm
Dynal Biotech		Magnetic beads
Eiffel Technologies	Drug delivery	Reducing size of the drug particles to 50–100 nm

Company	Major area of activity	Technology
EnviroSystems, Inc.	Surface desinfectsant	Nanoemulsions
Evident Technologies	Luminescent biomarkers	Semiconductor quantum dots with amine or carboxyl groups on the surface, emission from 350 to 2500 nm
Immunicon	Tarcking and separation of different cell types	magnetic core surrounded by a polymeric layer coated with antibodies for capturing cells
KES Science and Technology, Inc.	AiroCide filters	Nano-TiO2 to destroy airborne pathogens
NanoBio Cortporation	Pharmaceutical	Antimicrobal nano-emulsions
NanoCarrier Co., Ltd	Drug delivery	Micellar nanoparticles for encapsulation of drugs, proteins, DNA
NanoPharm AG	Drug delivery	Polybutilcyanoacrylate nanoparticles are coated with drugs and then with surfactant, can go across the blood-brain barrier
Nanoplex Technologies, Inc	Nanobarcodes for bioanalysis	
Nanoprobes,	Gold nanoparticles for	Gold nanoparticles bio-conjugates for TEM and/or

Company	Major area of activity	Technology
Inc.	biological markers	fluorescent microscopy
Nanoshpere, Inc.	Gold biomarkers	DNA barcode attached to each nanoprobe for identification purposes, PCR is used to amplify the signal; also catalytic silver deposition to amplify the signal using surface plasmon resonance
NanoMed Pharmaceutical, Inc.	Drug delivery	Nanoparticles for drug delivery
Oxonica Ltd	Sunscreens	Doped transparent nanoparticles to effectively absorb harmful UV and convert it into heat
PSiVida Ltd	Tissue engineering, implants, drugs and gene delivery, bio-filtration	Exploiting material properties of nanostructured porous silicone
Smith & Nephew	Acticoat bandages	Nanocrystal silver is highly toxic to pathogenes
QuantumDot Corporation	Luminescent biomarkers	Bioconjugated semiconductor quantum dots

FUTURE DIRECTIONS

As it stands now, the majority of commercial nanoparticle applications in medicine are geared towards drug delivery. In biosciences, nanoparticles are replacing organic dyes in the applications that require high photo-stability as well as high multiplexing capabilities. There are some developments in directing and remotely controlling the functions of nano-probes, for example driving magnetic nanoparticles to the tumour and then making them either to release the drug load or just heating them in order to destroy the surrounding tissue. The major trend in further development of nanomaterials is to make them multifunctional and controllable by external signals or by local environment thus essentially turning them into nano-devices.

REFERENCES

1. Feynman R: There's plenty of room at the bottom. Science. 1991, 254: 1300-1301.

2. Murray CB, Kagan CR, Bawendi MG: Synthesis and characterisation of monodisperse nanocrystals and close-packed nanocrystal assemblies. Annu Rev Mater Sci. 2000, 30: 545-610. 10.1146/annurev.matsci.30.1.545.

3. Mazzola L: Commercializing nanotechnology. Nature Biotechnology. 2003, 21: 1137-1143. 10.1038/nbt1003-1137.

4. Paull R, Wolfe J, Hebert P, Sinkula M: Investing in nanotechnology. Nature Biotechnology. 2003, 21: 1134-1147. 10.1038/nbt1003-1144.

5. Taton TA: Nanostructures as tailored biological probes. Trends Biotechnol. 2002, 20: 277-279. 10.1016/S0167-7799(02)01973-X.

6. Whitesides GM: The 'right' size in Nanobiotechnology. Nature Biotechnology. 2003, 21: 1161-1165. 10.1038/nbt872.

7. Parak WJ, Gerion D, Pellegrino T, Zanchet D, Micheel C, Williams CS, Boudreau R, Le Gros MA, Larabell CA, Alivisatos AP: Biological applications of colloidal nanocrystals. Nanotechnology. 2003, 14: R15-R27. 10.1088/0957-4484/14/7/201.

8. Pankhurst QA, Connolly J, Jones SK, Dobson J: Applications of magnetic nanoparticles in biomedicine. J Phys D: Appl Phys. 2003, 36: R167-R181. 10.1088/0022-3727/36/13/201.

9. Yan H, Park SH, Finkelstein G, Reif JH, LaBean TH: DNA-templated self-assembly of protein arrays and highly conductive nanowires. Science. 2003, 301: 1882-1884. 10.1126/science.1089389.

10. Keren K, Berman RS, Buchstab E, Sivan U, Braun E: DNA-templated carbon nanotube field-effect transistor. Science. 2003, 302: 1380-1382. 10.1126/science.1091022.

11. Bruchez M, Moronne M, Gin P, Weiss S, Alivisatos AP: Semiconductor nanocrystals as fluorescent biological labels. Science. 1998, 281: 2013-2016. 10.1126/science.281.5385.2013.

12. Chan WCW, Nie SM: Quantum dot bioconjugates for ultrasensitive nonisotopic detection. Science. 1998, 281: 2016-2018. 10.1126/science.281.5385.2016.

13. Wang S, Mamedova N, Kotov NA, Chen W, Studer J: Antigen/antibody immunocomplex from CdTe nanoparticle bioconjugates. Nano Letters. 2002, 2: 817-822. 10.1021/nl0255193.

14. Mah C, Zolotukhin I, Fraites TJ, Dobson J, Batich C, Byrne BJ: Microsphere-mediated delivery of recombinant AAV vectors *in vitro* and *in vivo*. Mol Therapy. 2000, 1: S239-10.1006/mthe.2000.0174.

15. Panatarotto D, Prtidos CD, Hoebeke J, Brown F, Kramer E, Briand JP, Muller S, Prato M, Bianco A: Immunization with peptide-functionalized carbon nanotubes enhances virus-specific neutralizing antibody responses. Chemistry&Biology. 2003, 10: 961-966.

16. Edelstein RL, Tamanaha CR, Sheehan PE, Miller MM, Baselt DR, Whitman LJ, Colton RJ: The BARC biosensor applied to the detection of biological warfare agents. Biosensors Bioelectron. 2000, 14: 805-813. 10.1016/S0956-5663(99)00054-8.

17. Nam JM, Thaxton CC, Mirkin CA: Nanoparticles-based bio-bar codes for the ultrasensitive detection of proteins. Science. 2003, 301: 1884-1886. 10.1126/science.1088755.

18. Mahtab R, Rogers JP, Murphy CJ: Protein-sized quantum dot luminescence can distinguish between "straight", "bent", and "kinked" oligonucleotides. J Am Chem Soc. 1995, 117: 9099-9100.

19. Ma J, Wong H, Kong LB, Peng KW: Biomimetic processing of nanocrystallite bioactive apatite coating on titanium. Nanotechnology. 2003, 14: 619-623. 10.1088/0957-4484/14/6/310.

20. de la Isla A, Brostow W, Bujard B, Estevez M, Rodriguez JR, Vargas S, Castano VM: Nanohybrid scratch resistant coating for teeth and bone viscoelasticity manifested in tribology. Mat Resr Innovat. 2003, 7: 110-114.

21. Yoshida J, Kobayashi T: Intracellular hyperthermia for cancer using magnetite cationic liposomes. J Magn Magn Mater. 1999, 194: 176-184.

22. Molday RS, MacKenzie D: Immunospecific ferromagnetic iron dextran reagents for the labeling and magnetic separation of cells. J Immunol Methods. 1982, 52: 353-367. 10.1016/0022-1759(82)90007-2.

23. Weissleder R, Elizondo G, Wittenburg J, Rabito CA, Bengele HH, Josephson L: Ultrasmall superparamagnetic iron oxide: characterization of a new class of contrast agents for MR imaging. Radiology. 1990, 175: 489-493.

24. Parak WJ, Boudreau R, Gros ML, Gerion D, Zanchet D, Micheel CM, Williams SC, Alivisatos AP, Larabell CA: Cell motility and metastatic potential studies based on quantum dot imaging of phagokinetic tracks. Adv Mater. 2002, 14: 882-885. 10.1002/1521-4095(20020618)14:12<882::AID-ADMA882>3.0.CO;2-Y.

25. Sinani VA, Koktysh DS, Yun BG, Matts RL, Pappas TC, Motamedi M, Thomas SN, Kotov NA: Collagen coating promotes biocompatibility of semiconductor nanoparticles in stratified LBL films. Nano Letters. 2003, 3: 1177-1182. 10.1021/nl0255045.

26. Zhang Y, Kohler N, Zhang M: Surface modification of superparamagnetic magnetite nanoparticles and their intracellular uptake. Biomaterials. 2002, 23: 1553-1561. 10.1016/S0142-9612(01)00267-8.

27. Gutwein LG, Webster TJ: Affects of alumina and titania nanoparticulates on bone cell function. American Ceramic Society 26 th Annual Meeting Conference Proceedings. 2003,

28. Roy I, Ohulchanskyy TY, Pudavar HE, Bergey EJ, Oseroff AR, Morgan J, Dougherty TJ, Prasad PN: Ceramic-based nanoparticles entrapping water-insoluble photosensitizing anticancer drugs: a novel drug-carrier system for photodynamic therapy. J Am Chem Soc. 2003, 125: 7860-7865. 10.1021/ja0343095.

29. Han M, Gao X, Su JZ, Nie S: Quantum-dot-tagged microbeads for multiplexed optical coding of biomolecules. Nature Biotechnology. 2001, 19: 631-635. 10.1038/90228.

30. Reich DH, Tanase M, Hultgren A, Bauer LA, Chen CS, Meyer GJ: Biological applications of multifunctional magnetic nanowires. J Appl Phys. 2003, 93: 7275-7280. 10.1063/1.1558672.

31. Cao YC, Jin R, Nam JM, Thaxton CS, Mirkin CA: Raman dye-labeled nanoparticle probes for proteins. JACS. 2003, 125: 14676-14677. 10.1021/ja0366235.

3

DAMAGE ASSESSMENT OF SHORT GLASS FIBER REINFORCED POLYESTER COMPOSITES: A COMPARATIVE STUDY

Amar Patnaik[1], Sandhyarani Biswas[2], Ritesh Kaundal[1] and Alok Satapathy[4]

[1] Department of Mechanical Engineering, National Institute of Technology, Hamirpur, India

[2] Department of Mechanical Engineering, National Institute of Technology, Rourkela, India

1. INTRODUCTION

The main goals of many challenges to the utilization of by-product cement kiln dusts (CKDs) as partial replacement of Portland cement (PC) environmental protection agencies are to seek ways to minimize the dual problems of disposal and health hazards of these by-products. For many years, by-products such as fly-ash, silica fume and slag were considered as waste materials. These by-products have been successfully used in the construction industry as a portland cement substitute (Krazowski & Emery, 1981; Al-jabri et al., 2006). However, with the growing of industries the new and different types of by-products are being generated by various industries, which could have being a promising future for partial replacement of portland cement. CKDs typically fine powders that are generated during the cement manufacturing process, and then carried off in the flue gases then subsequently collected in electrostatic precipitators. The portion of CKDs that are not returned back to the cement manufacture process, or otherwise used beneficially, are placed in landfills. A limited number of studies have carried out the use of CKD as industrial wastes in fabrication of polymer composites. The effect of CKD on the compressive strength of cement paste was studied and the corrosion behavior of embedded reinforcement and observed that up to 5wt% of CKD of cement had less effect on cement paste strength and on reinforcement (El-Sayed et al., 1991). It has also reported that when CKD and blast furnace slag are added in proper ratio to ordinary cement, the compressive strength and the corrosion-resistance of the mix increase dramatically (Batis et al., 2002). It has also been reported that the use of CKD in concrete as a partial replacement of cement (Sri Ravindrarajah, 1982). The percentages of cement replacement by weight were in different percentage i.e. from 0, 25, 50, 75, and 100 in cement paste and 0, 15, 25, 35, and 45 in both 1:1.5:3 and 1:2:4 concretes. As CKD is a cementitious material and it showed good strength as compared with general cement. Therefore, particulate played a vital role in the development of commercially feasible fiber reinforced polymers. Not only they provide significant cost reductions but also certain fillers may improve processing, physical properties and improved the mechanical strength (Rothon, 2002). There are few, if any, publications exploring the effect of hard ceramic particles on the wear resistance (especially for erosion resistance) of composites. Further, there are few publications discussing the erosion wear mechanisms of elastomers and their composites. Based on the above study, computational simulation has provided an effective and economical approach to study wear problems. During past decades, there were a number of computational models proposed to simulate wear processes at macro/meso and atomic/nano-scales (Ling & Pan, 1986; Ludema & Bayer, 1991; Hsu et al., 1997; Komvopoulos & Choi, 1992). The most widely used macro/meso-scopic model is FEM (finite element method), which can be used to deal with elastic/plastic contact problems and failure processes in the contact region. The two-dimensional (2D) water drop impact model with DYNA3D code was established (Adler, 1995). The calculation did not involve any solid erodent. Despite this, to date there are no economically viable and environmentally acceptable solutions for effective utilization of the high-residue volumes that have developed. Current best practice is to contain the material within specially constructed landfill sites, called red mud ponds/stacks or bauxite residue disposal areas (BRDAs). Recently, the potential utilization of red mud filled glass/bamboo epoxy composites was explored and investigated its significant effect on the erosion wear rate of the resulting composites (Biswas and Satapathy, 2009, 2010). Few researchers also studied the solid particle erosion behavior of particulate filled polymer matrix composites and revealed that parameters such as impact velocity, impingement angle, fiber loading/filler content, erodent size, erodent temperature and stand-off distance etc. largely depends on the erosion behavior of polymer composites (Patnaik et al., 2010a, 2010b, 2010c, 2010d,2010e, 2010f, 2010g).

In view of the above literature in this work, a non-linear finite element model has been proposed and compared the simulated results with experimental results. The utility of the new formulation is proven through its application onto a short glass fiber reinforced polyester based composites filled with cement by-pass dust as reinforcing particulates. This model is simple and flexible, which helps to gain insight into various wear processes. The eroded surfaces of these composites are analyzed with scanning electron microscopy (SEM), and the erosion wear mechanisms of the composites are investigated.

2. EXPERIMENTAL

2.1. Preparation Of Composites

Short glass fibers (elastic modulus of 72.5 GPa and possess a density of 2.59 gm/cc) of 6mm length are used to prepare the composites. The unsaturated isophthalic polyester resin (Elastic modulus 3.25GPa, density 1.35gm/cc) is manufactured by Ciba Geigy and locally supplied by Northern Polymers Ltd. New Delhi, India. The composite fabricated in two different sets. One having different fiber loading, with varying the weight fraction of fibers from 10wt% to 50wt% at an increment of 10wt% and the designation of individual composites are reported in Table 1. Secondly, cement by-pass dust (CBPD) particulate is mixed with short glass fiber reinforced polyester resin with three different percentages (0wt%, 10wt% and 20wt% of CBPD) and the mixture is poured into various moulds conforming to the requirements of various testing conditions and characterization standards. The entrapped air bubbles (if any) are removed carefully with a sliding roller and the mould is closed for curing at a temperature of 30^{0}C for 24 h at a constant pressure of 10 kg/cm^2.

2.2. Air-Jet Erosion Tester

The erosion experiments are carried out as per ASTM G76 on the erosive wear test rig. The equipment was designed to feed abrasive particles into a high velocity air stream, which propelled the particles against the specimen surface (Strzepa et al., 1993; Routbort et al., 1981). The erodent particles entrained in a stream of compressed air and accelerated down to a 65mm long brass nozzle with 3mm inside diameter to impact on a target material mounted on an angle fixture. The steady state erosion rates are measured as a function of the erodent particle size and the amount of erodent particle used in each experiment was measured with the help of double disc method. The steady state erosion rate was determined by weighing the sample after the end of each test. While the impingement angles ranges from 30° to 90° and the test duration was 20min for each run. The erodent used for this test was river silica sand particle of three different sizes, i.e. 250, 350 and 450μm. The sample was cleaned with a blast of compressed air before each weighing to remove all loosely adhering debris. The mass loss from the target was measured with an analytical balance of ±0.01mg accuracy. The erosion rate was measured in terms of mass of material removed per unit mass of erodent impacted. The process is repeated every 10 min till the erosion rate attains a constant value called steady-state-erosion-rate.

Table 1. Designation and composition of unfilled short glass fiber reinforced composites

Sl. No.	Designation of composites	Composition of composites
1.	PGFL-1	Polyester resin + 10wt% Short glass fiber
2.	PGFL-2	Polyester resin + 20wt% Short glass fiber
3.	PGFL-3	Polyester resin + 30wt% Short glass fiber
4.	PGFL-4	Polyester resin + 40wt% Short glass fiber
5.	PGFL-5	Polyester resin + 50wt% Short glass fiber

2.3. Finite Element Model

The erosive processes are simulated using an explicit dynamic code ANSYS/AUTO-DYN. The eight-node brick hexahedral elements with one integration point are used in the 3D simulation. The mesh is refined to standard cubic element at the impacted area in order to calculate the erosion rate. It has reported in literature that simulating a single particle was not sufficient and three or more particles were needed to simulate the erosion process (ElTobgy et al., 2005). In this study, 100 spherical particles were used so as to ensure the accuracy of the model. All of the particles strike the target center area at random locations. The "simultaneous group" was employed for computational efficiency and analytical organization. There are 10 groups which contain 100 particles aggregately in the proposed model. Every group has 10 particles which would impact the surface simultaneously and be followed by another simultaneous particles group, and so on. According to the researchers, the distance between any two particles' centers in the same group is no less than 0.6r (r is the radius of the particles) to avoid the damage interaction (Woytowitz and Richman, 1999). Figure 1 shows the schematic diagram of the prepared composites by using ANSYS/AUTO-DYN. The finite element model of the target material and simulated nozzle is shown in Figure 2. There are 164, 765 nodes and 147, 840 finite elements in the model. All of the bottom nodes of the target materials are fixed. For the particles, the rotation degrees of freedom are constrained. Generally, the erosion rate (gm/gm) was used to characterize the erosion performance of the target materials. The solution time is set to 1.3t10, where t10 is the time when the last particles group contacted to the target. In this study, the impact angle varies from 30° to 90° in increments of 15°. The impact velocity varies from 43 to 65 m/s.

2.4. Plan Of Experiments (Taguchi's Techniques)

For the elaboration of experiments plan, the method of Taguchi for five control factors at three levels is used, being understood by levels taken by the factors. Table 2 represents the factors to be studied and the assignment of the corresponding levels. The array chosen is the L_{27} (3^{13}) which has 27 rows corresponding to the number of tests (22 degrees of freedom) with 13 columns at three levels. The factors and the interactions are assigned to the columns. In practice, these factors can be assigned arbitrarily to any of the arrays columns, provided that all combinations must be included as per orthogonal array design. After assigning appropriate level settings, the S/N analysis (S/N: signal-to-noise ratio) is needed to evaluate the proposed experiment results. In S/N analysis, the greater the S/N, the better the experimental results:

$$\eta = -10\log(M.S.D.) \tag{1}$$

where M.S.D. is the mean-square deviation for the output characteristic (erosive wear rate).

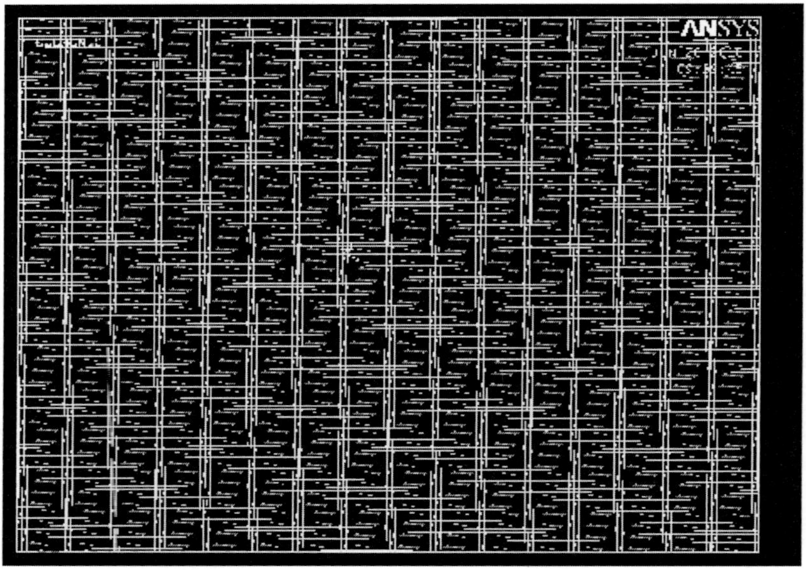

Figure 1. Schematic diagram of the particulate filled composite

As mentioned earlier, there are three categories of quality characteristics, i.e. lower-the-better, higher-the-better, and nominal-the-better. To obtain optimal performance, lower-the-better characteristic for erosion rate must be taken. The mean-square deviation (M.S.D.) for the lower-the-better characteristic can be expressed as (Glen, 1993):

$$M.S.D = \frac{1}{m}\sum_{i=1}^{m} T_i^{2} \tag{2}$$

where m is the number of tests and T_i is the value of experimental result of the ith test.

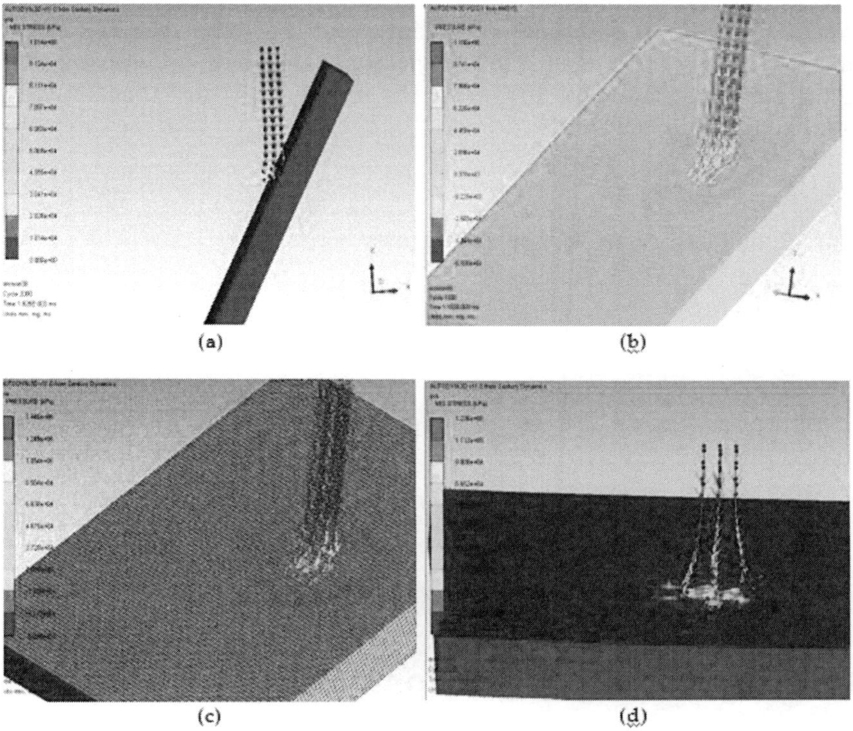

Figure 2. Schematic diagram of target composite mate rial and nozzle (a: 30° impingement angle, b: 45° impingement angle, c: 60° impingement angle and d: 90° impingement angle)

The plan of the experiments is as follows: the first column is assigned to impact velocity (A), the second column to CBPD content (B), the fifth column to impingement angle (C), the ninth column to stand-off distance (D) and the tenth column to erodent size (E), the third and fourth column are assigned to $(AB)_1$ and $(AB)_2$ respectively to estimate interaction between impact velocity (A) and CBPD percentage (B), the sixth and seventh column are assigned to $(BC)_1$ and $(BC)_2$ respectively to estimate interaction between the CBPD content (B) and impingement angle (C), the eight and eleventh column are assigned to $(AC)_1$ and $(AC)_2$ respectively to estimate interaction between the impact velocity (A) and impingement angle (C) and the remaining columns are used to estimate experimental errors. The output to be studied is erosion rate (E_r) and the tests are repeated twice corresponding to 54 tests. Furthermore, a statistical analysis of variance (ANOVA) is performed to identify the process parameters that are statistically significant. With the S/N and ANOVA analyses, the optimal combination of the process parameters can be predicted to a useful level of accuracy. Finally, a confirmation experiment is conducted to verify the optimal process parameters obtained from the parameter design.

Table 2. Levels for various control factors

Control factor	Level			
	I	II	III	Units
A:Velocity of impact	43	54	65	m/sec
B:CBPD content	0	10	20	%
C:Impingement angle	30	60	90	degree
D:Stand-off distance	65	75	85	mm
E:Erodent size	250	350	450	μm

3. RESULTS AND DISCUSSION

3.1. Erosive Wear Of The Composites

The erosive wear output characteristics are systematically determined by the Taguchi's technique as a multi-parametric optimization model, since the erosive wear may be a functional variable of five independent control factors and their interactions. Using this technique (a) firstly the steady state erosion characteristics of the composites are determined for selected level of optimally controlled operating variables and (b) subsequently control factors have been identified by Taguchi experimental techniques.

3.1.1. Effect Of Impingement Angle On Erosion Rate

Solid-particle erosion is a complex wear mechanism, influenced by a number of factors such as impact velocity, impingement angle, erodent size, erodent temperature, stand-off distance, materials properties, particles geometry, environment temperature etc. Among these, angle of impingement is the one of the most important factor and widely studied parameter in the erosion study of materials (Hutchings, 1992; Tsuda et al., 2006). When the erosion rate is measured as a function of impingement angle, two types of material behavior generally observed in the target material i.e. ductile and brittle nature. The ductile behavior of materials is characterized by maximum erosion at acute angle (15-30°) and for brittle behavior of materials, the maximum erosion is observed at normal impingement angle (90°). But as far as polymer composites are concerned the composite materials show versatile in nature depending upon the fabrication techniques and the erosion rate show slightly different behavior. The reinforced composites show a semi-ductile behavior having the maximum erosion rate in the range of 45-60° (Hutchings, 1992), unlike the above two categories. This classification, however, is not absolute as the erosion of material has a strong dependence on erosion conditions such as the properties of target material. In the present study, Figure 3 shows that maximum erosion rate appears at 60°, indicating semi-ductile erosion behavior irrespective of fiber loading.

3.1.2. Surface Morphology

Generally surface morphology of eroded surfaces indicates whether solid particle erosive wear has been occurred either by a ductile or brittle mechanism. Hence, scanning electron microscopy (SEM) studies have been done to ascertain the wear mechanism at 30°, 60° and 90° impingement angles.Figure 4a show micrographs of eroded surfaces of 10wt% short glass fiber reinforced polyester composites at 30° and 90° impingement angles at constant impact velocity of 43m/s, stand-off distance 75mm and erodent size 450µm. It is evident from the micrograph (Figure 4a) that the material removal in the composite is dominated by micro-ploughing, micro-cutting and plastic deformation. The plastically deformed material subsequently removed from the surface by micro-cutting leads to maximum erosion rate at 60° impingement angle; most material is lost when a maximum strain in the target is exceeded and formation of micro-cracks and embedment of fragments of sand particles is evident from the micrograph (Figure 4b).

Micrograph (Figure 4c) of eroded surface of 20wt% glass fiber based composite shows deeper micro-cracks on the composite surface due to brittle nature of polyester resin. However, the normal impact did not result in higher erosion rate like brittle materials. During normal impact the largest part of the initial energy is converted into heat and hence matrix is softened which resulted in embedment of sand particles (Figure 4d). The embedded sand particles control the further erosion of the target surface.Figures 4e and 4f shows micrographs of eroded surfaces of 30wt% and 40 wt% glass fiber reinforced polyester composite respectively. At oblique impingement angle (60°), micrographs (Figures 4e and4f) show matrix is plastically deformed and amount of deformation is proportional to impact velocity of particles. At 43 m/sec impact velocity removal of matrix along the length of the fiber and subsequently getting removed can be seen from the micrograph (Figure 4e). At same impact velocity for 40wt% glass fiber based composite little high degree of plastic deformation and removal of matrix and hence fibers protruded out from the matrix can be seen in the micrograph (Figure 4f). Similarly, for 50wt% glass fiber based composites (Figures 4g, h)show similar behaviour as shown in Figures 4e and 4f but the target surface is completely eroded out and exposure of glass fibers at 60° impingement angle. The exposed fibers are broken into small fragments resulting in excessive wear.

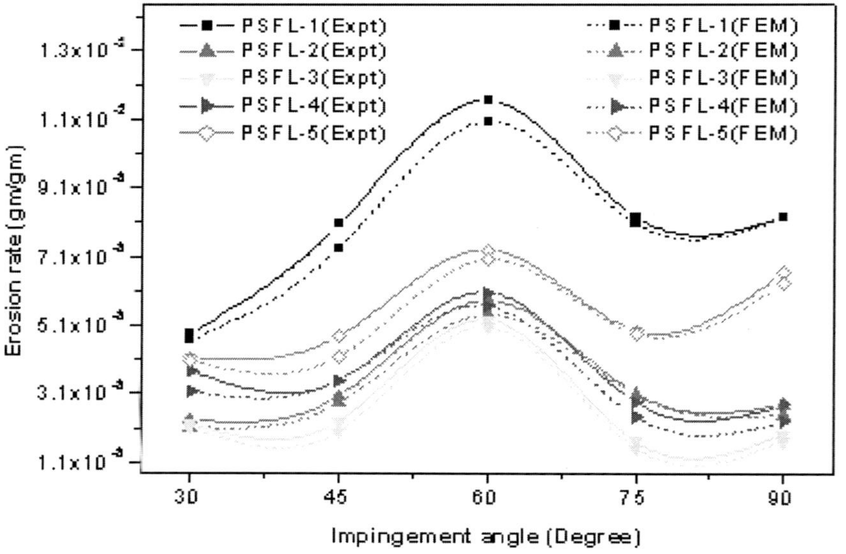

Figure 3. Influence of impingement angle on erosion rates of unfilled glass fiber-polyester composites (Impact velocity: 43m/sec, stand-off distance: 75mm and erodent size: 450μm).

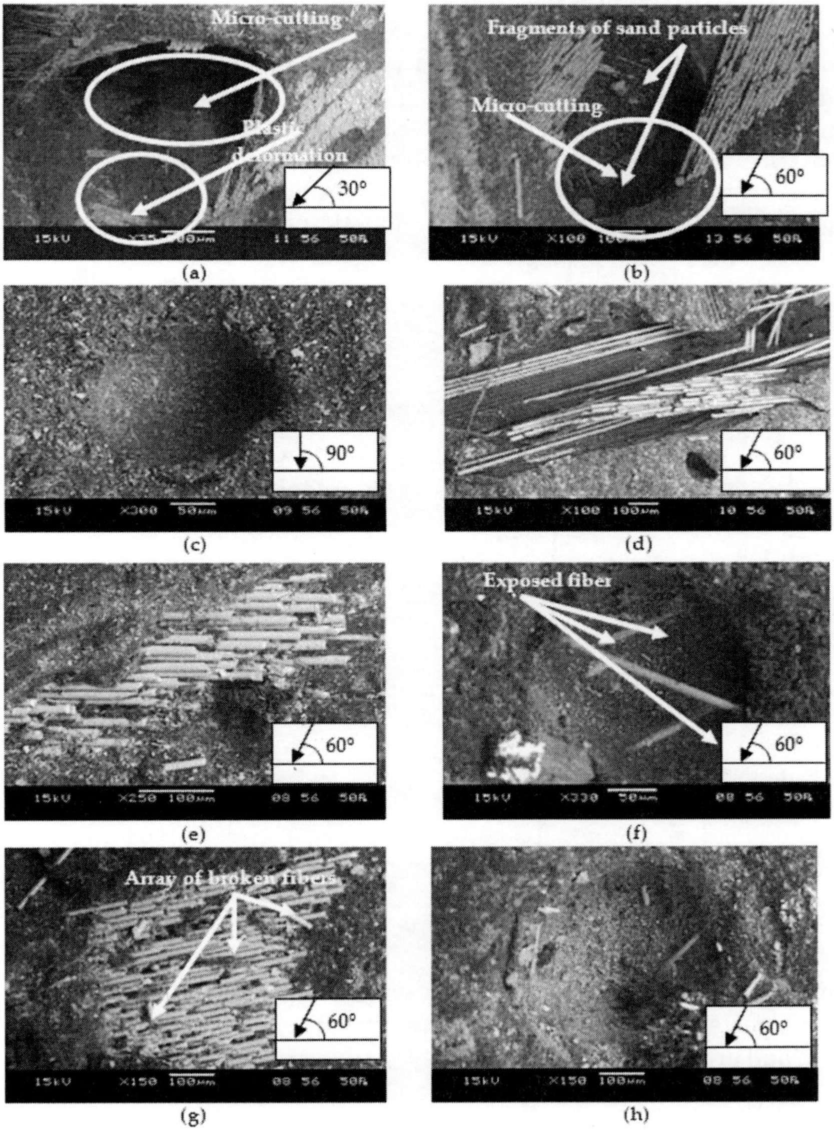

Figure 4. SEM micrographs of the eroded surfaces of the unfilled glass fiber-polyester composites

3.1.3. Influence Of Impact Velocity

In order to study the effect of impact velocity on erosion rate, erosion tests are performed on the short glass fiber reinforced polyester composites by varying the impact velocity from 43 to 65 m/s. As seen in Figure 5, erosive wear rates of the samples are remarkably higher with the increase in impact velocity conforming to the theoretical expectations. Particles have a higher kinetic energy at higher impact velocity, which results in greater impingement effect and results in increased erosion rate.

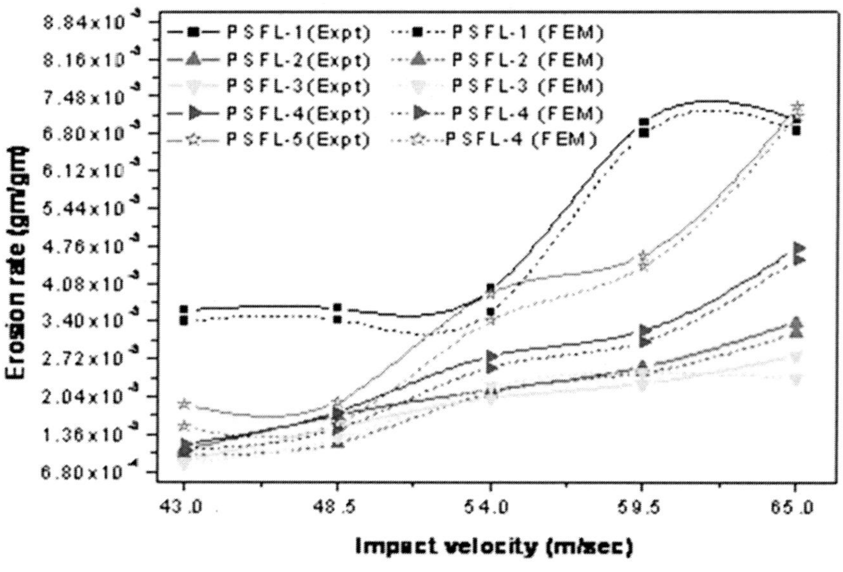

Figure 5. Influence of impact velocity on erosion rates of unfilled glass fiber-polyester composites (Impingement angle: 60^0, stand-off distance: 75mm and erodent size: 450µm).

It is clear from Figure 5 that steady-state erosion rate of all the composites increases with increase in impact velocity. Similarly, in computationally the finite element simulated results seems to be parallel to the experimental results as shown in Figure 5. It was observed that the composite with 30 wt% of short glass fiber reinforced composite is most resistant to erosion rate followed by the composites with 20, 50 and 40 wt% fiber contents in a decreasing order. Thus it is clearly from Figure that velocity of the erosive particles has a very strong significant effect on erosion rate however, their real mechanistic aspects may need further investigation.

3.2. Steady State Erosion Rate (For Particulate Filled Composites)

3.2.1. Influence Of Impingement Angle

In case of CBPD filled short glass fiber-polyester composites, the erosion rate increases monotonically with the increase in impingement angle and reaches maximum at 60° impingement angle for both particulate filled composites and unfilled composite. However, particulate filled composites shows lesser erosion rate as compared to unfilled composites as shown in Figure 6.

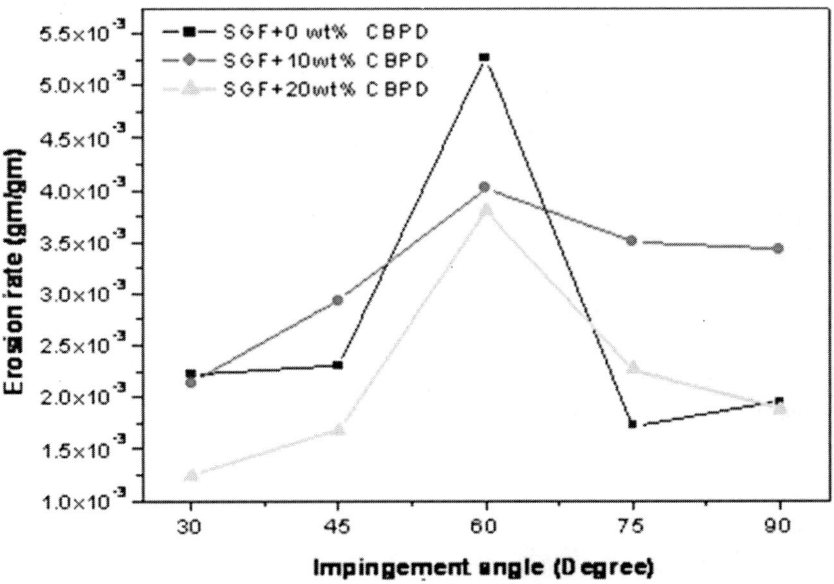

Figure 6. Influence of impingement angle on erosion rates of particulate filled composites (Impact velocity: 43 m/sec, stand-off distance: 75mm and erodent size: 450μm).

At low impingement angle the surface damage is seemed to be less and the erosion rate is also very low. Earlier studies have also reported that the crack length decreases with decreasing angle, i.e. at lower impingement angle it is small and attain maximum towards 90° (Suresh & Harsa, 2006). Longer cracks imply more rapid erosion rate taking place and the volume of material removal is approximately proportional to square of the surface crack length. This has only happened in case of normal impact velocity and hence the erosion rate shows stronger dependence on impact velocity and the angle of impingement. The maximum erosion rate for unfilled and particulate filled composites show at an angle of 60° as shown in Figure 6 under similar condition as studied in case of unfilled composites (Figure 4). But earlier studies have proved that two different processes contribute erosion on brittle materials, i.e. when particles impact at near normal incidence, lateral cracks induce mass loss, but the radial cracks do not contribute significantly to materials removal. For impact at oblique incidence, however, radial cracks can lead to materials removal (John Rajesh et al., 2001). In the present study, all the composites neither show brittle response nor does it behave ductile. It shows semi-brittle-ductile nature as shown in Figure 6.

3.2.2. Influence Of Impact Velocity

Similarly, the variation of erosion rate of unfilled and CBPD filled composites with impact velocity is shown in Figure 7. Erosion trials are conducted at five different impact velocities. It is seen, in this figure that for all the composite samples, the erosion rates gradually increases by the increase in impact velocity. The increase in erosion rate with impact velocity can be attributed to increased penetration of particles on impact as a result of dissipation of greater amount of particle thermal energy to the target surface. This leads to more surface damage, enhanced sub-critical crack growth etc. and consequently to the reduction in erosion resistance.

3.2.3. Taguchi Analysis and Response Optimization

As an evaluation tool for determining robustness, the signal-to-noise (S/N) ratio is the most important component of a parameter design. In the Taguchi method, the term 'signal' represents the desirable target i.e. erosion rate and the term 'noise' represents the undesirable value. The analysis is made using the computational software used for design of experiment applications (MINITAB 15). The response table (S/N ratio) for erosion rate is mentioned in Table 3. The overall mean for the S/N ratio of erosion rate is found to be 61.76db. Effect of control factors on erosion rate is shown in Figure 8. It is clear from Figure that factors A_1, B_3, C_1, D_3 and E_3 have been observed to be the most significant level which experimentally influences the erosion rate on the basis of smaller-the-better characteristics. The respective interaction plots are observed in the Figures 9a-c.

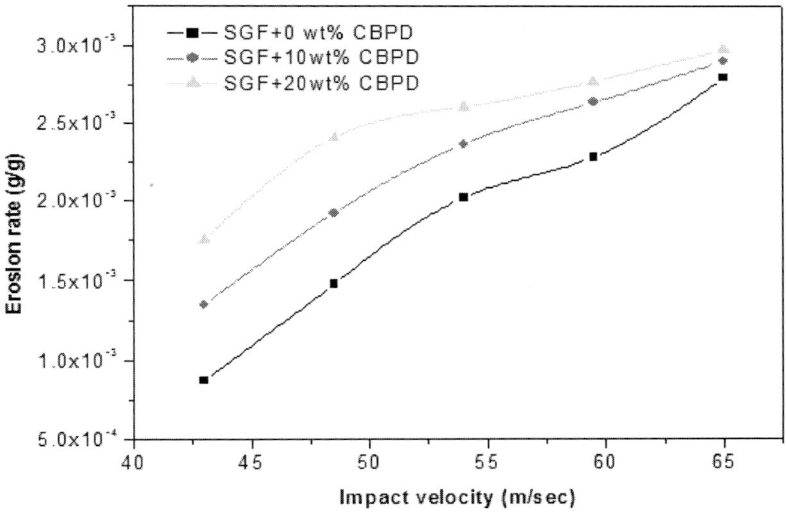

Figure 7. Influence of impact velocity on erosion rates of particulate filled composites (Impingement angle: 60°, stand-off distance: 75mm and erodent size: 450μm).

3.2.4. Surface Morphology Of Taguchi Experimental Results

To verify the erosion mechanisms of glass fiber reinforced polyester composites, eroded surfaces of the composites are observed as per Taguchi experimental orthogonal array design (see Table 3) by scanning electron microscopy (SEM). Figure 10a, 10b shows the SEM micrographs of eroded composite sample studied at lower impingement angle (see Table 3, Experiment 1) shows the random distribution of short glass fiber and removable of matrix material on the composite surface. Figures 10c and 10d show for 10wt% of CBPD filler composite that the erosion rate decreases (see Table 3, Experiment 6) with lower particle size (250μm). When the erodent particle size is more (450μm) (seeTable 3, Experiment 5), it has more energy to chip-off the target material as shown in Figure 10d. But in the case of a small particle size, it has insufficient energy to create the crack in the target material and thus lead to low erosion rate. In other hand, when the erodent particle strikes the composite surface, at 90° impingement angle the fiber damage resulting from recurring micro-ploughing in combination with the fibers getting slowly worn out from the matrix surfaces are observed. But it is not the case as tested same composite with 350μm particle size with impingement angle 60° (Figure 10e, f) shows more erosive surface with higher erosion rate (see Table 3, Experiment 4).

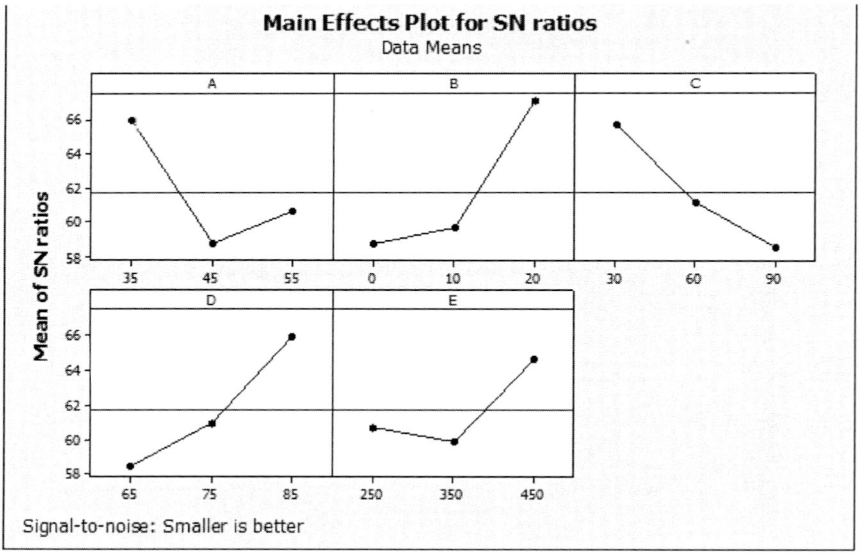

Figure 8. Effect of control factors on erosion rate

It has been discussed that the impact on brittle materials at an oblique angle produced radial cracks at an angle to the surface and they can contribute to only matrix material loss (Scattergood et al., 1981;Lawn, 1993). Radial cracks can also contribute to material loss when they drive through a relatively thin wall. In such a case, the material loss will occur without the formation of a lateral crack. Larger erodents produce deeper radial cracks. The tendency for material loss to occur from radial cracking should increase with erodent diameter (Lee etal., 2005; Milman et al., 1999). With the increasing of the CBPD particles in the composites from 10wt% to 20wt%, the wear rate of the composites increased gradually, reached a maximum and then declined gradually. The peak wear resistance for 20wt% CBPD particulate filled glass-polyester composites shows 60° impingement angle (see Figure 6). As shown from Figure 10f, the worn surface of particulate filled glass fiber-polyester composites when 350μm erodent particle stricks at an impact velocity 54m/sec the composite surface at lower impingement angle (30°) (see Table 3, Experiment 16) shows only removal of matrix material on the composite surface. Figure 10g shows a hole formed after an CBPD particle is removed from the surface. The inside surface of the hole seemed very smooth and clear which indicated that one CBPD particle wholly debonded from the matrix with the propagation of interfacial cracks (Figure 10h) due in part to the poor interfacial bond strength. However, on further increase in erodent particle from 250μm to 350μm size and impingement angle from 30° to 60° (seeTable 3, Experiment 26) the matrix materials removed from the surface along with exposure of glass fibers as shown in Figure 10i. The excessive amount of material loss in the composite is due to, too much CBPD particles were added (20 wt%), the interface area may have been too large and particles may have touched directly, so interfacial defects and cracks could largely emerge. Thus, CBPD particles could be more easily removed and could not improve the wear resistances. So the wear resistance of the composites declined with the increased content of the CBPD particles. In Figure 10j,an CBPD particle with a flat top surface protruding slightly over the worn surface, around whose back half interface distributed a clear gap (see Table 3, Experiment 25). The morphology showed that the top surface of the CBPD particle was heavily polished and became flat by the impact of the erodent particles with high velocity (65m/sec) at an angle of 60°, and the back part of the CBPD particle debonded and slipped from the matrix along the interface.

3.3. Anova and The Effects Of Factors

Table 4 shows the results of the ANOVA with the erosion rate (E_r) for CBPD filled short glass fiber reinforced polyester composites. The objective of ANOVA is to analyze the influence of impact velocity (A),

CBPD content (B), impingement angle (C), stand-off distance (D) and erodent size (E) on the total variance of the results. This analysis was undertaken for a level of significance of 5% that is for a level of confidence of 95%.

Table 3. Experimental design using L_{27} orthogonal array

Expt. No.	Impact Velociy (A)(m/s)	CBPD content (B) (%)	Impingement angle (C) (Degree)	Stand-off Distance (D)(mm)	Erodent size (E) (μm)	Erosion rate (Er) (gm/gm)	S/N Ratio (db)
1	43	0	30	65	250	0.0003303	69.6224
2	43	0	60	75	350	0.0002466	72.1588
3	43	0	90	85	450	0.0001246	78.0908
4	43	10	30	75	350	0.0008917	60.9959
5	43	10	60	85	450	0.0005550	65.1141
6	43	10	90	65	250	0.0047442	46.4768
7	43	20	30	85	450	0.0001267	77.9468
8	43	20	60	65	250	0.0005667	64.9334
9	43	20	90	75	350	0.0012350	58.1667
10	54	0	30	75	450	0.0014625	56.6981
11	54	0	60	85	250	0.0028121	51.0194
12	54	0	90	65	350	0.0027000	51.3727
13	54	10	30	85	250	0.0001375	77.2339
14	54	10	60	65	350	0.0033938	49.3864
15	54	10	90	75	450	0.0041087	47.7258
16	54	20	30	65	350	0.0008175	61.7502
17	54	20	60	75	450	0.0001844	74.6860
18	54	20	90	85	250	0.0011706	58.6316
19	65	0	30	85	350	0.0024783	52.1171
20	65	0	60	65	450	0.0045143	46.9082
21	65	0	90	75	250	0.0031857	49.9359
22	65	10	30	65	450	0.0005443	65.2835
23	65	10	60	75	250	0.0012014	58.4060
24	65	10	90	85	350	0.0005114	65.8243
25	65	20	30	75	250	0.0003300	69.6297
26	65	20	60	85	350	0.0004295	67.3403
27	65	20	90	65	450	0.0003138	70.0667

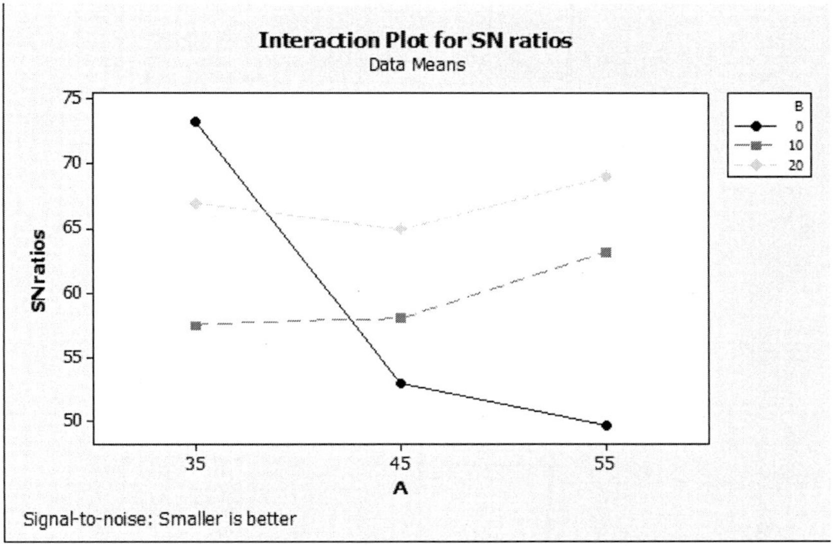

Figure 9a. Interaction graph between factor A and factor B (A×B) for erosion rate

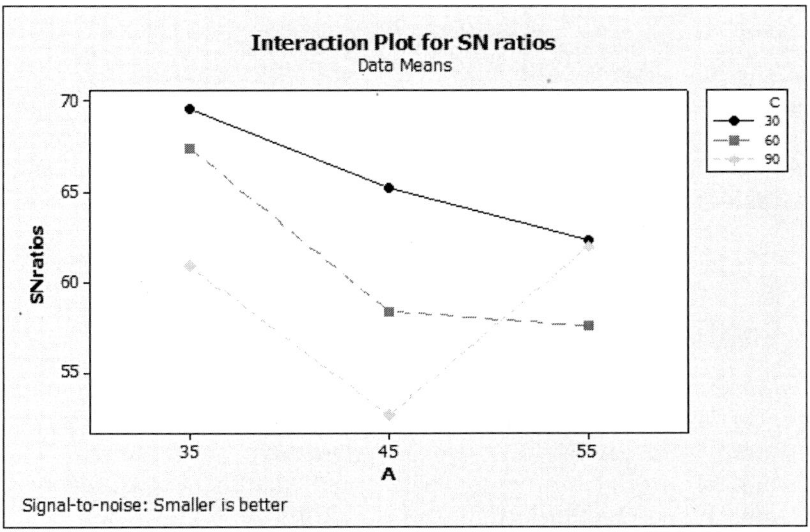

Figure 9b. Interaction graph between factor A and factor C (A×C) for erosion rate

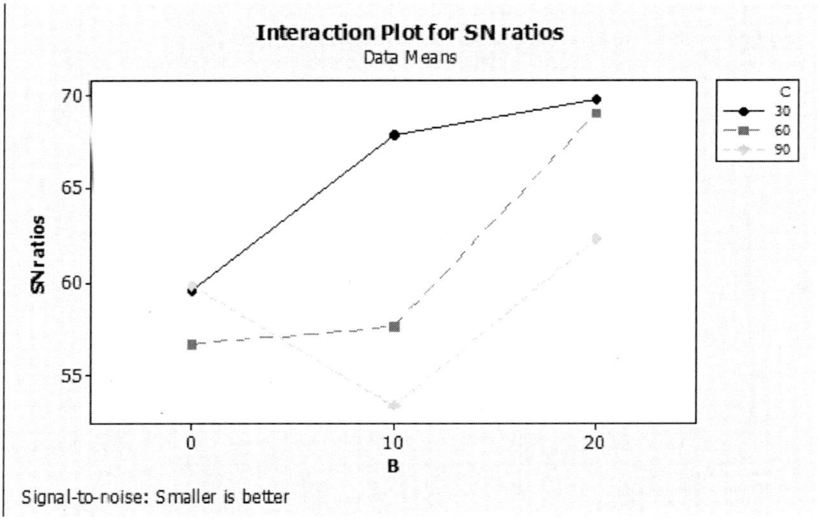

Figure 9c. Interaction graph between factor B and factor C (B×C) for erosion rate

From Table 4 it is observed that CBPD content (p = 0.120), stand-off distance (p = 0.187), impact velocity (p = 0.195) and impingement angle (p = 0.206) have great influence on the out put performance (erosion rate). The interactions of impact velocity and CBPD content (p = 0.102) has greater significant influence on erosion rate but the factor erodent size (p = 0.388), interaction between impact velocity and impingement angle (p = 0.575) and CBPD content and impingement angle (p = 0.479) have less significant contribution on erosion rate.

3.4. Confirmation Experiment

The confirmation experiment is the final step in the first iteration of the design of experiment process. The purpose of the confirmation experiment is to validate the conclusions drawn during the Taguchi experimental design process. The confirmation experiment is performed by conducting a new test condition in combination of the significant factors and their interaction levels on erosion rate as reported in Table 4. The final step is to predict and verify the improvement of the erosion resistance. The predictive value using the optimal levels of the input parameters can be calculated as:

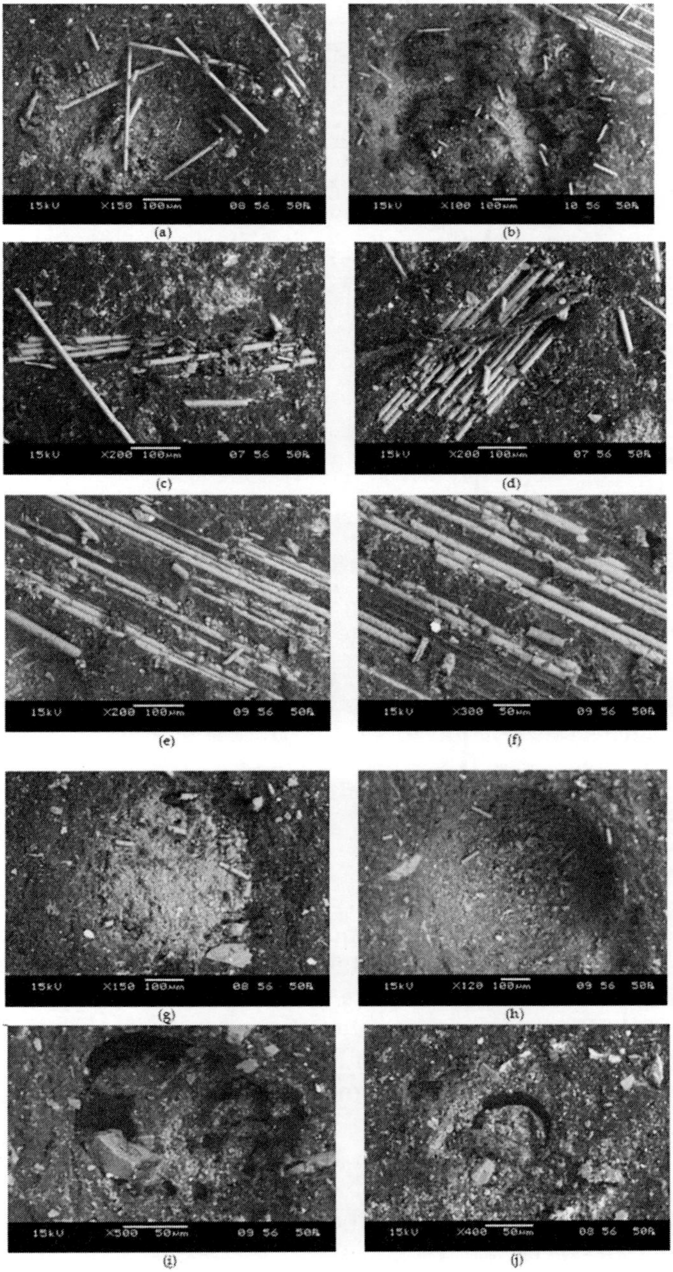

Figure 10. SEM micrographs of the eroded glass fiber-polyester composites filled with CBPD

Table 4. ANOVA table for erosion rate

Source	DF	Seq SS	Adj SS	Adj MS	F	P
A	2	252.51	252.51	126.26	2.53	0.195
B	2	377.09	377.09	188.54	3.77	0.120
C	2	240.40	240.40	120.20	2.41	0.206
D	2	262.48	262.48	131.24	2.63	0.187
E	2	121.18	121.18	60.59	1.21	0.388
A×B	4	809.70	809.70	202.42	4.05	0.102
A×C	4	163.40	163.40	40.85	0.82	0.575
B×C	4	211.27	211.27	52.82	1.06	0.479
Error	4	199.89	199.89	49.97		
Total	26	2637.93				

$$\bar{\eta}_1 = \bar{T} + (\bar{A}_2 - \bar{T}) + (\bar{B}_2 - \bar{T}) + \left[(\bar{A}_2\bar{B}_2 - \bar{T}) - (\bar{A}_2 - \bar{T}) - (\bar{B}_2 - \bar{T})\right] + (\bar{C}_3 - \bar{T}) + (\bar{D}_2 - \bar{T}) \quad (3)$$

$\eta 1$ Predicted average

$\bar{T}$ Overall experimental average

$\bar{A}_2$, $\bar{B}_2$, $\bar{C}_3$ and $\bar{D}_2$ Mean response for factors and interactions at designated levels.

By combining like terms, the equation reduces to

$$\bar{\eta}_1 = \bar{A}_2\bar{B}_2 + \bar{C}_3 + \bar{D}_2 - 2\bar{T} \quad (4)$$

After solving the above predictive equation the erosion rate is found to be and the experimental result is 51.54 dB. The resulting model seems to be capable of predicting wear rate to a reasonable level of accuracy. An error of 1.78% for the S/N ratio of wear rate is observed. This validates the development of the mathematical model for predicting the measures of performance based on knowledge of the input parameters.

4. CONCLUSIONS

1) The particle reinforced composites show good tribological properties. The erosion rate for unfilled composites 30wt% glass fiber polyester composites show better erosion resistance whereas,

for particulate filled composites 20wt% filled content shows superior erosion resistance as compared with the rest of the filled and unfilled composites.

2) The variation of erosion rate with impingement angles, the material loss is dictated mainly more at 60^0 impingement angles, in the steady state erosion rate with other control factors are remain constant. However, compared with the experimental results, the FE model (ANSYS/AUTO-DYN) is much closer to the experimental results. The major advantages of simulated results are during experimental study it is very difficult to analysis the flow direction and particularly at low impingement angle most of the erodent particles are sliding on the target material instead of reback of erodent particles. However, in finite element simulated model the above conditions can be measured easily including the residual stress and the depth of penetration which is difficult to determine by experimental method.

3) In eroded samples observed in scanning electron microscope shows mostly two types of wear mechanisms i.e. micro-cutting and micro-ploughing actions. It shows formation of lips and their flow along the direction of the erodent. At a medium angle of impingement (60°), the material loss is governs mostly by cutting action. It also depicts formation of coarser and larger lips. At a higher angle of impingement (90°), the material loss is dictated by micro-fracturing mechanism. However, CBPD particulates provide the protection against erosive wear in the composite.

ACKNOWLEDGEMENTS

The authors are grateful to the financial supports of the research project Ref. No. SR/FTP/ETA-49/08 by Department of Science and Technology, India.

REFERENCES

1. W. F. Adler, 1995Waterdrop Impact Modeling, Wear, 186-187 341 351

2. K. S. Al-Jabri, R. A. Taha, A. Al-Hashmi, A. S. Al-Harthy, 2006Effect of Copper Slag and Cement By-pass Dust addition on Mechanical Properties of Concrete, Construction and Building Materials. 20 322 331

3. G. Batis, E. Rakanta, E. Sideri, E. Chaniotakis, A. Papageorgiou, 2002Advantages of Simultaneous use of Cement Kiln Dust and Blast Furnace Slag, In: Proceedings of the International Conference on Challenges of Concrete Construction, University of Dundee, Dundee, UK.

4. S. Biswas, A. Satapathy, 2009Tribo-Performance Analysis of Red Mud Filled Glass-Epoxy Composites Using Taguchi Experimental Design, Materials & Design, 30 8 2841 2853

5. S. Biswas, A. Satapathy, 2010A Comparative Study on Erosion Characteristics of Red Mud Filled Bamboo-Epoxy and Glass-Epoxy Composites, Materials & Design, 31 4 1752 1767

6. H. A. El -Sayed, N. A. Gabr, S. Hanafi, M. A. Mohran, 1991Re-utilization of By-pass Kiln Dust in Cement Manufacture, In: Proceedings of the International Conference on Blended Cement in Construction, Sheffield, UK.

7. M. S. El Tobgy, E. Ng, M. A. Elbestawi, 2005Finite Element Modeling of Erosive Wear, International Journal of Machine Tools and Manufacture, 45 1337 1346

8. S. P. Glen, 1993 Taguchi Methods: A Hands-on Approach, Addison-Wesley, New York.

9. S. M. Hsu, M. C. Shen, A. W. Ruff, 1997Wear Prediction for Metals, Tribology International, 30 5 377 383

10. I. M. Hutchings, 1992Ductile brittle transitions and wear maps for the erosion and abrasion of brittle materials, Journal of Physics D: Applied Physics, 25 A212 A221

11. Rajesh. J. John, J. Bijwe, U. S. Tewari, B. Venkataraman, 2001Erosive Wear Behaviour of Various Polyamides, Wear, 249 702 714

12. K. Komvopoulos, D. H. Choi, 1992Elastic Finite Element Analysis of Multi-Asperity Contacts, ASME Journal of Tribology, 114 823 831

13. L. Krazowski, J. J. Emery, 1981Use of Cement Kiln Dust as a filler in Asphalt Mixes, In: Proceedings, ORF/CANMET Symposium on Mineral Fllers, Ontario Research Foundation, Toronto.

14. B. Lawn, 1993 Fracture in Brittle Solids, 2nd ed., Cambridge University Press, Cam-bridge, UK.

15. S. H. Lee, Y. I. Lee, Y. W. Kim, R. J. Xie, M. Mitomo, G. D. Zhan, 2005Mechanical Properties of Hot-Forged Silicon Carbide Ceramics, Scripta Materialia, 52 153 156

16. F. F. Ling, C. H. T. Pan, 1986Approaches to Modeling of Friction and Wear, Proceedings of the Workshop on the Use of Surface Deformation Models to Predict Tribology Behavior, Springer-Verlag, New York.

17. K. C. Ludema, R. G. Bayer, 1991Tribological Modeling for Mechanical Designers, ASTM STP1105, American Society for Testing and Materials, Philadelphia.

18. Y. V. Milman, S. I. Chugunova, I. V. Goncharova, T. Chudoba, W. Lojkowski, W. Gooch, 1999Temperature Dependence of Hardness in Silicon-Carbide Ceramics with Different Porosity, International Journal of Refractory Metals and Hard Materials, 17 5 361 368

19. A. Patnaik, R. Kaundal, A. Satapathy, S. Biswas, K. Pradeep, 2010aSolid Particle Erosion of Particulate Filled Short Glass Fiber Reinforced Polyester Resin Composites, Advanced Materials Research, 123-125 213 216

20. A. Patnaik, Abdulla, A. Satapathy, S. Biswas, B. K. Satapathy, 2010bA Study on Possible Correlation between Thermal Conductivity and Wear Resistance of Particulate Filled Polymer Composites, Materials and Design, 31 837 849

21. A. Patnaik, A. Satatpathy, N. Chand, N. M. Barkoula, S. Biswas, 2010cSolid Particle Erosion Wear Characteristics of Fiber and Particulate Filled Polymer Composites: A Review, Wear, 68 1-2 249 263

22. A. Patnaik, A. Satatpathy, M. Dwivedy, S. Biswas, 2010dWear behaviour of Plant-fiber (Pine-Bark) and Cement-kiln-dust Reinforced Polyester Composites using Taguchi Experimental Model, Journal of Composite Materials, 44 5 559 574

23. A. Patnaik, S. Tejyan, A. Rawal, 2010eSolid Particle Erosion Behavior of Needlepunched Nonwoven Reinforced Composites, Research Journal of Textile and Apparel, 14 3 12 22

24. A. Patnaik, A. Satatpathy, S. S. Mahapatra, R. R. Dash, 2010fModified Erosion Wear Characteristics of Glass-Polyester Composites by Silicon Carbide Filling: A Parametric Study using Taguchi Technique, International Journal of Materials and Product Technology (Materials Processing Technology), 38 2-3 131 152

25. A. Patnaik, S. S. Mahapatra, R. R. Dash, 2010gFlyash Filling for Improved Wear Resistance of Glass-Polyester Composites : An Experimental Study, The Journal of Solid Waste Technology and Management, 36 1 26 37

26. R. N. Rothon, 2002Particlulate Fillers for Polymers, RAPRA Review Reports, 12 9 164

27. J. L. Routbort, M. E. Gulden, E. Marshall, 1981Particle Size Distribution Effects on the Solid Particle Erosion of Brittle Materials, Wear, 71 363 373

28. R. O. Scattergood, J. L. Routbort, A. P. L. Turner, 1981Velocity and Size Dependence of the Erosion Rates in Silicon, Wear, 67 227 232

29. Ravindrarajah. R. Sri, 1982Usage of Cement Kiln Dust in Concrete, International Journal of Cement Composites and Lightweight Concrete, 4 2 95 102

30. P. Strzepa, E. J. Zamirowski, J. B. Kupperman, K. C. Goretta, J. L. Routbort, 1993Indentation, Erosion and Strength Degradation of Silicon-Alloyed Pyrolytic Carbon, Journal of Materials Science, 28 5917 5921

31. A. Suresh, A. P. Harsha, 2006Study of Erosion Efficiency of Polymers and Polymer Composites, Polymer Testing, 25 188 196

32. K. Tsuda, M. Kubouchi, T. Sakai, A. H. Saputra, N. Mitomo, 2006General Method for Predicting the Sand Erosion Rate of GFRP, Wear, 260 1045 1052

33. P. J. Woytowitz, R. H. Richman, 1999Modeling of Damage from Multiple Impacts by Spherical Particles, Wear, 233-235 120 133

4

COMPOSITES OF POLYMER HYDROGELS AND NANOPARTICULATE SYSTEMS FOR BIOMEDICAL AND PHARMACEUTICAL APPLICATIONS

Fuli Zhao †, Dan Yao †, Ruiwei Guo, Liandong Deng, Anjie Dong and Jianhua Zhang *

Department of Polymer Science and Engineering, School of Chemical Engineering and Technology, Tianjin University, Tianjin 300072, China

ABSTRACT

Due to their unique structures and properties, three-dimensional hydrogels and nanostructured particles have been widely studied and shown a very high potential for medical, therapeutic and diagnostic applications. However, hydrogels and nanoparticulate systems have respective disadvantages that limit their widespread applications. Recently, the incorporation of nanostructured fillers into hydrogels has been developed as an innovative means for the creation of novel materials with diverse functionality in order to meet new challenges. In this review, the fundamentals of hydrogels and nanoparticles (NPs) were briefly discussed, and then we comprehensively summarized recent advances in the design, synthesis, functionalization and application of nanocomposite hydrogels with enhanced mechanical, biological and physicochemical properties. Moreover, the current challenges and future opportunities for the use of these promising materials in the biomedical sector, especially the nanocomposite hydrogels produced from hydrogels and polymeric NPs, are discussed.

KEYWORDS

hydrogels; nanoparticles; nanocomposite hydrogels; hybrid hydrogels; nanocomposites; biomedical applications

1. HYDROGELS AND THEIR BIOMEDICAL AND PHARMACEUTICAL APPLICATIONS

Due to the remarkable characteristics, such as flexibility and versatility in fabrication, variety in composition, high tunability in the physical, chemical, and biological properties, high moldability in shape, especially their excellent biocompatibility and similarity to native extracellular matrix (ECM), hydrogels as promising materials are of greatest significance in the biomedical fields and have been most extensively studied in academic and industrial research [1,2,3,4,5,6,7,8]. Hydrogels are a class of materials that present a three-dimensional (3D) network formed from hydrophilic homopolymers or copolymers crosslinked to form insoluble polymer matrices, which are able to absorb a large amount of water or biological fluids. Moreover, some "intelligent" hydrogels could undergo reversible and significant changes in structure, shape, and/or property after being exposed to external physical/chemical/mechanical stimuli, such as pH value, temperature, ionic strength, enzymatic activity, glucose concentration, light, electric field, magnetic field, pressure and solvent composition or a combination of them [9,10,11,12]. Therefore, these stimuli-sensitive are expected to contribute significantly to the exploration and development of next generation of biomaterials for biological and biomedical applications, such as self-regulated and site-specific drug delivery systems, specialized separation systems or bioreactors.

Since Wichterle and Lim first reported a hydrogel for contact lens application in 1960 based on crosslinked poly-2-hydroxy-ethylmethacrylate (PHEMA) hydrogel by use of ethylene glycol dimethacrylate (EGDMA) [13], significant progress has been achieved and a diverse range of polymers have been used for the synthesis and fabrication of hydrogels for various applications. Besides the PHEMA, a great number of synthetic polymers were explored for preparation of hydrogels by polymerization or cross-linking, such as poly(ethylene glycol) (PEG), poly(vinyl alcohol) (PVA), polyacrylic acid (PAA), poly(2-hydroxypropyl methacrylamide) (PHPMA), polyvinyl pyrrolidone (PVP), polyurethanes (PU), polyacrylamide (PAM), poly(dimethylaminoethyl methacrylate) (PDMAEMA) and poly(*N*-isopropyl acrylamide) (PNIPAM) as well as polypeptides and polyesters. The polymeric hydrogels from synthetic sources are advantageous, because they usually possess controllable chemical structure and architecture, degradation rates, and mechanical strengths. However, most of them lack biological cues for biological applications, especially for proliferation of cells and tissue regeneration [2,3,4,6]. Natural hydrogels derived from natural sources, such as chitosan (CS), hyaluronic acid (HA), gelatin (GEL), collagen (COL), alginate (ALG), elastin (ELA), heparin (HEP) and chondroitin sulfate (CRS), are appealing for biological and biomedical applications because of their cell-interactive properties, cell adhesion and cell signaling, as well as biodegradability and excellent biocompatibility [2,3,4,8]. However, their uses are often restricted due to uncontrollable degradation and structure, low mechanical properties, potential immunogenicity and high batch-to-batch variations [2,3,14]. The well designed and tailored hydrogel materials can be obtained by the combination of natural and synthetic polymers, which have resulted in a shift towards the use of hydrogels based on semisynthetic polymers for biological and biomedical applications, such as P(PEG-co-peptides), ALG-g-(PEG-PPO-PEG), P(PLGA-co-serine), COL-acrylate, P(HPMA-g-peptide) and HA-g-PNIPAM [2,3,6,14].

Briefly, a hydrogel is a crosslinked hydrophilic polymeric network with high amount of water, and thus any techniques which are able to crosslink polymer chains can be applied to prepare a hydrogel. Gelation can take place either by chemical crosslinking (chemical gelation involving formation of permanent covalent bonds) or by physical crosslinking (physical gelation through the emergence of reversible and transient junctions). Chemically crosslinked hydrogels are commonly prepared by radical polymerization of a hydrophilic monomer in the presence of a multifunctional cross-linking agent, or condensation polymerization of monomers with multifunctionalities [3,6,15,16]. Chemical gelation can also occur through covalent bonding between polymeric chains through energy irradiation, chemical reactions or sulphur vulcanization by the addition of different chemicals (crosslinker) [3,6,15,16]. The 3D integrity of hydrogels in the swollen state can also be retained by physical crosslinking through molecular entanglements and/or secondary forces, such as hydrogen bonding, ionic or hydrophobic interactions [3,6,15,17]. Recently, the emergence of bio-orthogonal and supramolecular chemistries provides great opportunities for scientists to create multifunctional physically cross-linked hydrogel matrices [15,17]. All of these interactions are reversible, and thus they can be disrupted by changing physical conditions (pH, temperature, and ionic interactions), as shown in Figure 1. Generally speaking, chemically crosslinked gels often have much higher mechanical properties than their physically crosslinked counterparts. However, the residual organic solvents, chemical crosslinkers, and/or unreacted monomers could leach out from the hydrogels continuously and cause cytotoxicity.

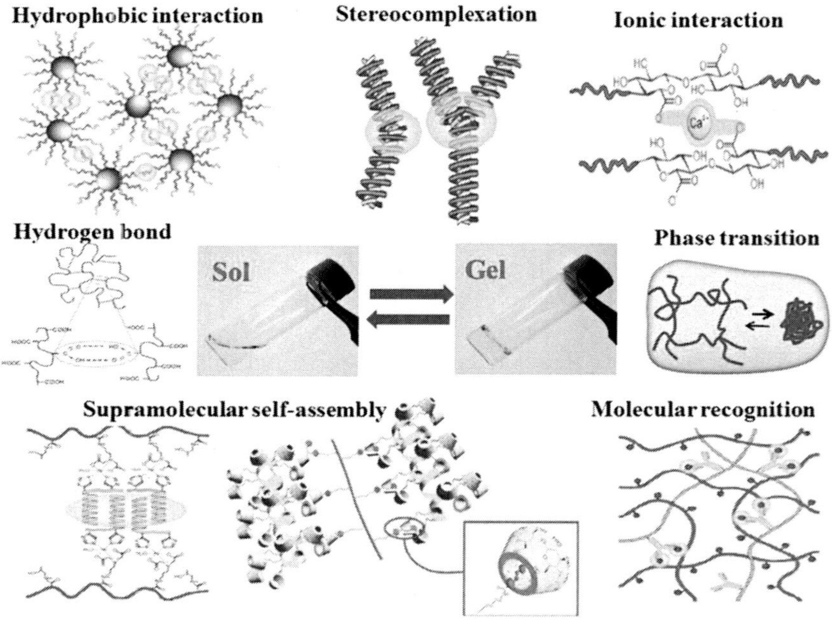

Figure 1. Schematic illustration of methods for formation of physically crosslinked hydrogels.

Hydrogels have a variety of properties [3,5,6,17], such as absorption capacity, swelling behavior, stability and degradation, bioadhesion and bioactivity, permeability, mechanical properties optical and surface properties, which make them promising materials for diverse applications. In theory, most of features can be engineered into a hydrogel system [5,17]. Biodegradability of hydrogels is essential for biomedical applications that require controlled resorption *in vivo* or local dissolution to favor cell activities and facilitate tissues reparation. Most of hydrogels are able to undergo local or bulk dissolution by hydrolysis, proteolysis or disentanglement [5,17]. One of the most remarkable characteristics of hydrogels is their highly porous structure, allow for a depot maintaining a high local concentration of drug at the targeted tissues. Therefore, the transportation capability of hydrophobic/hydrophilic molecules is very important for a hydrogel-based drug delivery system. The drug diffusion and release behavior through the polymer network can easily be regulated by manipulating the crosslinking density in the gel matrix and thus controlling the porosity (mesh size) of hydrogels [1]. For their use as scaffolds for tissue regeneration or as tissue adhesives in surgical reparation, the bioadhesion ability of hydrogels is also a significant property that allows tissues and cells to adhere to hydrogels. The hydrogels can be modified by using linker molecules that enable non-covalent or covalent molecular interactions between the gel matrix and its surroundings to improve the cell/tissue-adhesive properties [4,5]. The design of hydrogels with a good mechanical performance is of crucial importance in many biomedical application areas. In addition, the water content of a swollen hydrogel plays a key role in the biomedical applications of hydrogels, as it can significantly influence the solute diffusion and mechanical properties of the hydrogels [5,6]. In sum, the suitability and performance of hydrogels as unique materials in a biomedical application largely depend on their network architecture. The network structure of hydrogels can be characterized mainly by analyzing the polymer volume fraction in the swollen state, detecting the polymer chain length between two adjacent crosslinking points and calculating the corresponding mesh size [7,17]. The molecular structure of hydrogels can also be tuned by the selection of polymer, crosslinker and type of crosslinking as well as by the control of the crosslinking density, which offer great versatility in preparing hydrogels with tailorable network and mechanical strengths.

As mentioned above, due to the aforementioned remarkable properties, hydrogels as promising materials are highly suitable for different applications, especially for diagnostic and therapeutic applications in biomedical areas. They not only can serve as a vehicle to load and deliver drug or protein to tissues [1], but also can act as scaffolding materials to fabricate artificial tissues and organs for transplantation and serve as wound dressings, barriers or adhesives between material and tissue surfaces [4,7,8]. In addition, they are able to be applied to mimic the natural extracellular matrix (ECM) and thus offer support for cell migration, proliferation and adhesion, which can offer not only a solution for large-scale cell manufacturing and protein production, but also a platform for stem cell research and cell therapy [5]. Moreover, hydrogels can be cast into practically any shape, size, or form. And especially the stimuli-responsive hydrogels can respond to their surrounding environments, which have opened a new avenue for biomedical research in numerous leading edge applications, such as medical and biological sensors and actuators, diagnostic imaging and devices [3,5,7].

Overall, hydrogels have become a popular choice of materials for biomedical and pharmaceutical applications. However, despite the many favorable characteristics of hydrogels, problems still exist in many aspects, limiting their broad clinical applications. For example, for the hydrogel-based drug delivery system, the homogeneity and quantity of drug loading into hydrogels is often very difficult, particularly in the case of hydrophobic drugs [1]. The large pore sizes and high water content of most hydrogels often lead to relatively rapid drug release [1]. The premature drug release outside the targeted site not only decreases the therapeutic efficacy, but also causes serious local or systemic toxicity [18]. Besides, some hydrophobic drug crystal precipitation tends to occur in the hydrogels, generally causing poor formulation stability and a low drug-loading capacity [19]. Hydrogels are generally prepared from a single polymer network, which often does not hold all the required mechanical and biological properties for tissue engineering applications. They commonly tend to be too flexible to meet the mechanical demands of certain applications, leading to premature dissolution or flow away of the hydrogels from a targeted local site [1,3,6,7]. Moreover, the current hydrogels used in tissue engineering applications are still unable to replicate the complexities of the native environment, and thus often leading to unsatisfactory results. There is an increasing need for functional hydrogels to be used as biosensors, actuators and biologic devices, which are generally based on the measurement of hydrogels' changes in response to magnetic, optical and/or electrical stimuli. However, most of hydrogels made from polymers are completely inert and non-responsive under the magnetic and electrical conditions [3,7]. Apparently, each of these issues significantly restricts the practical use of hydrogels in the clinic. As a result, it is imperative to develop new materials and novel strategies for the introduction of physicochemical, mechanical, and biological functionality into hydrogels to meet the rigorous and diverse demands.

One of the efficient and simple approaches to improve the hydrogels properties and to include multiple functionalities is the incorporation of different entities into hydrogels to fabricate composite hydrogel matrices [3,20,21,22]. For example, the strategy of introducing the secondary polymers into hydrogels by physical blending or copolymerization to form hybrid hydrogels or interpenetrating network structure (IPN) hydrogels [3]. A variety of nanoparticles (NPs) have distinct physicochemical properties and favorable functions that are not commonly found in polymers. Recently, considerable efforts have been made to incorporate various kinds of NPs into gel matrices to create novel nanocomposite hydrogels with superior properties and tailored functionalities [20,22,23,24,25,26,27]. This review will summarize the most recent developments in the domain of nanocomposite hydrogels in biomedical and pharmaceutical applications, specifically highlighting some interesting discoveries in composites of hydrogel and polymeric NPs for biomedical applications.

2. NANOPARTICLES AND THEIR BIOMEDICAL AND PHARMACEUTICAL APPLICATIONS

Parallel to the aforementioned developments in hydrogels, nanoparticulate systems have gained a great amount of attention in the past several decades as one of the most promising biomedical materials, due to their unique physicochemical properties, nano-sized characteristics, controlled shape and versatile modification possibilities as well as well-defined multifunctionalities [28,29,30,31,32,33,34]. Significant progress has been achieved in preparation and functionalization of various nanoparticles for a diverse array of biomedical applications. Researchers have developed and employed a wide variety of nanomaterials, such as carbon-based NPs [35,36] (e.g., carbon nanotube, graphene, fullerene), silica-based NPs [37,38], and other inorganic NPs [30,39], semiconductors NPs [40,41], metal and metal oxide NPs [33,42] (e.g., Au [43,44,45], Ag [46] and iron oxides [47,48]), as well as liposomes [49,50,51], and polymer-based NPs (e.g., polymeric NPs, micelles and nanogels, polymersomes, dendrimer, polymer-drug conjugations, emulsion particles and protein NPs) [28,32,34]. As shown in Figure 2, these nanomaterials provide a powerful platform for site-specific and controllable delivery of drugs, gene, proteins and other bioactive agents; some of them exhibited noticeable antibacterial, antiviral and antifungal activities. Some other inorganic nanoparticles, such as gold NPs, semiconductor NPs and magnetic NPs, can be used for photoacoustic, photothermal or photodynamic as well as hyperthermal therapy, due to their unique physical properties. In addition, some functional NPs can find applications for a new generation of intelligent biosensing, bioseparation, bioimaging, cell labeling and diagnosing as well as monitoring of cells and tissues.

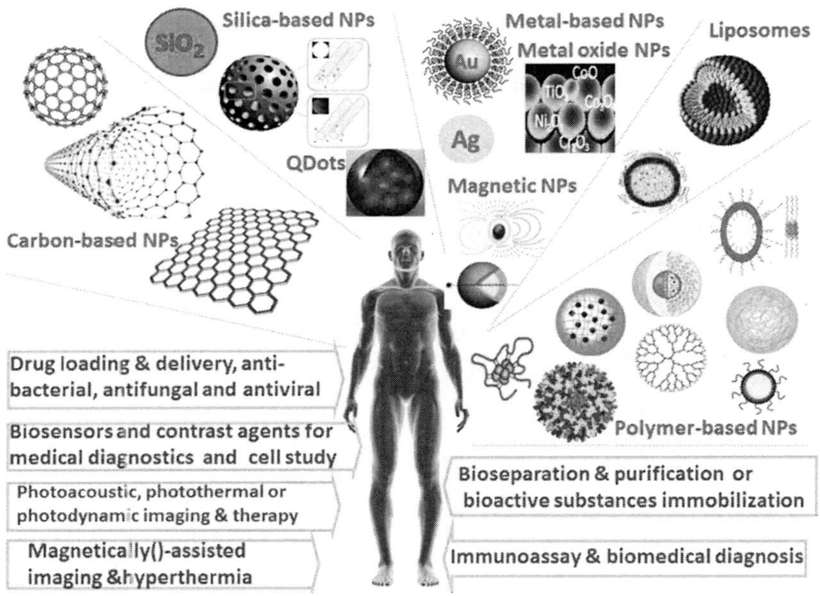

Figure 2. Examples of typical nanoparticles and their applications in biomedical fields.

NPs enter the human body by injection, inhalation, ingestion or through the skin. Once invading a biological milieu, NPs will inevitably contact with a huge variety of biomolecules in body fluids or blood, such as sugars, proteins, and lipids, which could immediately coat the NP surfaces leading to the formation of the so-called "protein corona" [52,53,54]. If the NPs can avoid the clearance by the reticuloendothelial system (RES) and/or mononuclear phagocytic system (MPS) and remain in circulation, they will reach the target tissues or cells and finally be internalized by targeted cells. Undoubtedly, NPs' transport behavior in blood, uptake and clearance by macrophages, consequent biodistribution and metabolism, as well as their internalization by targeted cells are critical for their therapeutic and diagnostic applications and are dictated by the physicochemical properties of NPs, such as hydrophilic–lipophilic property, surface feature and particle size, particle shape [52,53,54,55,56,57,58].

Generally, the biological use of NPs requires hydrophilicity for good dispersion in water or serum to impede aggregation and opsonization, and thus NPs are usually coated with hydrophilic polymers, surfactants, and biodegradable copolymers, e.g., PEG. However, hydrophobicity is also required to promote the interaction of NPs with the cellular membrane to improve uptake into cells. As a result, the anchoring of the hydrophilic ligands to the NPs surface can prolong the circulation time of NPs, but they also can dramatically suppress the interactions between NPs and cells, thus leading to a minimal internalization [32,45,53,54,59]. It was also worth pointing out that both corona formations between NPs with proteins in body fluids and interactions between NPs-bound ligands with cell membranes are influenced by the ligand type, ligand structure and ligand packing density on the surface of NPs.

Surface charge is another important factor that can definitely affect the fate of NPs administered in biological systems. Because the cell membranes have a slight negative charge and cellular uptakes are driven by electrostatic attractions, positively charged NPs, compared with NPs with a neutral or negative charge, have higher affinity with negatively charged cell membranes, leading to a faster and more efficient cellular uptake [53,60]. However, positive NPs always have serious aggregation and rapid clearance after injection due to the nonspecific interactions with blood components.

In addition of surface properties of NPs, nanomaterial size and shape also contribute significantly to their biodistribution in circulation and the clearance from the body, as well as interaction with subcellular organelles, cells, tissues, and organisms [54,55,61]. NPs, are often defined as solid particles with a size in the range of 10–1000 nm. However, the size of NPs suitable for systemic administration for therapeutic applications might be in the range between 2 and 200 nm [52,54,55,61]. Firstly, the size of NPs can affect their permeability out of the vasculature, the clearance from the body and biodistribution *in vivo*. Some previous studies have indicated that the smaller the size of NPs will lead to the less the interactions with endothelial cells and the higher the concentration in a blood vessel, as well as the longer the blood circulation time [53,54,62]. NPs with size of less than 20 nm showed wide spread in various organs by crossing the tight endothelial junctions. They were found in almost all organs, including the blood, heart, lungs liver, kidneys, spleen, testes, thymus, and brain. In general, small particles are less likely to be taken up by macrophages than large ones. However, NPs with size of less than 20 nm are rapidly excreted through the glomeruli of kidney. Large NPs tended to aggregate under physiological conditions and be retained mechanically by capillaries, which were mainly detected only in the blood, liver and spleen [53,54,62]. Generally, the ideal size of NPs for long circulation should be in the range of 20–200 nm. The size of NPs should be larger than 20 nm in diameter in order to avoid filtration by the kidney and smaller than 200 nm to avoid specific sequestration by fenestra of liver and sinusoids in spleen. In the case of accumulation of NPs in tumor tissues, the pores or fenestrations in tumor tissues are much larger (>100 nm) than the tight endothelial junctions of normal vessels (<10 nm), allowing NPs to passively accumulate inside the tumor (termed enhanced permeation and retention (EPR) effect) after extravasation. Researches have demonstrated that the capacity of NPs to navigate between the tumor interstitium decreased with increasing size [58]. Large NPs are very difficult to extravasate far beyond the blood vessel, whereas small NPs easily enter the tumor parenchyme through tumor vasculatures, but they remain there only transiently due to the rapid secretion from the tumor. The size of NPs is also one of the major factors determining the efficiency and mechanism of cellular uptake. NPs are internalized into the cell through various pathways, such as pinocytosis and phagocytosis. Large particles are most likely to be engulfed by macropinocytosis. Internalization of NPs with a diameter of less than 200 nm involved a clathrin-mediated endocytosis process. With size increasing, the caveolae-mediated internalization becomes more predominant [53,54,55]. The shape of NPs also plays a very important role in their *in vivo* behavior and interaction with cells. Some studies suggested that spherical NPs of similar size were uptaken significantly more efficiently than non-spherical NPs, due to greater membrane wrapping time required for the elongated NPs. However, non-spherical NPs exhibit superior properties to their spherical counterparts in terms of escaping from phagocytosis, circulating in blood and binding to the target tissues [53,54,55,58,61]. In sum, prior to the widespread applications of NPs in biomedical and pharmaceutical field, it should be very important to demonstrate how the physicochemical properties of NPs interact with subcellular organelles, cells, tissues, and organisms. And the effect of physicochemical properties of NPs on biological systems should be fully taken into consideration toward the development of the next generation of therapeutic and diagnostic NPs.

Despite the many advantages making NPs a great candidate for various biomedical applications, there remains a lot of pressing challenges for us to face in the future. In certain applications such as the development of drug delivering systems, for example, "Soft" drug carriers (including liposomes, micelles, dendrimers and polymeric NPs) often suffer from the premature release of drugs under harsh environmental conditions and incontrollable drug release rate *in vivo*. These bioorganic NPs, such as liposomes and polymeric micelles, are generally formed by spontaneous aggregations, which are dynamically unstable system owing to the weak nature of the cohesive interactions. They often displayed low resistance to dilution-induced disassembly *in vivo* and high instability during storage as well as relatively small loading capacity [63]. Compared to the above soft organic NPs, "Hard" nanomaterials (e.g., NPs originated from carbon, silica, metal and metal oxides, quantum dots (QDots), *et al.*) have many fundamental advantages. These various inorganic NPs possess plenty of intrinsic functionalities, because of the optical, magnetic, electrical and physical properties. They could be made into integrated and multifunctional systems, including stimuli-responsive release, specific tissue/cell type targeting, *in vivo*imaging for diagnosis, photothermal treatment, and drug delivery monitoring, which are normally unable to be found in the above "Soft" nanomaterials [30,35,43,46,47]. Besides, "Hard" inorganic NPs are generally consisted of a three-dimensional arrangement with linked atoms mainly via covalent or metallic bonds. And therefore these particles are highly ordered and extremely stable in circulation and during storage [30,35,43,46,47]. However, compared with the polymeric NPs, these inorganic NPs often suffer from a very low biocompatibility and a very high foreign body response or inflammatory response [39,43,45,59]. In fact, as for the biomedical applications of most of nanomaterials, the toxicity toward normal tissues or non-targeted cells is the most formidable obstacle to be overcome. In addition, to exert its pharmacological effects, NPs administrated must be remained in the circulation long enough and reach the required concentration. Therefore, another main challenge toward designing NPs for biomedical applications is how to avoid rapid clearance from circulation by MPS or RES after administration. It was well known that, once entering blood circulation, NPs will be rapidly coated with serum proteins and then cleared through renal systems or sequestered in the liver or spleen [53,54,58,62]. In order to enhance the biocompatibility and evade the opsonization process, various biomolecules and ligands were anchored by physical adsorption or chemisorption on the surface of NPs, especially for inorganic NPs. The functionalized NPs displayed higher colloidal stability, improved biocompatibility as well as anti-opsonization activity under physiological conditions [45,59]. However, the physicochemical characteristics of the particle can drastically change with even the slightest changes in the morphology and surface of NPs arising from functionalization process, significantly impairing their functionality and performance the high cost. Moreover, the complicated synthesis steps and low stability as well as harsh reaction condition and tedious purification process limit their large-scale production. Therefore, there is still an urgent need to develop a couple of simple and efficient strategies to functionalize NPs in order to foster the successful applications of NPs for diagnostic and therapeutic purposes.

3. COMPOSITES OF HYDROGELS AND NANOPARTICLES

3.1. Nanocomposite Hydrogels: A Brief Overview

As mentioned above, because of their unique structures and properties, the three-dimensional hydrogels and NPs have shown a very high potential for medical therapeutic and diagnostic applications. However, some inherent limitations of hydrogels and NPs inhibited their widespread applications. For example, with the rapid development of biomedicine, there is an increasing demand for multifunctional hydrogel with enhanced mechanical property. However, the limited functionalities and inferior mechanical properties of hydrogels have hindered their wide applications. As for NPs, they have plenty of unique multifunctionalities for biomedical applications, but the high instability, low biocompatibility and opsonization activity still remain significant clinical challenges. To meet increasingly rigorous requirements for biomedical use, various strategies have been explored to surmount the drawbacks but meanwhile maintain the advantages of hydrogels and NPs. Among them, the design and fabrication of novel biomaterials with desirable functions and excellent performances by innovative combination of these two entirely different types of materials has gained enormous attention [20,22,23,24,25,26,27,64].

Composite materials are materials which consisted of two or more constituent materials with significantly different physicochemical properties. When two or more components are combined in a single entity where the individual components remain distinct and separate, the resultant composite materials can exhibit markedly different characteristics from the individual components. Originally, the advantages of fabrication of composite materials by incorporating a second component into the main matrix are usually for the purpose of strength enhancement and reinforcement. For example, mud and straw were combined to form bricks several thousand years ago. Concretes are actually a composite of stones in a matrix of cement, which have been widely used for buildings. Recently, to address the specific biological and clinical challenges, the incorporation of nanostructured fillers into hydrogels has been developed as a popular means for the creation of novel materials with diverse functionality. The resulting nanocomposite hydrogels, also known as hybrid hydrogels, can be defined as composite material that physically or covalently incorporates nanosized particles or nanostructures into a matrix of cross-linked polymer networks. These nanocomposite hydrogels have found fascinating interest for various biomedical applications due to combining the characteristics of a nanoparticulate material (e.g., very small size and high surface area) with hydrogel system (e.g., hydrophilicity and extremely high water content) [64]. As shown in Figure 3, the rigid NPs and flexible polymers can be looked as bricks and mortar. The polymer mortar connects the bricks to maintain the structural integrity and offer stability to the NPs; meanwhile NPs bricks render new functionalities to the composites. As a result, the challenges in the application of hydrogels and NPs, as mentioned above, could potentially be overcome by the fabrication of nanocomposite hydrogels. For example, the main aim of incorporation of NPs into hydrogel networks in the early attempts was to improve mechanical properties of hydrogel [65,66]. Usuki *et al.* significantly improved the tensile strength of hydrogels by incorporating montmorillonite, a type of natural silicate mineral ("clay") NPs, into nylon-6 network [65]. The inorganic clay NPs are also rendered hydrophilicity and solubility when they are dispersed into 3D polymer networks, which can meet the essential requirements for the reasonable wound dressing with some desirable characteristics [67]. In recent reports, nanocomposite hydrogels prepared by incorporation of silica NPs with hydrogels showed remarkable improvements in mechanical stiffness, tissue stickiness and bioactivity compared with unmodified hydrogels [68,69]. In addition, the incorporation of metal NPs into hydrogel networks will impart superior physicochemical properties absent in the individual components. For example, incorporation of gold or silver NPs into a polymer matrix will render excellent antimicrobial property [70,71]. Similarly, the entrapment of magnetic NPs into polymer networks can make hydrogels exhibit magnetic character. Meanwhile, the NPs embedded in hydrogel matrix are endowed with particular surface properties, which can avoid aggregation and sedimentation [72]. In sum, the innovative combinations of hydrogels and NPs generate not only structural diversity but also a plurality of property enhancements [64].

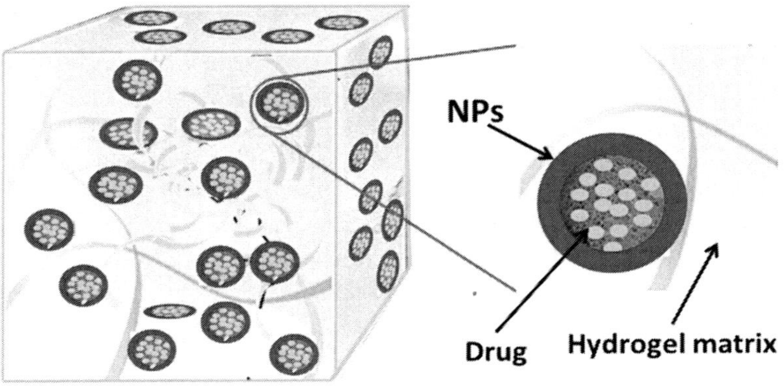

Figure 3. Schematic illustration of typical nanocomposite hydrogels from hydrogels and drug-loaded nanoparticles (NPs). NPs were non-covalently or covalently immobilized in a hydrogel matrix.

Apparently, the fabrication of composite materials of hydrogels and NPs provides an efficient route to improve their performances and expand their application scopes. A range of NPs such as carbon-based nanomaterials (carbon nanotubes (CNTs), graphene, fullerene), inorganic/ceramic NPs (silica, silicates, clay, hydroxyapatite, calcium phosphate, quantum dots (QDots)), metal/metal-oxide nanoparticles (gold, silver, copper, platinum, iron oxides and titanium oxides) and organic NPs (polymer NPs, micelles, polymersomes, dendrimers and liposomes), have been integrated within the hydrogel networks to obtain nanocomposite hydrogels with tailored functionality and superior physicochemical and biological properties. As the definition suggested, nanocomposite hydrogels can be made through the combination of NPs and hydrogels via various mechanisms. Recently, Thoniyot *et al.* [27] summarized the main approaches to prepare nanoparticle-hydrogel composites. The simplest method to prepare a nanocomposite hydrogel is the gelation of a suspension of pre-ready NPs in a solution of hydrogel-forming monomer. For example, Sershen *et al.* report a gold (Au) NPs- acrylamide hydrogel composites can be prepared by adding pre-formed Au NPs into a solution of acrylamide monomers and then initiating the polymerization by addition of the gelation initiator [73]. This approach has been widely used to prepare a variety of nanocomposite hydrogels containing various NPs. However, some drawbacks limit its wide applications, such as the aggregation of NPs in monomer solution before and during the gelation process, and the leaching of NPs out of the hydrogel matrix if the cross link density is low [74,75]. Another common approach to obtain nanocomposite hydrogels is to physically introduce NPs into a hydrogel networks after gelation. Some hydrogels, such as PAM hydrogel can highly swell in water but dramatically shrink in an aprotic solvent acetone. NPs can be introduced into this kind of hydrogels in the repeated swelling-shrinking process [76,77]. Typically, the suspended NPs in water will be embedded in the swelling hydrogel when the shrunken gel in acetone was placed in an aqueous solution of NPs. After washing out the weakly surface-adsorbed NPs, the nanocomposite hydrogel was obtained [76,77]. The NPs can be introduced into hydrogel matrix and remain attached within the gel, possibly owing to the physical entanglement and/or hydrogen bonding interactions between the polymer chains and NPs surface. Alternatively, the incorporation of NPs into pre-formed gels could also be achieved by repeated heating, centrifugation and redispersion according to similar mechanism [78]. The NPs can also be created *in situ* after gelation of the hydrogel and subsequent conversion of the nanoparticle precursors suspended in the gel [79]. In fact, the porous structures of hydrogel networks are very suitable for *in situ* formation of NPs, as the free space among the gel network provided a nanoscopic pot for formation and growth of NPs without aggregations [20,27]. Apparently, this method involves loading NPs precursors into the hydrogel networks, rather than preformed NPs. In a typical procedure, Au (III) or Ag (I) ions were embedded PNIPAM or PAA hydrogel networks and then the Au or Ag ions functionalized polymer hydrogels were reduced or hydrolyzed to yield metal NPs-hydrogel composites [79]. One particularly innovative approach to yield NPs-hydrogel composites is the use of crosslinker groups on the outer surface of NPs to form the hydrogel matrix [80,81]. This concept was well demonstrated by Rose *et al.* for adhesion between two hydrogels [81]. A rapid and strong adhesion between two hydrogels can be achieved by adding a droplet at room temperature of silica NPs solution on the surface of the hydrogels, as shown in Figure 4. This method was dependent on the ability of NPs to act as a connector between polymer chains or to be adsorbed onto the polymer gels. Finally, the nanocomposite material can also be obtained by using the interactions among NPs, polymers, and distinct gelator molecules to form positively charged polymers into NPs-hydrogels composites [82]. These strategies for the fabrication of nanocomposite hydrogels bearing tailored functionality have opened up new possibilities in fabricating advanced biomaterials for various biotechnological and biomedical and applications. Different types of NPs-hydrogel composites and their associated properties and applications will be described below.

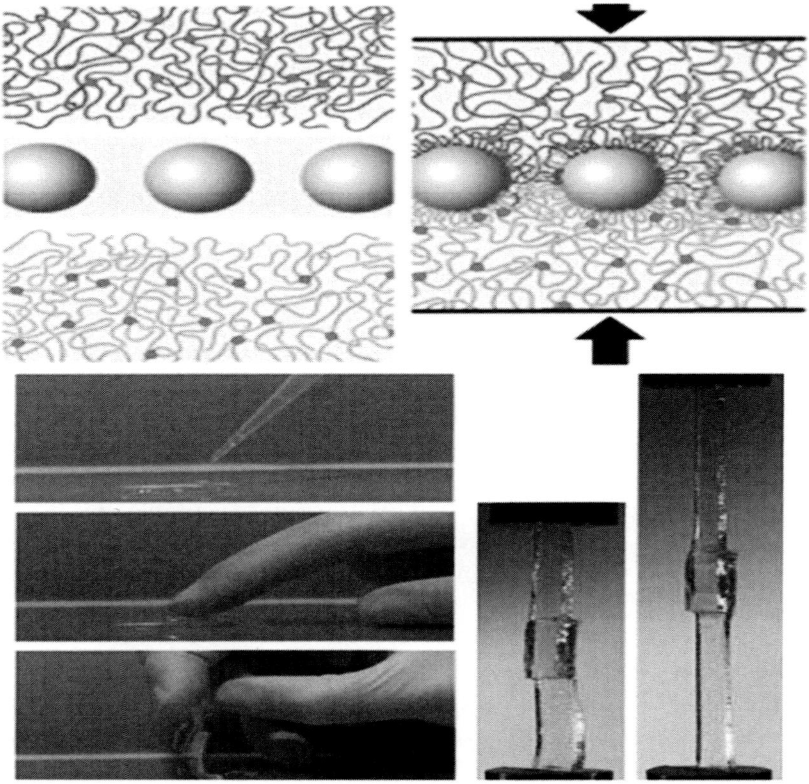

Figure 4. Schematic illustration of the concept of gluing swollen polymer networks together using silica NPs, reproduced with permission from [81]. Copyright Nature Publishing Group, 2014.

3.2. Nanocomposite Hydrogels from Hydrogels and Carbon-based NPs

Hydrogels, generally prepared from a single or multiply polymer networks, often lack all the required mechanical properties for tissue engineering applications. And they are also short of functions (such as electrical, optical and thermal conductivity) for the applications of controlled drug delivery or biosensing and bioimaging. Therefore, some strategies have been employed to improve the properties of hydrogels, mainly by inclusion different entities into hydrogel matrices to create composite hydrogels [3,24,25,27]. In the past two decades, carbon-based NPs, such as carbon nanotubes (CNTs), graphene and fullerene (C_{60}) have gained worldwide popularity, due to their excellent multifunctional feature and unique physicochemical properties (*i.e.*, mechanical strength, corrosion resistance, low cost and friendly to the environment, especially thermal, electrical, and optical conductivity). As a result, extensive research efforts are reported to utilize the favorable and unique properties of carbon-based NPs to create composite materials for various applications, including sensors and actuators, photothermal and photodynamic therapy, diagnostic imaging and drug delivery devices [36,83,84]. For example, due to their remarkable mechanical properties, carbon-based NPs have been widely used to incorporate into hydrogels to improve their physical and mechanical properties for a variety of tissue applications [85,86,87,88,89, 90,91,92,93,94]. A great number of studies have demonstrated that a small amount of carbon-based NPs, such as CNTs or graphene, in the polymer matrix can markedly change its mechanical, electrical and thermal properties [87,88,89,90]. However, the key problem of the practical use of CNTs is their poor dispersibility in water, organic solvents and polymer matrix [87,88,89,90,95,96]. Generally, some pretreatments, such as functionalization with strong acid or surface modification with surfactants, polymers or proteins, are needed to improve the dispersion of CNTs [25]. Among various carbon-based NPs, graphene-based materials (including graphene, graphene oxide (GO) and reduced graphene oxide (rGO)) have been extensively explored as a novel reinforcing agent to support polymers for biomedical applications, due to their distinctive properties, such as large surface area (2620 m^2/g), high aspect ratio, outstanding mechanical properties (Young's modulus of 1 TPa and intrinsic strength of 130 GPa), excellent thermal and electrical conductivity as well as good optical transparency [36,84,97,98]. However, graphene is neither lipophilic nor hydrophilic, often causing a high difficulty in their processing and low compatibility with polymer matrices [97]. In contrast, GO, a single monomolecular layer of graphite with various oxygen containing groups such as epoxide, carbonyl, carboxyl and hydroxyl groups, is amphiphilic and thus can be dispersed in water and alcohol through hydrogen bonding. This will significantly facilitate molecular dispersion of GO in polymeric matrix and simplify the processing of GO-based composites. As a result,

GO as a reinforcement agent can be directly introduced to into polymer matrices by solvent blending method. Nevertheless, as shown in Figure 5, the stable inclusion of graphene in a polymer matrix often requires pretreating graphene (such as surface and structural modifications) or post-treating hybrid hydrogel (such as crosslinking) [91,97,99]. Undoubtedly, the incorporation of graphene-based materials into hydrogel has proved extremely valuable for the improvement of the mechanical and thermal properties of the hydrogels. However, it was worth pointing out that the embedment of graphene-based materials into a hydrogel matrix is often indispensable, especially when these graphene-based materials were used in biomedical and biological fields. Normally, the graphene sheets have sharp edges, which can assist penetration through cell membranes. However, the sharp edges will also markedly enhance the cytotoxicity [100]. The confinement of graphene in a water-rich environment of hydrogels can change its interaction with cells and thus offer a solution to prevent this effect [91].

In addition, hydrogels commonly made from non-conductive polymers are not conductive towards optical, thermal and electrical stimuli. However, there is an increasing need for conductive hydrogels in the biomedical applications, such as biosensors, actuators and neural tissue engineering. Therefore, carbon-based NPs with excellent optical, thermal and electrical conductivity are considered as highly promising materials to engineer conductive stimuli-responsive hydrogels [3,83,84,95]. A variety of optical or electrical-responsive nanocomposite hydrogels as actuator, biosensor or tissue engineering scaffold have been developed by incorporating with carbon-based NPs [91,92, 101,102]. Overall, due to the fact that the carbon-based NPs often possess highly desirable functions that cannot be commonly found in polymers, the strategies of integrating carbon-based NPs into polymer hydrogels to develop composite hydrogels with enhanced mechanical strength and specialized properties (e.g., electrical, thermal and optical conductivity) are expected to be extensively explored for biomedical applications. And these and related topics have been well covered in several excellent reviews [3,85,95,97]. And therefore, as shown in Table 1, we just summarized the latest advances of experimental research on nanocomposite hydrogels from hydrogels and carbon-based NPs for biomedical and pharmaceutical applicants.

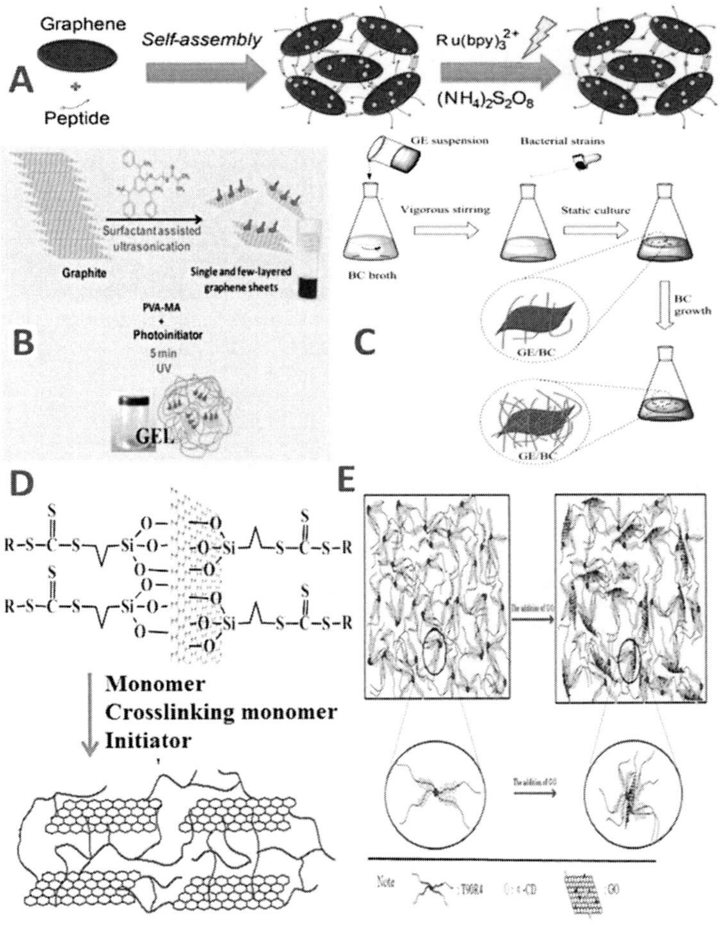

Figure 5. Schematic illustrations of incorporating graphene-based materials into hydrogel matrixes; The formation of graphene-hydrogel composite materials by (**A**): self-assembly and photo-crosslinking between graphene and peptide, reproduced with permission from [99]. Copyright Royal Society of Chemistry, 2015; (**B**): photopolymerization of vinyl moieties present on surface of graphene and polymer chains, reproduced with permission from [91]. Copyright American Chemical Society, 2015; (**C**): *in situ* biosynthesis, reproduced with permission from [103]. Copyright Royal Society of Chemistry, 2014; (**D**): surface-grafted polymerization, reproduced with permission from [104]. Copyright American Chemical Society, 2012; (**E**): supramolecular self-assembly by hydrogen bonding, reproduced with permission from [102]. Copyright Royal Society of Chemistry, 2015.

Table 1. Summary of the most recent advances of nanocomposite hydrogels from hydrogels and carbon-based NPs for biomedical applications.

Fillers	Hydrogels	Applications and Functions	Ref.
Carbon Nanotubes (CNTs)	PVA	Drug delivery; Incorporation of CNTs caused a significant increase in tensile strength, decrease in elongation at break, and increase in Young's modulus of the PVA hydrogels; Improved the electrical conductivity and antibacterial activities of the hydrogel systems; CNTs imparted to hydrogels the ability of electrical stimulation-controlled drug release	[105,106]
	PHEMA	Tissue engineering; CNTs were added to improve electrical conductivity and mechanical properties; CNTs makes hydrogel more suitable as nerve conduits	[87]
	PNIPAM	Tissue engineering; Incorporation of CNTs could improve the bioactivities, especially supportive adhesion of PNIPAM to encapsulated cells and favor their efficacy in myocardial repair	[88]
	PAM	Biomaterials; CNTs can significantly improve the mechanical properties and swelling behavior of PAM hydrogels	[89]
	P(AM-co-EBA)	Drug delivery; CNTs was covalently included into films and acted as a functional element for both improving the electro-responsivity and modulating the release of anionic and cationic drugs	[90]
	P(EGDMA-co-HEMA)	Tissue engineering; CNTs can improve the swelling and mechanical properties	[107]
	P(VPB-co-DMA)	Biomaterials; CNTs could increase storage moduli of the hydrogel and endow the hydrogel with self-healing property	[96]
	MPEG/α-CD	Drug delivery; CNTs improved the hydrophobic drug loading amount; MPEG not only dispersed CNTs in aqueous solution, but also functioned as hydrogel matrix by interacting with α-CD	[108]
	GG-g-AA	Drug delivery; CNTs can increase the hydrophobicity of hydrogel, ensure the high environmental stability and drug retention properties of hydrogel	[109]
	BC/SA	Drug delivery; CNTs can be used to obtain the composite hydrogel with the highest electric sensitivity; Drug release from hybrid hydrogels showed a significant pulsatile characteristic	[110]
	DNA	Tissue engineering; Incorporation of CNTs significantly strengthens the mechanical properties of DNA hydrogel; DNA improved the biocompatibility of CNTs and the resulting hybrid hydrogel	[111]
	PEI-Fc	Drug delivery; CNTs can improve the electrical sensitivity properties of hydrogel and can be used to tune the stiffness and degree of swelling	[112]
	GEL	Tissue engineering; CNTs supported the contractile property of engineered cardiac tissues, improved the formation of gap junction and enhanced the excitation-contraction coupling	[113]
	CS	Antimicrobial biomaterials; The addition of CNTs in the composite hydrogel scaffold significantly reduced the water uptake ability, increased the antimicrobial activity and decreased the toxicity	[114]

Fillers	Hydrogels	Applications and Functions	Ref.
Carbon Nanotubes (CNTs)	COL	Tissue engineering; CNTs enhanced the proliferative potential of the cells; Hydrogel offers excellent 3-D conditions for the growth of cells; CNTs within the hydrogel significantly stimulate cell secretion of neurotrophic factors and expression of neural markers	[115]
	Ambidextrous gelators	Biomaterials; The inclusion of SWNTs within the ambidextrous gels improved the mechanical rigidity of the resulting soft nanocomposites	[116]
	Peptide	Tissue engineering; Peptide enables the dispersion of the CNTs in aqueous media; CNTs make hybrid hydrogels have superior physical and mechanical properties	[117]
	Lysozyme	Drug delivery and tissue engineering; CNTs contributed to gelation, improved mechanical properties and improved the biological activity of hydrogels	[118]
Graphene	PAA	Drug delivery; Graphene/PAA hydrogel was prepared with pH-sensitive, good thermal, strong mechanical and excellent elastic properties	[119]
	PMAA	Drug delivery; The graphene hybrid hydrogels with enhanced mechanical, electrical, and thermal properties were prepared by *in situ* free radical polymerization, which exhibited controlled drug release in a pulsatile fashion upon the ON/OFF application of low electrical voltages	[120]
	PVA/P(MA-co-NIPAM)	Drug delivery; Small amounts of embedded graphene affect the behavior of mechanical and thermal properties of the hydrogel without altering its hydration state; The entrapment of graphene increase the possibility for functionalization and its dispersibility in water	[91]
	Bacterial cellulose	Tissue engineering; Graphene was used to reinforce various hydrogels, leading to excellent mechanical, electrical and biological properties; Graphene also changed the crystallinity of bacterial cellulose	[103]
	CS/GEL	Neural tissue engineering; Graphene improved the electrical conductivity and mechanical properties, but decreased the porosity, swelling ratio and *in vitro* biodegradability	[92]
	β-CD /PDMA	Drug delivery; β-CD endowed the self-healing property of hydrogel as anticancer drug carrier, more drug loading and better drug release behavior	[121]
Graphene Oxide (GO) & reduced graphene oxide (rGO)	PEG	Drug delivery; PEG endowed GO high aqueous solubility and stability in physiological solutions; GO imparted to hydrogel superior loading capacity for hydrophobic drug	[93]
	PLA–PEG–PLA	Drug delivery; Hydrogel can serve as a sustained delivery depot of the drugs. GO can increase drug loading capacity	[94]
	PEGDGE	Drug delivery; Enhanced mechanical and electrical properties were observed with increased rGO content; Hydrogel can greatly improve the stability of rGO dispersions	[101]

Fillers	Hydrogels	Applications and Functions	Ref.
Graphene Oxide (GO) & reduced graphene oxide (rGO)	PAA/GEL	Tissue engineering; GO worked as reinforcement agent can significantly increase the tensile strength (71%) and elongation at break (26%) of composite hydrogels	[122]
	GEL	Drug delivery; GO worked as physical crosslinker, and the multiple crosslinking sites between gelatin chain and GO nanosheet rendered GO–gelatin hybrid hydrogels high mechanical performance; Drug release behaviors from these hydrogels could be manipulated by controlling the crosslinking density and pore sizes of the hydrogels	[123]
	α-CD/PEO-PPO	Drug delivery; The incorporation of GO enhanced the mechanical strength and stability of the hybrid hydrogels enormously; Hybrid hydrogels can really exhibit a sustained drug release effect	[102,124]
	Peptide	Drug delivery; Mechanical stability, syringe-injectability, low erosion rate, and near-infrared light triggered release functionalities	[99]
	KGM/SA	Drug delivery; GO as a drug-binding effector for anticancer drug loading and release; SA as the pH sensitive agent	[125]
	CS	Blood toxins adsorption; The incorporation of graphene oxide into the chitosan matrix improved both the Young's modulus and the compressive strength of the hybrid hydrogel, as well as its adsorption capacity for bilirubin	[126]
	clay-PDMA	Biomaterial dressing; GO sheets in the hybrid hydrogels acted as not only a collaborative cross-linking agent but also as a NIR absorber to absorb the NIR irradiation energy and transform it to thermal energy rapidly and efficiently, resulting in a rapid temperature increase of the GO containing gels.	[127]
Fullerene (C_{60})	PNIPAM	Drug delivery; The incorporation C_{60} improved the mechanical strength and decreased the swelling ratio and reduced the lower critical solution temperature of PNIPAM	[128]
	PHEMA	Drug delivery; Equilibrium swelling ratio is lower due to the presence of hydrophobic fullerene moieties; The glass transition temperature and thermo-stability were increased with increasing fullerene content	[129]
	GC-g-DMA	Tumor therapy; The incorporation C_{60} endowed the hydrogel an advanced property for endosomal pH targeting and an elevated photodynamic tumor cell ablation at pH 5.0 through an enhanced singlet oxygen generation at pH 5.0	[130]

Poly(acrylamide-co-*N,N'*-ethylene bisacrylamide) P(AM-co-EBA); Poly(methacrylic acid) PMAA; Glycol chitosan grafted with 2,3-dimethylmaleic acid (GC-g-DMA); Poly(*N,N*-dimethylacrylamide-4-((4-vinylbenzyloxy)carbonyl)phenylboronic acid) Poly(VPB-co-DMA); Polyethylene glycol monomethyl ether (PGME); Acrylic acid grafted guar gum (GG-g-AA); Bacterial cellulose and sodium alginate (BC/SA); Ferrocene-grafted polyethylenimine (PEI-Fc); Poly(methacryloyl-co-*N*-isopropylacrylammide) P(MA-co-NIPAM); Polylactide-poly(ethylene glycol)-polylactide (PLA-PEG-PLA); Konjac glucomannan/sodium alginate (KGM/SA); PDMAA: Poly(*N,N*-dimethylacrylamide) (PDMA); Glycol chitosan grafted with 2,3-dimethylmaleic acid (GC-g-DMA); Polyethylene glycol diglycidyl ether (PEGDGE).

3.3. Nanocomposite Hydrogels from Hydrogels and Inorganic NPs or Semiconductor NPs

In addition of carbon-based nanomaterials, some other inorganic NPs, such as silicon-based NPs and quantum dots (QDots) have been widely used to combine with the polymeric network to create composite materials with unique properties and functions. Especially, silicon-based NPs, including silica NPs (SiO_2), mesoporous silica NPs (MSNs), hydroxyapatite (HAP) and bioactive glasses as well as various kinds of clay (such as montmorillonite (MMT), laponite (LAP), attapulgite and halloysite) are of greatest significance and have been utilized as the nanofillers in nanocomposite hydrogels, largely due to their excellent biocompatibility, chemical inertia, high thermal and mechanical stability, facile surface modification and easy functionalization. Most of these silicon-based NPs or nanoclays are already present in the body or are necessary for the normal functioning of human tissues [131]. Silicon is very important in skeletal development. Silicon promotes collagen type I synthesis and also stimulates the osteogenic differentiation in human stem cells [131]. HAP is a major mineral component and an essential ingredient of normal bone and teeth and HAP used biomaterials for bone regeneration can promote new bone ingrowth through osteoconduction mechanism without causing any local or systemic toxicity, inflammation or foreign body response [132,133]. Some studies demonstrated that some nanoclays, such as MMT and LAP, can induce osteogenic differentiation of human mesenchymal stem cells in the absence of exogenous growth factors [131,134]. It was well-known that many of the currently available polymer hydrogels suffer from poor mechanical properties and inability to induce mineralization. These unique bioactive properties and high mechanical strength of the silicon-based NPs and nanoclays will be of great interest to develop nanocomposite hydrogels with excellent mechanical and biological properties for the repair and regeneration of human tissues and body functions [3]. Especially, the anisotropic and plate-like, high aspect-ratio morphology of the nanoclays, lead to the formation of high non-covalent surface interactions between the nanoclays and polymer chains. These characteristics could therefore be used to create functional hybrid hydrogels via sol-gel process, which are very suitable for minimally invasive therapies, such as injectable tissue repair matrices or bone-related tissue engineering scaffolds [135,136]. In addition, due to their large specific surface area, controllable particle morphology and pore volume or interlayer space, excellent adsorption ability, ionic exchange capacity and outstanding adhesive ability, some silicon-based NPs, especially MSNs, MMT and LAP, have superb drug-carrying capability.

They have been widely used to integrate into hydrogel drug delivery systems to increase drug-loading, manipulate drug pharmacological profiles, minimize drug degradation and decrease detrimental drug side effects [137,138,139]. We summarized the advances of nanocomposite hydrogels from hydrogels and silicon-based NPs for biomedical applications. As shown in Table 2, various kinds of silicon-based NPs have been incorporated in different synthetic and natural polymers via physical or chemical approaches to obtain bioactive nanocomposite hydrogels. These resulting hydrogels with enhanced mechanical properties, good biocompatibility and biodegradability, tunable cell and tissue adhesion, or improved drug loading and drug release profiles, have exhibited plenty of important applications in biomedical and pharmaceutical fields, such as injectable or implantable sustained drug delivery system, controlled cell and tissue adhesion surfaces, antimicrobial films or wound dressings, tissue engineering and cell based therapies. For example, in our previous study, a novel organic-inorganic silica NPs-based bioadhesive hydrogel as an alternative vehicle for transdermal drug delivery system was developed [69]. As shown in Figure 6, the direct formation of silica NPs in PVA hydrogel was observed due to the reaction between PVA and the hydrolyzed silanol of γ-(glycidyloxypropyl)trimethoxysilane (GPTMS) in the sol-gel process. The silica NPs dispersed in the matrix form hard segments in the hydrogel, which can not only significantly improve mechanical strength, skin adhesion properties and film-forming properties of PVA gels, but also can reduce the crystalline regions of PVA and hence facilitate the diffusion of drug and water vapor.

In another interesting study, Ayyub *et al.* [140] developed a nanocomposite hydrogel respond to proteolytic degradation with an increase in stiffness through the formation of physical cross-links. The nanocomposite hydrogel was consisted of a colloidal array of silica NPs distributed within a PEG-peptide hydrogel, as shown in Figure 7. The negatively charged silica NPs at a sufficiently high concentration within the hydrogel can form a regular periodic structure as a photonic bandgap to reflect visible wavelengths of light. The hybrid hydrogels would collapse in response to diverse biological stimuli, causing a 1200% increase in elastic storage and loss modulus. Moreover, the material provided threshold responses, requiring a certain extent of proteolytic activity before the transition occurred. This allowed for the fabrication of Boolean logic gates responding to a specific assortment of proteases. The study demonstrated that incorporating silica NPs at a high density within a polymer hydrogel can not only result in mechanically strong and tissue-adhesive nanocomposite hydrogels, but also fabricate soft materials with a unique and counterintuitive response to environment-stimuli. The protease responsive hydrogel nanocomposites reported in this study could offer avenues in degradation-stiffening and collapsing materials for a variety of biomaterial applications.

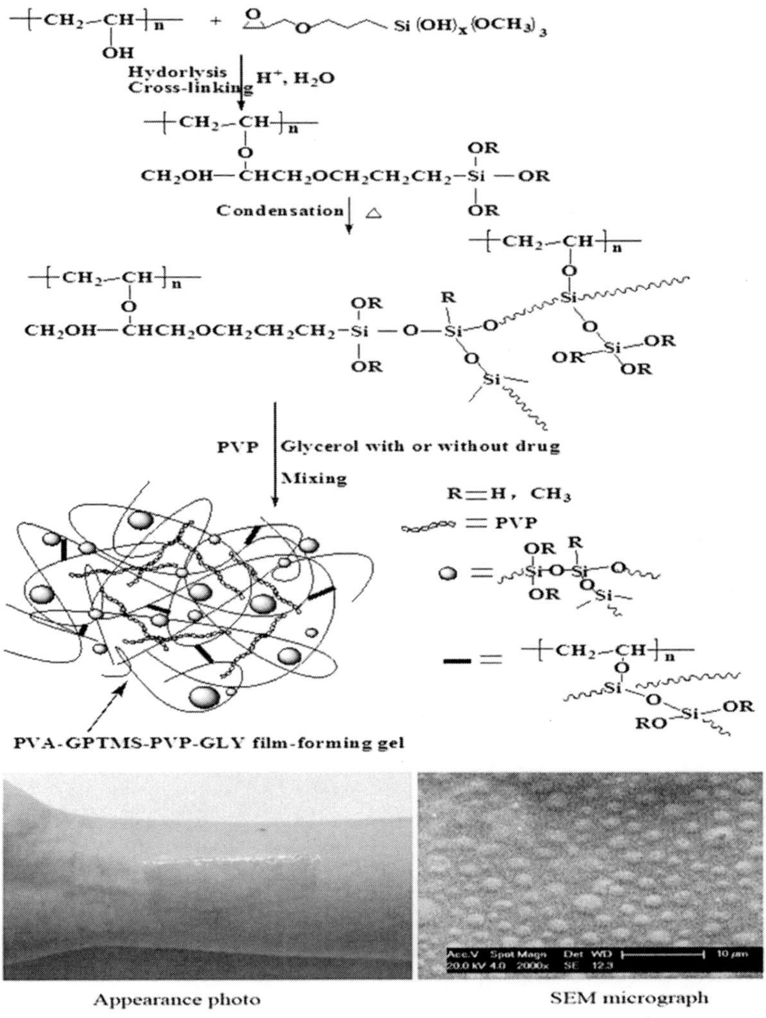

Figure 6. Illustration of the preparation of PVA-GPTMS-PVP-GLY film-forming gel and the appearance photo and SEM micrograph of the resultant gel, reproduced with permission from [69]. Copyright Elsevier, 2011.

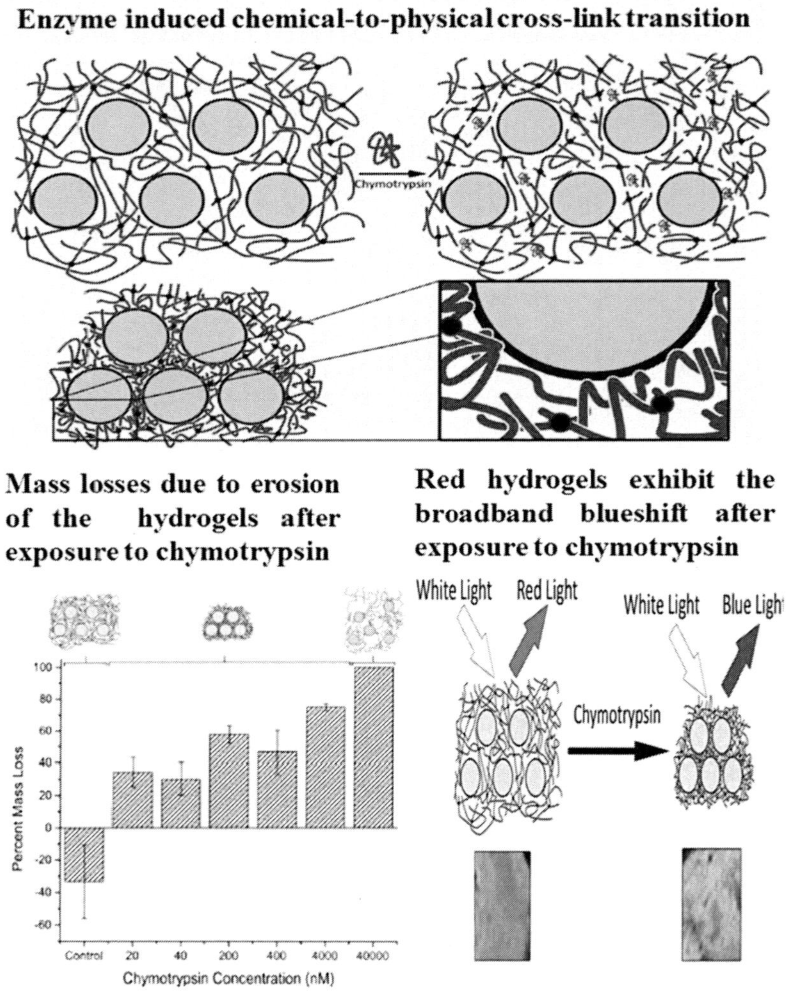

Figure 7. Depiction of the enzyme induced chemical-to-physical cross-link transition and subsequent mass loss and blue shift of nanocomposite hydrogels after exposure to chymotrypsin, reproduced with permission from [140]. Copyright American Chemical Society, 2015.

Among various inorganic NPs, semiconductor quantum dots (QDots), such as CdTe, CdS, carbon and silicon QDots, have attracted tremendous interest due to their unique photoluminescence behavior, including high quantum yield, large absorptivity, broad absorption spectrum, improved photo-stability, concentration and size-tunable emission [141,142,143,144,145,146,147]. These fluorescent QDots have been extensively studied and applied, such as imaging fixed and live cells all the way, high resolution localization of proteins and cellular structures, bioimaging *in vivo*, light-triggered drug delivery, immunoassays and innovative diagnostic methodologies [148,149]. For example, cells labeled with QDots with unique photoluminescence can be injected in small animals, which can be used to follow a particular pathway in the organism, such as the carcinogenesis process. However, due to high free energy of the surface and a large surface-volume ratio, QDots tend to easily aggregate in aqueous medium, resulting in fluorescence quenching. Therefore, the dispersity and stability of QDots are extremely important for their biomedical applications, such as biosensing, bioimaging, biolabeling and drug delivery. Many techniques, especially the surface functionalization by the attachment of ionic groups, polymers, lipids and proteins, have been employed to stabilize the colloidal QDs, but these approaches suffer from the complicated synthesis steps, harsh reaction condition and tedious purification process [29,141]. On the other hand, as described above, hydrogels have exhibited great potential in biomedical applications due to their unique physicochemical properties and porous network structures. In fact, hydrogels can also provide an ideal hydrated environment for the host of QDots. Moreover, the integration of fluorescent QDots into hydrogel matrix can not only provide protection to QDots against different harsh environments, but also render to the resulting composite hydrogels with unique photo-electronic properties that are highly suitable for therapeutic and diagnostic applications in biomedical fields [142,145,148,149,150,151,152]. Fabrication of fluorescent hydrogel by incorporating QDots in the polymer hydrogel matrix has achieved promising success. For example, QDots-loaded PNIPAM nanocomposite hydrogels have been used as light-controlled drug delivery system, thermo-sensitive biochemical monitor and sensor [144,150,153]. Generally, there are two distinct strategies for the incorporation of QDots into hydrogels, based on the covalent and noncovalent interactions between polymer chains and QDots. The former approach to QDs encapsulation involves the formation of covalent bonding, which is commonly realized by copolymerization or crosslinking of QDots carrying polymerizable ligands with proper monomers [141,146, 154,155]. This strategy can guarantee the firm immobilization of QDots in the hydrogel matrix. However, this approach usually involves tedious polymerization or synthetic process, and usually requires that the hydrogel polymers are able to be chemically tailored. An alternative method involves encapsulating the QDots by the diffusion of the QDots into the preformed hydrogel or the mixture of the QDots with the precursor solution of hydrogel [147,156,157,158]. The entrapment processes are mainly dependent on the electrostatic interactions, hydrophobic interactions, hydrogen bonding, and host–guest interactions. These noncovalent bonding interactions, obviously, can drive the free QDots to be mostly inhaled into the hydrogels, due to the large surface area of hydrogels. However, for the case of physical embedding, the weak noncovalent interactions between the networks and the QDs may result in squeezing of the trapped QDs from the networks and a loss of luminescence [141,147,157]. A wide variety of polymer hydrogels, such as PAM, PEG, PVA, PVP, COL and AGL, have been used to entrap QDs into the hydrogel matrices by means of the above strategies. Typically, Liu *et al.* [141] combined a physical embedment and a subsequent covalent crosslinking to firmly encapsulate QDots into agarose hydrogels. Cysteamine-capped QDots can be entrapped into the pores of agarose hydrogels by hydrogen bonding between hydroxyl groups of agarose and the amino groups of cysteamine. Subsequently, oxalaldehyde and polyethylenimine were then incorporated to produce a covalently crosslinked network to further stabilize the encapsulation. In another study, Zhang *et al.* [146] developed a sunlight self-initiated polymerization approach to prepare multifunctional nanocomposite hydrogels by using QDots as the initiator and introducing clay as additional non-covalent crosslinking points, as shown in Figure 8. The resulting hydrogels exhibit excellent mechanical strength and high elasticity. Moreover, the self-immobilized NPs within the functional nanocomposite gels exhibit inherent physicochemical properties, such as large absorptivity, high photoluminescence and size-tunable emission, which are comparable to those of their original solution.

3.4. Nanocomposite Hydrogels from Hydrogels and Metal NPs or Metal Oxide NPs

Hydrogels and metal or metal oxide NPs are two completely distinguished kinds of materials. Hydrogels are commonly made of natural and synthetic polymers, which possess numerous advantages over inorganic and metal materials, such as lower material cost and easier fabrication, better biocompatibility and biodegradability, and more versatile processability and functionalization. However, polymers usually display poor mechanical and thermal stability, and especially they lack some special functions, such as thermal, magnetic, optical and electrical conductivity [3]. With the increase of popularity of hydrogels in various biomedical applications including biosensing, bioimaging and biotracking/labeling, there is a growing need to engineer hydrogels with magnetic, electrical and optical properties. Metal and metal oxide NPs possess unique characters that are not commonly found in polymer materials, which have great potential as reinforcing elements to prepare composite hydrogels with unique characteristics and tunable properties [3,251]. Various types of metal and metal oxide NPs have been integrated into polymer hydrogels through a covalent or non-covalent fashion to fabricate nanocomposite hydrogels. Generally speaking, the incorporation of such NPs into hydrogel matrix by a covalent manner will lead to a relatively strong polymer-NPs interaction and thus change markedly the mechanical properties and swelling behaviors of hydrogels. On the other hand, there will be little impact on the viscoelastic and mechanical properties of the resultant hybrid hydrogels if the metallic NPs were incorporated into hydrogel networks by the weak non-covalent manner [251]. In this case, only the properties of the NPs, such as improved magnetic, optical, electrical conductivity and enhanced antimicrobial properties are added into the hydrogels. For example, the physical entrapment of Au NPs within the HA polymeric network did not improve the mechanical properties of hydrogel, due to the presence of only the weak interactions between polymers and NPs [252]. Nevertheless, significant increases in the stiffness were observed when surface-functionalized Au NPs can form covalent bonds to the polymer chains and crosslink with the polymeric networks [79,252,253].

A variety of metallic NPs, such as gold (Au), silver (Ag), platinum (Pt), cobalt (Co), nickel (Ni) and copper (Cu), and a series of metal oxide NPs including iron oxide (Fe_3O_4, Fe_2O_3), titania (TiO_2), zinc oxide (ZnO), cupric oxide (CuO) as well as metal alloys and salts with desirable physicochemical properties were combined with natural and synthetic polymer networks and then were designed for diverse biomedical and pharmaceutical applications, including biological sensing, *in vivo* bioimaging, cell separation, drug delivery, conductive scaffolds, switchable electronics, and disease treatment [3,251]. We summarized the biomedical and biological applications of nanocomposite hydrogels from hydrogels and metal or metal oxide NPs, as shown in Table 3. Due to their unique functions and relatively simple preparation process, Ag, Au and magnetic iron oxide Fe_3O_4 NPs are the most widely investigated class of NPs for the fabrication of functional nanocomposite hydrogels.

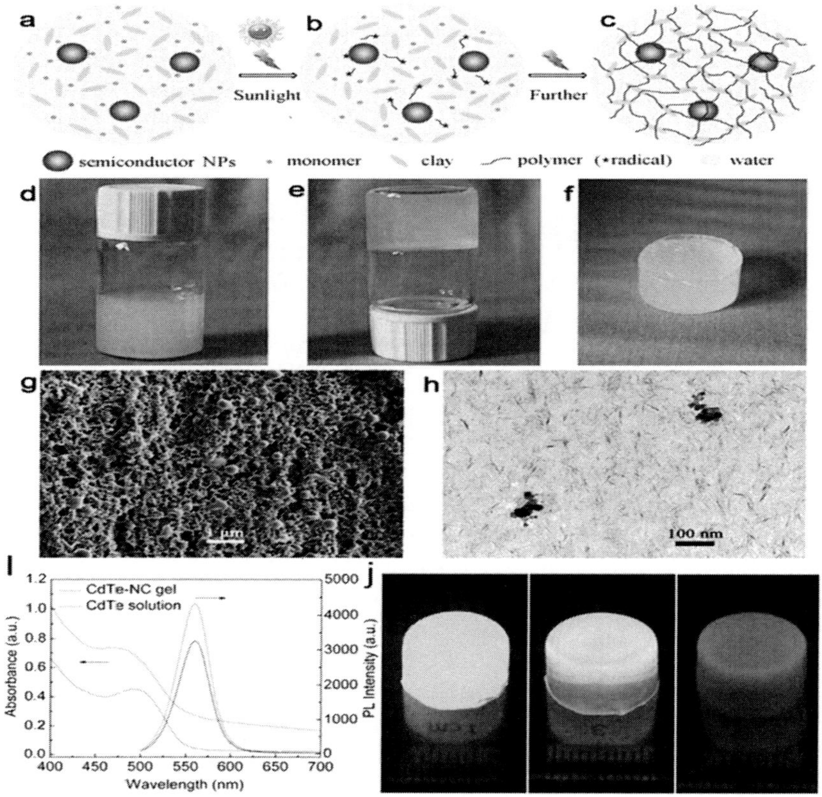

Figure 8. Proposed mechanism (**a**–**c**), photographs (**d**–**f**), and SEM image (**g**) and TEM image (**h**), UV-Vis absorbance and images under an ultraviolet lamp of the functional nanocomposite hydrogels prepared by a QDots-initiated polymerization approach under sunlight, reproduced with permission from [146]. Copyright Nature Publishing Group, 2013.

Recently, Ag NPs with unique optical, electronic, and antibacterial properties have attracted much attention and have found many applications in field of medicine and pharmaceutics. One of the most attractive features of Ag NPs is their remarkably strong broad-spectrum antimicrobial activity in combination with a fairly low toxicity against human tissue [46,254,255,256,257,258]. However, because of strong dipole-dipole attractions between particles and high surface energy, the lack of sufficient physicochemical and dispersion stability in aqueous media limited their applications. As described above, hydrogels with large free space among the crosslinked networks can not only act as a reservoir for massive NPs loading, but also serve as a nanoreactor template for the nucleation and growth of NPs [259]. Embedding Ag NPs in crosslinked polymer networks to obtain Ag nanocomposite hydrogels has proved to be a most promising solution for stabilization and distribution problem, due to the excellent capacity of stabilizing NPs. Normally, Ag NPs can be introduced into the matrix of polymer hydrogel by simply blending the NPs with the preformed hydrogel, entrapping NPs during the swelling process or adding the NPs during the gelation process [260]. And Ag NPs can also be generated *in situ* in the swollen hydrogel, which can achieve a more uniform distribution of Ag-NPs [261,262]. As described in Table 3, the incorporation of Ag NPs into hydrogels will render them with high antibacterial activity and usually change their mechanical toughness, swelling ratio and stimuli responsiveness. Hydrogels work as an efficient stabilizer of Ag NPs and control the release of the Ag NPs. More importantly, because the biocompatible hydrogels can maintain a moist environment at the wound interface, act as a barrier to microorganisms, allow sufficient air and water vapor to permeate through, and absorb or remove excess exudates, as well as attach at the target site and be easily removed without trauma and pain, Ag NPs-hydrogel composites have great potential to provide functional coatings for various applications, such as wound dressings, catheters and bone cement [263,264,265]. It was worth pointing out that the inclusion NPs into natural polymer hydrogels, such as CS, GEL, COL and peptides can produce biodegradable composite materials, which can be employed as implantable dressings without the need of removal [266,267,268]. In addition, Abdel-Halim *et al.* [269] reported an electrically conducting nanocomposite hydrogel consisted of silver NPs and guar gum/poly(acrylic acid)hydrogel. The nanocomposite hydrogel's electrical conductivity was found to be affected by the content of Ag NPs and the swelling ratio of the hydrogel. Endo *et al.* [256] developed a surface plasmon resonance (LSPR)-based optical glucose concentration sensor by incorporation of Ag NPs into pH-responsive hydrogels. The interparticle distances of the silver NPs in the hydrogel would be increased if glucose was present on the surface of this biosensor. And thus the absorbance strength of the LSPR was decreased, which can be used to specifically determine glucose concentrations.

Table 2. Summary of nanocomposite hydrogels from hydrogels and silicon-based NPs for biomedical applications.

Fillers	Hydrogels	Applications and Functions	Ref.
Silica (SiO_2) particles or NPs	HA/PNIPAM	Tissue engineering; SiO_2 NPs improve mechanical properties; HA can improve biocompatibility; The stability and the rigidity increase with the amount of PNIPAM	[159]
	PNIPAM	Drug delivery and tissue engineering; SiO_2 NPs improve the thermal and mechanical properties but decrease the swelling ratio	[160–168]
	PAM	Compression strength and elastic modulus of the composite hydrogels were significantly improved by adding SiO_2 NPs compared with pure PAM hydrogels	[169–173]
	Silicone	Implantable medical devices; SiO_2 can increase swelling behavior and improve stiffness of the hydrogel; Hybrid hydrogel can be used to decrease the protein absorption and the growth of bacteria	[174]
	PHEA	Drug delivery; The silica particles were well dispersed in hydrogels; The release rate of aspirin decreased with the increasing content of silica	[175]
	PVA	Biomaterials; The tensile strength of PVA hydrogel increased with SiO_2 NPs content. Proper content of SiO_2 NPs could also enhance the permeability and swelling property, resulting in the improvement of the capillary water absorption capacity of PVA hydrogel	[176]
	CS	Bone tissue engineering; SiO_2 NPs have a positive effect on cell viability and induce the occurrence of mineralization not only at the surface of the material but also in its entire volume; CS as bioactive injectable systems for bone tissue repair can undergo *in situ* gelation under physiological temperatures	[177,178]
	Peptide	Biosensor; PEG/peptide can form enzyme responsive hydrogel and collapse in response to diverse biological stimuli; The negatively charged silica NPs used are also at a sufficiently high concentration to form a regular periodic structure within the hydrogel, acting as a photonic bandgap which can reflect visible wavelengths of light	[138,140]
	Cellulose	Tissue engineering and cell cultures; SiO_2 preserves extracellular matrix-like materials and cellular proteins; The intact 3D spheroids can be recovered from the hydrogel by a cellulase enzyme for downstream applications	[179]
	PMAA or PAA	SiO_2 as adhesive fillers that interacted with PMAA chains can improved viscoelastic moduli (up to 8.7 times) and enhanced elasticity	[180,181]
	ALG or COL	Tissue engineering; SiO_2 can improve the mechanical properties, including surface roughness and hardness, of the hydrogel	[182–184]
	PEG	Drug delivery; Hydrogel can form sustained-release depot systems; Incorporation of SiO_2 NPs can regulate the pores size, gelation time and viscosity of the hydrogel; The drug loading capacity and drug release profiles could be tuned	[185,186]

Fillers	Hydrogels	Applications and Functions	Ref.
MSNs	CS	Drug delivery; MSNs facilitate the drug loading and their subsequent release; CS enables pH-triggered drug release and improves the endocytosis of target cells and cell biocompatibility	[187]
MSNs	CS	Drug delivery; The introduction of MSNs into CS will facilitate the gelation process at body temperature and also promote the elastic modulus; MSNs can load small drug molecules and then steadily release them for a long time period	[178]
MSNs	PClAETA	Biomaterials; Introduction of non-functionalized MSNs can improve the water absorption of the hydrogel	[188]
MSNs	Poly(aspartic acid)	Biomaterials; MSNs as crosslinker can optimize the pore morphology, improve thermal stability increase swelling ratio and enhance salt tolerance of the composite; MSNs can be uniformly and stably dispersed in the structure of the hydrogel	[189]
MSNs	PNIPAM	Biomaterials; MSNs can be used as effective "topological crosslinkers" to reinforce the PNIPAM hydrogels	[190]
HAP and nBG	GEL	Bone tissue regeneration; HAP significantly improves the stiffness of GEL hydrogels, while it maintains their structural integrity and swelling ratio; Introduction of HAP results in a lower swelling ratio, higher mechanical moduli, and better biocompatibility and promotes cell functional expression for osteon biofabrication; Gelatin hydrogels also provide natural cell binding motifs, making them amenable for 3D cell encapsulation; GEL provide cell-responsive characteristics, cell adhesion sites, and proteolytic degradability	[191–193]
HAP and nBG	Silk fibroin	Bone tissue regeneration; HAP can increase compression modulus and mechanical properties, decrease the water uptake ability, improve metabolic and alkaline phosphatase activities of osteoblastic cells; Hydrogel plays regulatory role in oriented nucleation and growth of HAP crystals	[133,194–197]
HAP and nBG	PAM	Tissue engineering; SiO_2 can improve the mechanical properties, including surface roughness and hardness	[198]
HAP and nBG	CS	Bone tissue regeneration; HAP enhances swelling, protein adsorption, exogenous biomineralization and osteoblast differentiation and also accelerates bone formation; Zn possessing excellent antimicrobial properties can tackle implant-associated microbial infections; CS can minimize immune response	[199–202]
HAP and nBG	PVA	Bone tissue regeneration; Hydrogels are employed as a matrix gel in a mineralization solution; HAP was formed in the polymer matrix under mild conditions, mimicking mineralization in natural bone formation	[132,203,204]
HAP and nBG	Pullulan	Bone tissue regeneration; Addition of HAP can improve compressive modulus of the scaffold, provide sites for cell adhesion, and render them osteoconductive *in vitro*; Hydrogels work as scaffolds for mineralization	[205]
HAP and nBG	PECE	Tissue engineering; PECE endowed hydrogel good thermosensitivity and injectability; HAP can improve mechanical properties of hydrogel	[206]

Fillers	Hydrogels	Applications and Functions	Ref.
HAP and nBG	PEG	Bone substitute material; The formation of HAP in hydrogel matrices enable the acquisition of bioactive composites materials with desired shapes.	[207,208]
HAP and nBG	ALG	Bone tissue engineering; Incorporation of nBG into hydrogel can combine excellent cellular adhesion, proliferation and differentiation properties, good biocompatibility and predictable degradation rates; ALG can increase lactate dehydrogenase and mitochondrial activity	[135,209–211]
HAP and nBG	CHS	Bone tissue engineering; nBG shows an excellent stimulatory effect on bone formation; CHS improves integration of the nBG to prevent particle migration and promotes bone regeneration; The composite can encapsulate bone marrow to form a mechanically stable construct	[212]
HAP and nBG	COL	Tissue engineering; Incorporation of nBG NPs improves mechanical stability and enhanced the proliferation rate and osteogenic differentiation	[136,213–215]
LAP	GEL/PAM	Tissue engineering; Incorporation of LAP can enhance thermal stability, tensile and stretching properties; GEL can significantly improve the hydrogel's pH-responsive properties and enhance the antithrombogenicity but decrease the degree of hemolysis of the gels	[216]
LAP	PEG/GEL	Tissue engineering; Hydrogel remain stable and provide a cell supportive microenvironment under normal cell culture conditions. LAP can preferentially induce osteogenic differentiation of human mesenchymal stem cells	[217]
LAP	PEG	Tissue engineering; Incorporation of LAP significantly reduced the cure time while enhancing the adhesive and bulk mechanical properties of the hydrogel; PEG adsorbed onto LAP, forms a compact layer of mostly loops and trains on top of the nanoparticle and large loops around the edge of the particles; PEG elicits minimal inflammatory response and exhibits an enhanced level of cellular infiltration	[218,219]
LAP	PAM	Tissue engineering; The incorporation of catechol on the PAM exhibits a strong affinity toward LAP and enhances stiffness and a viscous dissipation property	[220–222]
LAP	P(MEOMA-OEGMA)	Drug delivery and tissue engineering; LAP as physical cross-linker had significant influence on the microstructure and swelling/deswelling behaviors of hydrogels.; OEGMA can increase equilibrium swelling ratio and water retention; Hydrogel can provide thermosensitivity and the excellent biocompatibility	[223]
LAP	P(AM-DMAEMA)	Tissue engineering, cell culture substrates and biosensors; LAP serves as a physical cross-linker and can change the mechanical strength of the hydrogel under direct-current electric field	[224]
LAP	PNIPAM	Tissue engineering; LAP can influence the polymeric chain arrangement and increase the mechanical toughness and thermal stability	[225,226]
LAP	Pluronic F127	Tissue engineering; The interactions between LAP and Pluronic F127 may contribute to rearrangements of network structures at high deformations, leading to high elongations and improved toughness	[227]

Fillers	Hydrogels	Applications and Functions	Ref.
LAP	GMA	Biomaterials; LAP can reinforce mechanical toughness and elasticity; GMA regulates equilibrium swelling ratio and improves water content; Integrating LAP and GMA can improve self-standing ability and rheological, compression, and tensile properties	[228]
MMT	CS	Drug delivery; MMT can enhance the loading of positively charged drug and affect the hydrogel's drug release mechanism and swelling properties; The hydrogel can be used for controlled- release of drugs	[137,229]
	CS-g-PAM	Superabsorbent polymer composites; Incorporation of MMT can optimize their absorption capacity, improve their swelling rate and salt-resistant ability; Composites exhibit moderate antibacterial activity in acidic medium	[230]
	PAM	Drug delivery; MMT effects the equilibrium swelling and drug release behavior of the composite; MMT can improve the barrier property of nanocomposite hydrogels and decrease the burse effect. MA effects the pH-responsivity on equilibrium swelling and release of drug	[231–233]
	P(ATC-AM)	Drug delivery; MMT serves as chemical cross-linker to enhance strength and toughness and decrease the swelling degree; ATC is cationic monomer and exist cation- exchange reaction with MMT	[234]
	GEL or COL	Drug delivery and wound dressing; Drug intercalation results in changes in MMT layered space; Integration of drug loaded MMT and gelatin creates biodegradable composite hydrogels with controlled drug release property, improved mechanical and thermal properties	[235,236]
	PAA or PMA	Biomaterial implants; MMT is used for adsorbing drug, achieving high drug loading; PAA is used for pH/bacteria- responsive releasing	[237]
	Polysaccharide	MMT can enhance strength and toughness and decrease the swelling degree	[238]
	PNIPAM	Tissue engineering; MMT can influence the polymeric chain arrangement and the pore size and increase the complex viscosity and adhesion strength as well as thermal stability	[239–243]
Other silicate particles	P(AM-IA)	Drug delivery; Incorporation of hectorite offers the hydrogel with suitable water absorbency, shear-resistance, high gel strength and good thermal stability; IA provides pH-responsivity on drug release	[244–246]
	ALG	Drug delivery; Halloysite nanotubes improve complex surface topography and structural integrity and achieve a sustained release of the growth factor; The composites enhance repair and regeneration in damaged or diseased tissues	[139]
	PEG/P(AA-VP)	Drug delivery; Phyllosilicate enhances the water uptake with a desirable strength; Hydrogels are used for pH responsible and controlled release	[247]
	PCL-PEG-PCL	Bone regeneration; Mesoporous magnesium silicate can enhance the compressive strength, elastic modulus, and hydrophilicity of hydrogel, and promote cell attachment and proliferation and increase the degradability of the hydrogel	[248]

Fillers	Hydrogels	Applications and Functions	Ref.
Other silicate particles	PAM	Tissue engineering; Attapulgite nanorods grafted with vinyl groups serve as macro-crosslinkers, which can significantly increase the modulus, strength, and toughness of hydrogels;	[249]
	P(AA-NIPAM)	Tissue engineering; The presence of imogolite nanotubes produced strong hydrogels that exhibit thermo-responsive volume transition because of the coil/globule transition of PNIPAM chains.	[250]

Mesoporous silica nanoparticles (MSNs); Hydroxyapatite (HAP); Poly[(2-acryloyloxyethyl) trimethylammonium chloride] (PClAETA); Zinc-doped chitosan (Zn-CS); nano-scaled bioactive Glass (nBG); Chondroitin sulfate (CHS); Laponite (LAP); 2-(2-methoxyethoxy) ethyl methacrylate (MEOMA); oligo(ethylene glycol) methacrylate (OEGMA); Guanidinium-pendant methacrylamide (GMA); Montmorillonite (MMT); Poly(acrylamide-co-maleic acid) (P(AM-MA)); (3-acrylamidopropyl) trimethylammonium chloride (ATC); Poly(ε-caprolactone)-poly(ethylene glycol)–poly(ε-caprolactone) copolymer (PCEC); Poly(acrylamide-co-itaconic acid) (P(AM-IA)); Poly(acrylic acid-co-vinyl pyrrolidone) (P(AA-co-VP)); Poly(ε-caprolactone)-poly(ethylene glycol)-poly(ε-caprolactone) (PCL-PEG-PCL).

Similarly, the inclusion of Au NPs within the hydrogel matrix has also emerged as a promising way to form a novel class of soft nanocomposites for various applications in the biomedical fields [270,271,272,273]. Au NPs with excellent optical and LSPR properties from the visible regime to the near infrared (NIR) have received intensive attention in biomedical applications. NIR light is very useful for biomedical applications because these wavelengths can penetrate deeply into biological tissues with relatively little attenuation and minimal damage, which has been wide used as a stimulus for photothermal therapy (PTT), photodynamic therapy (PDT), bioimaging and remote-controlled smart drug delivery [31,274,275,276]. One of the most remarkable properties of Ag NPs is that they can effectively absorb and scatter light due to LSPR and then efficiently convert optical energy into thermal energy [43,44,45]. Due to their very high photothermal conversion efficiency, Au NPs-hydrogel composites in combination with NIR laser irradiation have aroused tremendous interest as therapeutic agents for cancer therapy, especially for the PTT and PDT. Compared to other cancer therapies, PTT and PDT are both excellent minimally invasive therapeutic treatments [270,271,273,275,276]. For example, Wang *et al.* [275] reported a composite hydrogel that contained Au NPs, spinach extract, and PEGDA via *in situ* photopolymerization. Spinach extract can act as a photoinitiator to initiate the formation of the PEGDA hydrogel and serve as an excellent photosensitizer to generate cytotoxic singlet oxygen ($^1O_2$) with oxygen to kill tumor cells. Au NPs as a photoabsorbing agent can generate heat from optical energy to induce tumor tissue hyperthermia and accelerate the rate of $^1O_2$ generation. The composite hydrogel shell on tumor cells can prevent the photosensitizer from migrating to normal tissue and keeps a high concentration on lesions, thereby enhancing the curative effect. More interestingly, a remote-controlled smart drug delivery can be easily achieved by NIR irradiating Au NPs-integrated temperature-responsive gel matrix. The gel structure will quickly collapse if the temperature rises above the lower critical solution temperature of the gel matrix due to the NIR-triggered photothermal effect. The fast response leads to an on demand burst release of drugs, which should be contributed to heat generation within the gel matrix, rather than external to the matrix [270,273,277]. Sershen *et al.* [277] firstly developed a photothermally modulated hydrogel using photoactive Au NPs in combination with thermal-responsive PNIPAM-PAM copolymers, which exhibited a controlled pulsatile release of methylene blue and proteins [277]. As shown in Figure 9, a novel chemo-photothermal co-therapy system was developed to prevent local breast cancer recurrence [270]. The temperature of the hybrid hydrogel system will be increased due to the photothermal effect of Au NPs under NIR irradiation, thus leading to a fast shrinkage of hydrogel and a significant increase in the release rate of loaded drug. This functional hydrogel effectively avoided breast cancer recurrence after primary tumor resection in a mouse model, due to the controlled administration of combination chemo-photothermal therapy and the decreased systemic toxicity. Besides using Au NPs based nanocomposite hydrogels for drug delivery and cancer therapy, they are also explored for biosensing [278], cellular therapy such as cell labeling and cell-biology research [279], and tissue regeneration [280].

Unlike those noble metals NPs, some transitional metal, such as ferreous (Fe), cobalt (Co) and nickel (Ni) based NPs, exhibit excellent magnetic properties. Among them, pure iron oxides, such as maghemite (γ-Fe_2O_3) and magnetite (Fe_3O_4) NPs are most common magnetic nanomaterials for biomedical applications, due to their superior biocompatibility [47,48]. Moreover, the magnetic properties of these NPs can be tuned from ferromagnetic to superparamagnetic by adjusting their sizes, because of quantum size effects [47,48]. As shown in Table 3, the incorporations of these magnetic NPs into polymer networks also offer a high potential for several biomedical applications, such as magnetic field induced hyperthermia, magnetic resonance imaging (MRI), drug delivery, cell tracking and protein separation [47,48,281]. The typical biomedical applications of these magnetic hydrogel were shown in Figure 10. For example, the superparamagnetic property of iron oxide NPs can cause a temperature rise under alternating magnetic fields, and thus the use of magnetic induction hyperthermia for cancer therapy and cell-biology research has recently gained great interest. Samal *et al.* [282] introduced magnetic NPs into silk fibroin protein hydrogel scaffolds via dip-coating methods. The obtained magnetic scaffolds showed excellent hyperthermia properties achieving temperature increases up to 8 °C in about 100 s, improving the cell adhesion and proliferation. Magnetic induction hyperthermia also allows the on-off response of the system depending on temperature. Liu *et al.* [283] incorporated magnetic NPs into PVA matrix to create composite hydrogels, which have proven externally controlled on–off drug release from the hydrogels. The drug release behaviors can be remote-controlled by the use of a magnetic field, and the release profiles of drug were highly dependent on the particle size, strength of the magnetic field and the duration of the switching time. Composite hydrogels of thermo-sensitive PNIPAM with superparamagnetic NPs exhibited remotely-controlled thermal-responsive swelling behaviors in an alternating magnetic field [284]. The covalent conjugation of magnetic Fe_3O_4 NPs and bovine serum albumin (BSA) on the hydrogels allows these materials to be applied in magnetically assisted bioseparation processes [285].

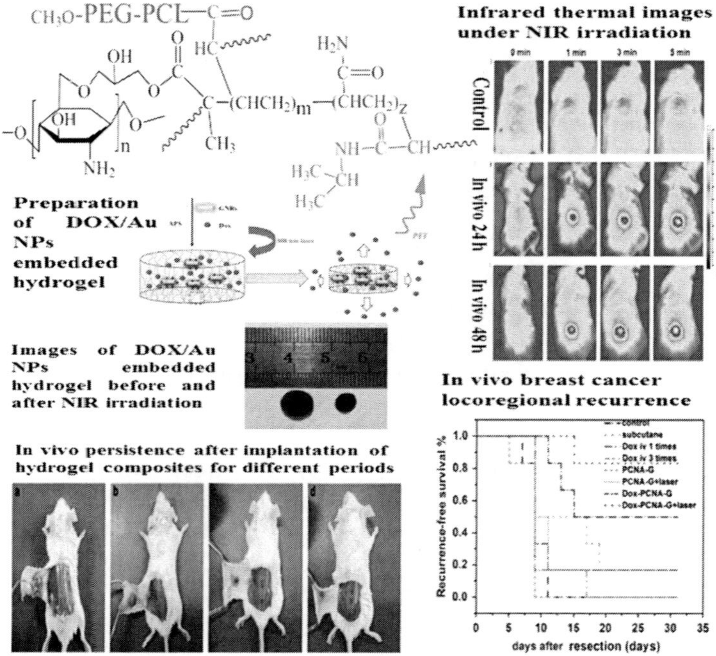

Figure 9. Schematic illustration of Doxorubicin (DOX)/Au NPs embedded hybrid hydrogels with the ability of photothermal therapy (PTT) and near infrared (NIR)-triggered thermo-responsive drug release, reproduced with permission from [270]. Copyright Nature Publishing Group, 2015.

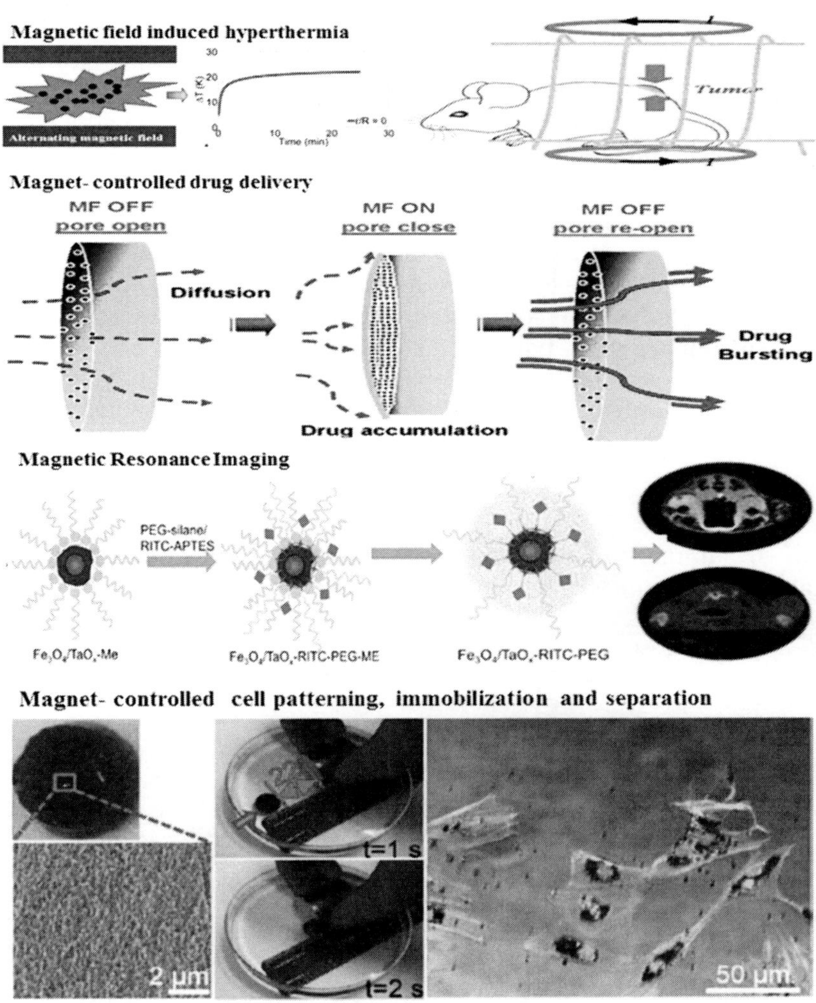

Figure 10. Schematic illustrations of typical biomedical applications of iron oxide NPs-hydrogel composites. Magnetic field induced hyperthermia, reproduced with permission from [286] Copyright Ivyspring International Publisher, 2012; Magnet-controlled drug delivery, reproduced with permission from [283]. Copyright American Chemical Society, 2006; Magnetic resonance imaging, reproduced with permission from [287]. Copyright American Chemical Society, 2012; Magnet-controlled cell patterning, immobilization and separation, reproduced with permission from [288]. Copyright American Chemical Society, 2015.

Several other metallic NPs, metal-oxide NPs and metal alloys NPs can also be used to integrate with polymer networks to enhance the bioactivity of hydrogels. For instance, magnetic Co or Ni NPs have also been introduced into hydrogels to form ultra-flexible and strong magnetic actuator [289,290]. Some titania NPs and alumina NPs were incorporated within polymeric matrix to enhanced cell adhesion and proliferation for tissue engineering applications [291,292]. However, very limited studies focused on the use of these NPs to fabricate hybrid hydrogels for biomedical applications because of their potential toxicity and rapid oxidation. Overall, various types metallic and metal-oxide colloidal particles were incorporated with polymer network structures, which can impart them with exclusive optical, electrical or magnetic properties. The resultant nanocomposite hydrogels are highly suitable for different biomedical applications, especially for diagnostic and therapeutic applications.

Table 3. Summary of nanocomposite hydrogels from hydrogels and metal or metal oxide NPs for biomedical applications.

Fillers	Hydrogels	Applications and Functions	Ref.
Ag NPs	PAM	Wound dressing; Ag NPs demonstrate excellent antibacterial activity; Polymer hydrogels were used to keep the NPs stable and contact with skin	[254,255,293–297]
	PAM	Biosensor; Ag NPs can change the strength of the localized surface plasm on resonance by regulating interparticle distances; PAM can be swelling with ionic strength change	[256]
	PVP or PVA	Wound dressing or tissue engineering; Ag NPs are antibacterial agents and enhance the mechanical and thermal strength of the hydrogel; Hydrogels work as an efficient stabilizer of Ag NPs and control the release of the Ag NPs	[257,258,298–302]
	PNIPAM	Antibiotic materials; Ag NPs work as antimicrobial agents; PNIPAM is temperature-responsive component and make Ag NPs stable and uniformly distributed	[259,303–306]
	PEGDMA	Contact lenses; Ag NPs work as antimicrobial agents	[307]
	Acrylate-based copolymers	Biosensors and drug delivery; Ag NPs work as antimicrobial agents; Copolymer hydrogel with different swelling properties is used to disperse Ag NPs	[262,266,308–310]
	Acrylate-based copolymers	Ophthalmic lenses; The addition of Ag NPs is associated with the reduction of UV-B transmittance and increase in tensile strength	[311]
	AMPS-Na	Burn wound dressing; Silver hydrogel is efficient at preventing bacterial colonization of wounds and has good inhibitory action against bacteria	[263,312]

Fillers	Hydrogels	Applications and Functions	Ref.
	NaDOC	Antibacterial materials; Ag NPs are antibacterial agents; NaDOC can immobilize Ag NPs via pH-induced self-assembly	[313]
	Graphene	Wound dressing; Ag NPs are antibacterial agents; Graphene is conductive to cellular adhesion and growth; The composite can significantly accelerate the healing rate of artificial wounds	[264]
	Silicone	Adhesion to catheters to prevent urinary tract infections; Ag NPs work as antimicrobial agents	[265]
	Lysine gelator	Nanomedicine; Gelator exhibits good gelation ability, low haemolytic activity and high biocompatibility to mammalian cells; Incorporation of Ag NPs exhibits excellent antibacterial activity and significant mechanical strength	[314]
Ag NPs	Peptides	Wound healing and sustained Ag NPs release; Ag NPs inhibit bacterial growth; Peptide fibers prevent aggregation of Ag NPs and improve the biocompatibility of the system	[267]
		Antibacterial materials and bioimaging; Fluorescent Ag NPs show excellent optical properties and antibacterial activity; Peptides make the Ag NPs stable	[315]
	PAA	Drug delivery; PAA is used to control the release of antibacterial drugs; Ag NPs can enhance the antibacterial activity	[316]
	Cellulose	Wound dressing and antimicrobial materials; Ag NPs show high antibacterial activity against Gram positive and Gram negative bacteria; Cellulose serves as a gelling agent as well as a reducing agent for $AgNO_3$ and a stabilizer for Ag NPs	[247,261,317]
	GEL or COL	Wound dressing; Ag NPs are antibacterial agents; Hydrogel is used for stabling Ag NPs	[268,318,319]
	GEL/CS	Drug delivery and tissue engineering; Benzotriazole maleimide functionalized Ag NPs work as cross-linkers, improve the storage modulus and decrease swelling ratios	[320]
	AG-CS	Biosensor; Ag NPs improve the conductivity and mechanical strength of the hydrogels and also have antimicrobial activity; Biopolymer moieties act as both reducing and stabilizing agents	[321]
	AG	Antibiotic materials; Hydrogels promote the Ag NPs protection and inhibit their aggregation; Ag NPs can produce Ag^+ to kill bacteria	[322,323]

Au NPs	PNIPAM	Drug delivery; Au NPs can be used as a photoabsorbing agent to generate heat from optical energy; PNIPAM is thermo-responsive to control the release of drugs	[270,324–326]
		Biosensor for the detection of DNA; Au NPs immobilize DNA strands	[278]
	PEG	Tissue engineering; RGD modified gold nanoarrays as scaffold to control cell localization and differentiation	[279]
		Drug delivery; Au NPs are utilized as building blocks; α-CD and PEG form temperature-responsive reversible supramolecular hydrogel through host–guest interaction	[271]
	PEGDA	Tumor therapy; Au NRs can be used as a photoabsorbing agent to generate heat from optical energy; The introduction of Au NRs is conducive to the formation of the hydrogel and accelerates the rate of cytotoxic singlet oxygen generation; PEGDA can stabilize Au NRs	[275]
	PPy	Clinical diagnosis; Au NPs can further increase the specific surface area to capture a large amount of antibodies as well as improve the capability of electron transfer	[327]
	PEDOT	Electroporation-assisted cell uptakes; Au NPs work as microelectrode; PEDOT modification is effective in preventing electrolysis during the electroporation	[328]

Fillers	Hydrogels	Applications and Functions	Ref.
Au NPs	CS	Drug delivery; Au NPs act as a role of physical cross-linking agents; This physical cross-linked hydrogel shows excellent drug loading ability	[272]
	GEL	Bone tissue engineering; The hydrogels loaded with Au NPs promote proliferation, differentiation, and alkaline phosphate activities of human adipose-derived stem cells and have a significant influence on new bone formation; GEL is a photo-curable hydrogel to embed Au NPs	[280]
	Peptide	Tissue engineering; Au NPs act as cross-linking agents and strengthen the prepared hydrogel	[329]
	DNA	Drug delivery and photothermal therapy; Au NPs can generate heat upon laser excitation and be used for photothermal therapy; DNA hydrogel can be fragmented by heat generations to release drugs and Au NPs	[273]
	DNAzyme	Detection of lead; Au NPs are used for visual detection; DNAzyme can be activated by lead and conducive to dissolve the hydrogel to release Au NPs	[330]
	Graphene	Biosensor for nitric oxide detection; Au NPs catalyze the electrochemical oxidation of NO; Graphene is used for uniformly depositing Au NPs	[331]
	κ-carrageenan	Drug delivery; Fe_3O_4 NPs offer the hydrogel with magnetic-responsiveness properties; κ-carrageenan is pH responsive	[332–335]

Fillers	Hydrogels	Applications and Functions	Ref.
Fe_3O_4 NPs	Starch	Drug delivery; Fe_3O_4 NPs endow the hydrogel with magnetic sensitivity; The composite also shows temperature/pH sensitivity	[336–338]
	Cellulose	Drug delivery; Magnetic nanoparticles are used as cross-linkers; The release of DOX is significantly enhanced at external magnetic field	[339,340]
	Fibrin	Tissue engineering; Fe_3O_4 NPs make the hydrogel show excellent hyperthermia properties under exposure to an alternating magnetic field; The presence of Fe_3O_4 NPs in the fibrin can improve cell adhesion and colonization of osteogenic cells	[282,341]
	HA	Magnetic resonance imaging and drug delivery; HA can be degraded by hyaluronidase and release hydrophobic drugs; Fe_3O_4 NPs serve as imaging agents to track the hydrogel degradation and degradation products *in vivo*	[342]
	CS	Drug delivery; The presence of Fe_3O_4 NPs can make the hydrogels have magnetic response property and deliver and release encapsulated anticancer agent at the tumor by the weak magnetic field	[343–345]
	ALG	Cell encapsulation; Alginate cell capsules with Fe_3O_4 NPs can be easily controlled and manipulated by external magnetic fields	[346]
Fillers	**Hydrogels**	**Applications and Functions**	**Ref.**
Fe_3O_4 NPs	COL-HA-PEG	Cartilage tissue engineering; The presence of Fe_3O_4 NPs can make the hydrogel travel to the tissue defect sites under remote magnetic guidance	[288]
	Guar gum	Drug delivery and theranostic; The NPs work as imaging agents and produce hyperthermia; Aminated guar gum can form the hydrogel without using toxic crosslinking agents	[347]
	PPZ	Drug delivery and biodetection; Ferrite nanoparticles is used for long-term magnetic response theragnosis; The composite can be used for sustainedly releasing cargos	[348]
	PAM	Drug delivery and protein separation; The presence of Fe_3O_4 NPs can make the hydrogel exhibit higher dielectric constants and magnetic responsiveness; The composite also has the properties of fast adsorption	[349–351]
		Drug delivery; Incorporation of Fe_3O_4 NPs endows the hydrogel with the ability of direction-dependent thermogenesis in an alternating magnetic field; The novel magnetic hydrogel shows a direction-dependent release of drugs that has a 3.4-fold difference between the two directions	[352]
	PNIPAM	Drug delivery; Incorporation of magnetic nanoparticles endows the hydrogel with on-demand pulsatile drug release triggered by alternating magnetic field	[353]
	PVP or PVA	Drug delivery; Fe_3O_4 NPs are used for magnetic drug targeting; The hydrogel is used for carry cancer therapeutic agent	[354–356]

Fillers	Hydrogels	Applications and Functions	Ref.
	PEGDA	Cell patterning; Incorporation of Fe_3O_4 NPs can manipulate space pattern of the hydrogel; PEGDA blocks are used as a stencil to define the area for cell adhesion and the second types of cells could be seeded after the magnetic block was removed to create heterotypic cell patterns	[357]
	Acrylate/acrylamide-based copolymers	Drug delivery and protein immobilization; Fe_3O_4 NPs endow the hydrogel with properties of magnetic stimuli-responsive; The copolymer hydrogel exhibits pH- and thermosensitive properties	[88,358–361]
	P(HEMA-GMA)-PANI	Immobilization of glucoamylase; Fe_3O_4 NPs show super-paramagnetism and make the hydrogel easily dispersed or separated from the medium without needing high magnetic intensity; Grafting of polyaniline on the hydrogel increases 3.4 times maximum adsorption capacity	[362]
Cu or CuO NPs	PVP/PVA	Bioimaging and drug delivery; Cu NPs with red emission properties are used for optical imaging as well as for flow cytometric probe of cellular uptake; Cu NPs also generate reactive oxygen species to enhance the efficiency of killing cancer cells; PVP is used for stabling the NPs and delivering drugs	[363]
	PEGDA	Antibacterial material; Cu NPs-modified hydrogel can be used for preventing bacterial infections	[364]
	Cellulose	Antibacterial material; CuO NPs endow the hydrogel with antibacterial activity	[365]
ZnO NPs	PAM	Wound dressing; ZnO NPs endow the hydrogel with antibacterial activity; PAM provides uniform distribution and binding of the NPs to the fiber surface and to prevent their agglomeration	[366]
	Cellulose	Antibacterial materials; ZnO NPs endow the hydrogel with antibacterial activity; The composite shows a pH and salt sensitive swelling behavior	[367,368]
	Chitin	Wound dressing; ZnO NPs endow the hydrogel with antibacterial activity and blood clotting ability; Incorporation of ZnO NPs improve healing rate and collagen deposition ability of composite bandages	[369]
ZnS NPs	PHEMA/PAA	Artificial cornea implants; ZnS NPs can improve the refractive index of the polymer	[370]
Pt NPs	Polyimide	Biosensor; Pt NPs work as microelectrades; The composite works as microsensor to detect glutamate *in vivo*	[371]
Ni NPs	Cellulose	Responsive biomaterials; Ni NPs offer the gel with magnetic response properties with can be changed by temperature and the size of the NPs	[289]
Co NPs	PHEMA	Tissue engineering; Co NPs enable the hydrogel that combine the use of saturation magnetizations and high particle loading, flexibility, and shape memory; PHEMA can reduce magnetic particle migration or loss	[290]

MnO_2 NPs	CS	Biosensor; MnO_2 NPs work as catalytic for H_2O_2; The composite can be used for detections of choline chloride	[372]
TiO_2 NPs	PAM/PNIPAM	Responsive biomaterials; Modified TiO_2 NPs serve as cross-linker and enhance the mechanical strength; The hydrogel bilayer is able to bend to opposite directions to form hydrogel circles in pure water and 20% NaCl solution	[373]

Poly(ethylene glycol diacrylate) (PEGDA); 2-acrylamido-2-methylpropane sulfonic acid sodium salt (AMPS-Na); Sodium deoxycholate (NaDOC); Agarose (AG); Polypyrrole (PPy); Poly(3,4-ethylenedioxythiophene) (PEDOT); Poly(organophosphazene) (PPZ); Polyaniline-grafted poly(2-hydroxyethylmethacrylate-co-glycidylmethacrylate) (P(HEMA-GMA)-PANI).

3.5. Nanocomposite Hydrogels from Hydrogels and Polymeric NPs or Liposomes

In comparison to conventional inorganic and low molecular weight organic materials, polymers exhibit unique properties, such as non-toxic and biocompatible nature, good physicochemical stability, excellent elasticity and ductility. Moreover, polymer chains are dynamic when polymers are conducted under usual conditions and particularly induced by an external environment, e.g., annealing, swelling and stretching, allowing for direct the self-assembly of NPs by use of polymers [32,34,374]. Especially because of the flexibility offered by macromolecular synthesis methods, the ability to respond to specific physiological or external stimuli, the almost infinite diversity of polymers in terms of property, architecture and composition, as well as their ease of functionalization, polymers have always been the first choice for the design and development of nanomaterials with tailored physicochemical and biological properties [32,34,374]. A wide variety of polymeric NPs, such as polymeric micelles, nanogels and particles, dendrimers as well as liposomes, have been developed and gained great attention in biomedical and pharmaceutical applications. The inclusion of these polymeric particles into a hydrogel system has been explored to improve the performance of hydrogels and polymer NPs [3]. This strategy strongly has provided the opportunity to obtain nanocomposite hydrogels with undoubtedly improved properties and performances. The advances in the fabrication of polymer NPs and hydrogels composites and their applications in drug delivery, tissue engineering, and other biorelated domains were summarized in Table 4. In addition, some important studies in this topic were highlighted in detail and shown as below.

Table 4. Summary of nanocomposite hydrogels from hydrogels and polymeric NPs or liposomes for biomedical applications.

Fillers	Hydrogels	Applications and Functions	Ref.
PLGA particles or NPs	PVA	Loading and delivery of biomacromolecular drugs; PLGA can increase drug loading and decrease burst release; Hydrogel can prevent the foreign body reaction(FBR), promote angiogenesis around subcutaneous implants and extend the lifetime of implantable biosensors	[375,376]
		Drug delivery; PVA hydrogel acts as a hydrophilic base to support the microspheres; PLGA microspheres serve as drug reservoirs to continuously drug release	[377–381]
	ALG	Drug delivery; Maintain consistent release of rhBMP-2; Improve bone formation and osseous integration	[382]
	PuraMatrixTM peptide	Drug delivery; PuraMatrixTM peptide hydrogel as vaccine adjuvants to recruit and activate immune cells	[383]
	F127	Drug delivery; Sustained release of protein drugs; Temper burst release and prolong delivery of drugs at the site of a spinal cord injury	[384,385]
	P407	Tissue engineering; Preserve protein structure and integrity, allowing a better and prolonged release profile and the maintenance of their biological activity. Hydrogel further protected it from the hydrophobic environment	[386]
	ALG	Cell-based tissue engineering; Offer a continuous and localized release of drug; Provide a physical support for microcapsules, facilitating administration, ensuring retention and recuperation and preventing dissemination; Reduce post-transplantation inflammation and foreign body reaction, thus prolonging the lifetime of the implant	[387]
	PEI-PEGDA	Drug delivery; Hydrogels are used for carrying dual or multi-molecular compounds and releasing them in a bimodal, sequential manner	[388]
	HAMC	Drug delivery; HAMC hydrogel is well-tolerated, having a minimal inflammatory response; NPs offer sustained release while the HAMC gel localizes the NPs at the site of injection	[389–395]
	PAM	Antivirulence treatment of local bacterial infection; RBC-coated NPs is an effective detoxification platform against bacterial infections; Hydrogels preserve the structural integrity and the functionalities of the contained NPs and offer additional engineering flexibility to improve the therapeutic efficacy	[396]

Fillers	Hydrogels	Applications and Functions	Ref.
PLGA particles or NPs	MC	Drug delivery; Particle localization, decreases initial burst, and further prolongs release	[397]
	GG	Sustained local delivery of drug; For the local treatment of osteoporosis and other bone tissue disorder; Improve the low bioavailability and decrease the high toxicity of sodium alendronate	[398]
	PNIPAM	Drug delivery; PLGA used for the isolation of the drug, a slower drug-release rate, and the achievement of different drug release profiles	[399]
	PAMAM	Drug delivery; The residence time of pilocarpine can be prolonged by hydrogel; PLGA nanoparticles are safe for delivery of ophthalmic agents and are capable of sustained delivery of antiglaucoma agents	[400]
	Fmoc-peptide	Drug delivery and tissue engineering; Peptide hydrogel will likely be more stable *in vivo*; Sustained release of drug	[401]
DBM-BHPEA NPs	Silicone	Protect the eyes from UV rays; Increase the pore size, allowing the particles to diffuse into the lenses.	[402]
HPMC-PEG NPs	Branched PAA	Drug delivery; Optimize drug therapies in which the release of hydrophobic compounds; Selectively direct to specific cell lines; Able to stay localized at the injection site	[403]
Silk fibroin particles	Silk fibroin	Tissue engineering; Offering well suited rheological features for injectability, and shape conformability into defect sites as well as controlled delivery rate	[404]
PHBV NPs	GG	Drug delivery; Minimally invasive administration and the controlled delivery of the active agent	[405]
PEG-PLA NPs	HPMC	Drug delivery; Shear-thinning and self-healing properties; Dual loading of a hydrophobic molecule into the NPs and a second hydrophilic molecule into the aqueous bulk of the gel, and thus enabled simultaneous release of both hydrophobic and hydrophilic drug *in vivo* from a single gel.	[406]
PPy NPs	ALG	Drug delivery and tissue engineering; PPy is stable in solution over the period of a month, and have good drug loading capacity; Localized depot releasing systems; Pulsatile releasing behaviors and maintain morphology and mechanical strength.	[407,408]
PHBHHx NPs	CS/GP	Drug delivery; Relatively strict control of long-term insulin release	[409]
PGT NPs	Silicone	Drug delivery in eye; Increase patient compliance and the bioavailability; Increase the release duration from the contact lenses; Establish safety and efficacy of glaucoma therapy by extended wear of nanoparticle loaded contact lenses	[410]
CS NPs	GX	Scaffold system for dual growth factor delivery in bone regeneration; Develop an *in situ* gelling biomaterial combining the viscoelasticity of natural polymers with the powerful antimicrobial properties of chitosan	[411]
PMMA NPs	Polysaccharide-PAA	Drug delivery; Hydrogels can remain localized at the site of injection, showing high biocompatibility and good ability to provide short term delivery. PMMA based NPs can be traced both *in vitro* and *in vivo* biological studies over a long period of time without side effects due to the biodegradation process	[412]

Fillers	Hydrogels	Applications and Functions	Ref.
PβAE NPs	PAMAM-DEX	siRNA delivery; High transfection efficiency and low cytotoxicity; Sustained delivery of the siRNA; Enhance the stability of the NPs	[413]
PLA particles	PECE	Drug delivery and tissue engineering; Increase the thickness of the corium; Increase cell adhesion of microspheres	[414]
PAMAM dendrimer	CS-PEG	Drug delivery; Dendrimer was used to increase the solubility, loading efficiency and homogeneity of hydrophobic drug; Hydrogel provide a local drug delivery depot for a prolonged drug release	[415]
	COL	Drug delivery; Dendrimer can improve the biostability and structural integrity of COL, which can make COL have higher denature temperature and resistance against collagenase digestion	[416]
	HA	Biofabrication; Form a fast cross-linking hydrogel; Improved the cell viability, proliferation, and attachment	[417]
	PEG-LA-DA	Tissue engineering; The multiple cross-linking sites present on the dendrimers can increase the cross-linking density at lower concentrations; The spherical dendrimers may provide discrete "molecular islands" in the network to limit swelling and improve mechanical properties; The multiple end-groups on the dendrimers facilitate the introduction of functional groups into the system at the nanoscale level	[418]
	Carbopol 980	Drug delivery; PAMAM dendrimers strongly affects their influence on the improvement of solubility and antifungal activity of drug	[419]
	PEG	Tissue engineering and drug delivery; Hydrogel crosslinked with dendrimers, showing improved cytocompatibility, controlled swelling and degradation	[420]
Solid lipid NPs	Poloxamer	Drug delivery; NPs can increase gel strength and mucoadhesive force; Easy to administer rectally; Gelled rapidly inside the body; Increased dissolution rate of the drug; Reduced initial burst effect	[421]
Liposomes	PAM	Drug delivery and antibacterial; Hydrogel can stabilize liposomes against fusion and preserves the structural integrity of the liposomes; The hydrogel formulation allows for controllable viscoeleasticity and tunable liposome release rate	[422]
	Peptide	Tissue regeneration; Liposomes can enhance binding of growth factors to peptide fibers of the gel matrix and achieve delayed release; Bimodal drug release	[423]
	HA	Drug delivery; Strengthen the network formed by HA chains, high efficiency encapsulation, easily injectable and less invasive, delay and control the release of drugs for local delivery	[424–427]
	PNIPAM	Drug delivery; The intact liposomes with their content can be controlled to release from hydrogel by changing the temperature, which can be used for temperature-triggered on-demand delivery	[428]
	PVA	Tissue engineering; Hydrogel can retain the structural integrity and contents of liposomes	[429]
	PVP-PAA-PBMA	Drug delivery, Hydrogel can ensure liposomes original vesicle structure; Liposomes can improve viscoelastic features and drug release profiles	[430]

Fillers	Hydrogels	Applications and Functions	Ref.
Micelles	PECA	Drug delivery; Improve docetaxel solubility and permeability; The pH-responsive hydrogel controlled the micelles diffusion excellently in intestinal environment, thus achieving the target delivery of drug-loaded micelles to small intestine and significantly improvement of the oral bioavailability of docetaxel	[431]
	PEG-PCL-PEG	Drug delivery and wound dressing; High drug loading and encapsulation efficiency; Improve combined curcumin solubility and permeability; Exhibit excellent wound healing activity and promote tissue reconstruction processes	[432]
	Agarose	Drug delivery; Achieve functional hydrogels capable of stimulus-triggered drug release	[433]
	PEG	Gene delivery; The incorporation of the nanosized micelles provided an excellent mean to tune physical properties of the hydrogels, such as increasing porosity and tunable mechanical property of the hydrogels and providing the best balance among hydrogel stiffness and porosity for cell survival	[434]
Nanogels or microgels	PEG	Tissue engineering; Hydrogel can control the degradation and release of nanogels; Possess relatively strong mechanical properties, biodegradability; Nanogels can trap proteins by hydrophobic interactions; Sustainedly release proteins in their native form	[435]
	PAM	Obtain macroscopic hydrogels with a fast response to external stimuli; Obtain strong hydrogel matrixes and composite gels with an internal structure	[436]
	PEG	Drug delivery; Multidrug delivery system; High drug loading and encapsulation efficiency; Encapsulate various hydrophobic substances	[437]
	DEX	Drug delivery; Both proteins and poorly water-soluble low-molecular-weight drugs can be efficiently encapsulated in the nanogels; Eliminating the initial burst release; Offer a maximum pharmacological efficiency at a minimum drug dose, reducing administration frequency and improving patient compliance	[438]
	Carbohydrate	Drug delivery; Improve loading or prolong delivery of a target drug; Prevent migration of the microgels from target sites; Provide an additional diffusive barrier for drug release; Reducing burst release effects and prolonging drug release kinetics	[439]
CHP nanogel	Hyaluronan	Drug delivery; Minimize denaturation by trapping the protein in a hydrated polymer network; Simultaneously achieved both simple drug loading and controlled release with no denaturation of the protein drugs	[440]
Vesicles	CS	Functional biomaterials; The vesicle serves as reversible "dynamic" crosslinks that hydrophobic chains can be either embedded into the bilayers of vesicle or pulled out; Vesicles can afford multi-action sites for hydrophobic interaction, indicating that the self-healing rate of such a hydrogel would be much faster	[441]
	PVA	Drug delivery; The vesicles can be evenly dispersed in PVA and are stable. The release can be triggered and the calcein diffuses afterwards quickly throughout the gel	[442]

Fillers	Hydrogels	Applications and Functions	Ref.
Vesicles	Xanthan	Drug delivery; The hydrogel shows a protective effect on vesicle integrity and leads to a slow release of the loaded model molecules from the polysaccharidic system; The vesicular structures may enhance the delivery of drug in the stratum corneum due to their specific constituents	[443]
Micro-droplet	ALG	Drug delivery and bone tissue engineering; Facilitate the regional regulation of encapsulated cell fate; *In situ* formation of localized, sustained bioactive molecule delivery to encapsulate stem cells for therapeutic applications	[444]
Virus	ALG	Regenerative medicine; Significant improvement in cell attachment; Mimic the biological niche of a functional tissue; Localization, delivery, and differentiation of stem cells	[445]

Poly(D,L-lactide-co-glycolide) (PLGA); Cholesterol-bearing pullulan (CHP); Gellan gum (GG); Poly(ethylene glycol) (PEI); Poly(ethylene glycol) dimethacrylate (PEGDMA); Hyaluronan/methyl cellulose (HAMC); Methylcellulose (MC); Polyamidoamine (PAMAM); Hydroxypropylmethyl cellulose derivatives (HPMC); 1,3-diphenylpropane-1,3-dione (dibenzoyl methane) (DBM); 2-(4-Benzoyl-3-hydroxyphenoxy)ethyl acrylate (BHPEA); *N*-acetylglucosamine(GlcNAc); Poly(hydroxybutyrate-co-hydroxyvalerate) (PHBV); Poly(-lactic acid) (PLA); Polypyrrole (PPy); Poly(3-hydroxybutyrate-co-3-hydroxyhexanoate) (PHBHHx); β-glycerophosphate disodium salt (GP); Propoxylated glyceryl triacylate (PET); Gellan xanthan (GX); Poly(methyl methacrylate) (PMMA); Oligopeptide-terminated poly(beta-aminoester) (PβAE); Poly(ethylene glycol)-poly(e-caprolactone)-poly(ethylene glycol) (PECE); Hydroxyethyl cellulose (HEC); Poly(lactic acid)-b-poly(ethylene glycol)-b-poly(lactic acid) with acrylate end-groups (PEG-LA-DA); Dual-reverse thermosensitive hydrogel (DRTH); Poly(2-vinyl pyridine)-b-poly(acrylic acid)-b-poly(n-butyl methacrylate) (PVP-PAA-PBMA); PEG-(Poly(ε-caprolactone-co-trimethylene carbonate)) (PEG-P(CL-co-TMC)); Polyethylene glycol dimethacrylate (PEGDMA); Poly(ε-caprolactone)-poly(ethylene glycol)-poly(ε-caprolactone) copolymer (PCEC); Poly(ethylene glycol)–poly(ε-caprolactone)-acryloyl chloride (PECA); Poly(ethylene oxide) (PEO)-b-poly(2-(*N*,*N*-diisopropylamino)ethyl methacrylate) (PEO-PDPAEMA); Polycarbonate (PC); Acryloyl group modified cholesterol-bearing pullulan (CHPOA); Dextrin (DEX); Cholesterol-bearing pullulan (CHP).

Polymeric NPs have obtained great attention in drug delivery applications because of their capability to entrap hydrophobic or hydrophilic drugs. Among the various polymer materials used in drug delivery and biomedical devices, poly(lactic-co-glycolic acid) (PLGA) has attracted most considerable attention because of its attractive properties, such as excellent biocompatibility and biodegradability, European Medicine Agency approval and FDA in drug delivery systems. Typically, PLGA-based polymeric NPs have been integrated into different hydrogels and the resulting nanocomposite hydrogels have shown promise to develop novel biomedical materials with useful functions. Lampe *et al.* [446] entrapped PLGA microparticles within a PEG-based hydrogel to locally release two types of neurothropic factors with different release profiles. PLGA microparticles incorporated hydrogels possessing coordinated drug delivery ability can release two different neurothropic factors in different regions of the brain, significantly reducing the microglial response relative to sham surgeries, and thus providing a useful tool for further development in combining drug delivery and tissue engineering. Hydrogels have been extensively investigated as a carrier for loading and local delivery of various hydrophilic molecular cells and compounds. To tailor the release rates of multiple drugs and also manipulate the phenotypic activities of co-cultured cells in a hydrogel system, Yonet-Tanyeri *et al.* [388] separately incorporated two isoforms of vascular endothelial growth factor (VEGF121 and VEGF165), into the poly(ethylene imine)-poly(ethylene glycol) diacrylate (PEI-PEGDA) gel and PLGA microparticles. And then the PEI-PEGDA gel and PLGA microparticles were embedded in the poly(ethylene glycol) dimethacrylate PEGDMA hydrogel. The PLGA microparticles were used as the VEGF165 carrier to limit initial VEGF165 burst. PEI-PEGDA gel was used as VEGF121 carrier. Apparently, a bimodal, sequential macromolecular release profile could be attained, including rapid VEGF121 release with the degradation of the PEI-PEGDA gel and sustained VEGF165 release from the PLGA microparticles. A novel polymer coating consisting of PLGA microspheres dispersed in PVA hydrogels was developed in combination with dummy sensors as a "smart" drug eluting coating with high biocompatibility for implantable biosensors to avoid the foreign body response, and thus improve sensor performance *in vivo* [376]. The polymeric microspheres slowly release tissue-modifying drugs at the implantation sites to control the inflammation and fibrous encapsulation, while the hydrogel allows rapid analyte diffusion to the sensing elements. Samanta *et al.* [407] reported a generalized pH-sensitive drug delivery system based on polypyrrole nanoparticles (PPy NPs) with good drug loading capacity, which can be immobilized in a calcium alginate hydrogel, allowing for sustained release of the incorporated drug for more than 21 days. Luo *et al.* [408] synthesized a kind of NIR laser responsive material from agarose/alginate double network hydrogels composited with PPy NPs. As shown in Figure 11, the combination of alginate and agarose was synergistic. The agarose/PPy network provides the hydrogel with capability of responding to light. On the other hand, non-thermal-responsive alginate keeps the mechanical properties of the hydrogel during irradiation. Moreover the agarose hydrogel network was able to form a supporting template for Ca^{2+} diffusion control, which regulated alginate gelation processes and forms objects with shapes. As a result, the hybrid hydrogel exhibited pulsatile releasing behaviors according to the laser switching while maintaining its morphology and mechanical strength.

Wang *et al.* [396] reported an advanced hybrid hydrogel that integrates toxin nanosponges (red blood cell (RBC) membrane-coated PLGA nanoparticle system) with PAM hydrogel for local treatment of bacterial infection, as shown inFigure 12. The hydrogel can effectively withhold the nanosponges in its matrix without compromising toxin transport for neutralization. After subcutaneous injection to mice, the hybrid hydrogel can also effectively anchor the nanosponges at the injection sites (a property facilitating toxin absorption), showing significant antivirulence therapeutic efficacy. This study demonstrated the incorporation of the nanosponges into hydrogel provide a new and effective detoxification strategy for the treatment of localized bacterial infection. In addition of PLGA particles, a wide variety of polymeric NPs or microparticles have been used to combine with hydrogel system to form nanocomposite hydrogels with unique physicochemical and biological properties, as shown in Table 4. These approaches have provided a promising novel platform for drug delivery and tissue engineering, especially for sustained delivery of multiple drugs at respective rate.

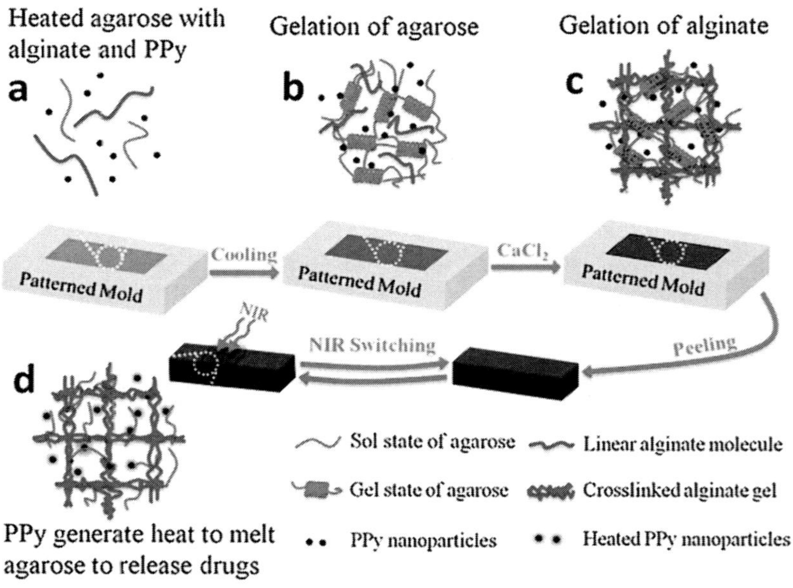

Figure 11. Schematic illustration of a NIR laser responsive deformation-free hydrogel containing embedded polypyrrole (PPy) NPs: (**a**) pre-heated agarose solution was mixed with alginate and PPy NPs and injected into patterned mold; (**b**) cooling to induce gelation of agarose, determining shape of the hydrogel and served as a template for alginate gelation; (**c**) $CaCl_2$ was added to form an alginate network, which provided the mechanical support during laser irradiation; (**d**) periodic switching of NIR laser to digitally control agarose melting for pulsatile drug release, reproduced with permission from [408]. Copyright Royal Society of Chemistry, 2014.

Polymer micelles are a typical type of nanoscale polymeric particles, which are generated from self-assembly of amphiphilic block-copolymers in water. They present a core-shell structure wherein the hydrophilic corona imparts colloidal stability against aggregation while the hydrophobic core can serve as a microenvironment for the incorporation of drugs. Due to their abundant chemical diversity, functionality and tunable physicochemical properties, micelles are one of the most versatile nanocarriers for the delivery of hydrophobic drugs [34]. The incorporation of the nanosized micelles into hydrogels provides not only a multiple drug delivery system, but also an excellent means to tune physical properties of the hydrogels. Jin *et al.* [433] developed pH-responsive block copolymer micelles as drug carriers for use in hydrogel scaffolds for neural repair, as shown in Figure 13. These micelles are formed from PEG-b-poly(2-(*N,N*-diisopropylamino)ethyl methacrylate) (PEG-b-PDPAEMA) with pH sensitivity due to the tertiary amine groups of PDPAEMA. The incorporation of these micelles into agarose hydrogels to obtain hybrid hydrogels with capable of stimulus-triggered drug release, which could allow the controlled delivery of hydrophobic neuroprotective agents to improve survival of transplanted cells in tune with signals from the surrounding pathological environment. Micelles have also been incorporated into hydrogels to enhance their mechanical properties. Xiao and coworkers [447] covalently incorporated polymer self-assembled micelles into PAM hydrogel. The mechanical properties of the micelle-linked PAM hydrogel can be regulated by adjusting the amount of micelles. 4-fold increase in Young's modulus and 2-fold increase in tensile stress were observed when the concentration of micelles was increased from 7.5 to 15 mg/mL.

Liposomes are a kind of micelles from the self-assembly of phospholipids dispersed in an aqueous environment. These vesicles contain an aqueous volume entirely enclosed by a lipid bilayer membrane. Manipulation of composition, size, charge and lamellarity of liposomes allow the entrapment of payload both in the aqueous compartment and within the membrane. Hydrogels embedded with liposomes may generate a drug delivery system with desirable features and superior performance. Weiner *et al.* [448] may be the first to introduce liposomes into collagen gel. They observed that the release rate of payloads encapsulated in liposomes decreased in the presence of the hydrogel. The use of liposomes for drug incorporation within the matrix of hydrogel contact lenses is especially promising due to their thermodynamic stability and high drug-loading capacity [449]. Especially, this kind of nanocomposite hydrogels enables the possibility of achieving a controlled and sustained drug release in the eye without compromising the optical performance. In other words, this strategy can avoid the major limitation of common dosage (eye drops), which is the extremely short residence time of drug in the cornea because of the rapid clearance and the dilutive effect of tears [449]. Gulsen *et al.* [450] encapsulated various drugs in liposomes consisting of dimyristoyl phosphatidylcholine (DMPC) followed by incorporation into the hydrogel contact lenses. It was found that these hydrogel can control drug release for up to 8 days. Wickremasinghe *et al.* [423] created a unique nanocomposite hydrogel by stepwise self-assembly of liposomes and multidomain peptide fibers, as shown in Figure 14A. The two-component system can achieve controlled release of desired growth factors and cytokines over multiple time scale. This bimodal release system would have great potential in the field of tissue engineering and regenerative medicine.

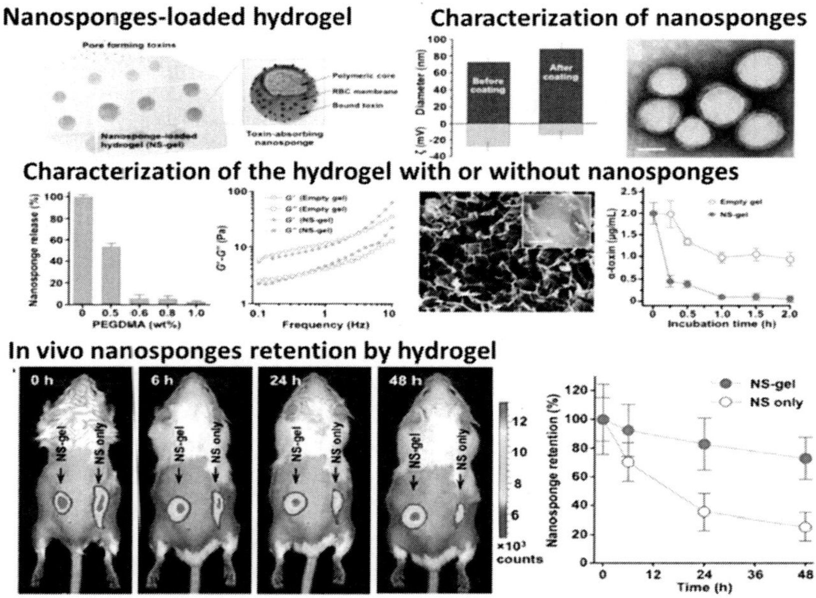

Figure 12. Schematic illustration of a hydrogel retaining toxin-absorbing nanosponges (NS-gel) for local treatment of bacterial infection. The toxin nanosponge was constructed with a poly(D,L-lactide-co-glycolide) (PLGA) core wrapped in natural red blood cell (RBC) bilayer membrane and was subsequently embedded into polyacrylamide (PAM) hydrogel, reproduced with permission from [396]. Copyright John Wiley and Sons, 2015.

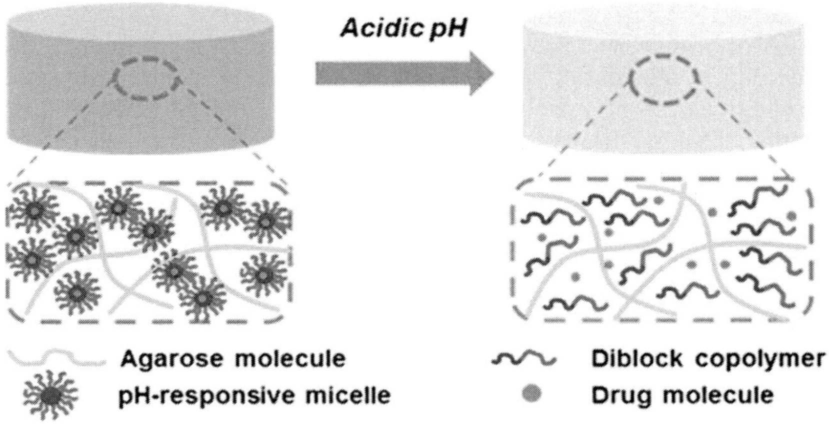

Figure 13. Schematic illustration of agarose hydrogels embedded with pH-responsive diblock copolymer micelles for triggered release of substances, reproduced with permission from [433]. Copyright American Chemical Society, 2013.

Polymer vesicles, usually called polymersomes, are sphere-like shell structures in which an aqueous compartment is enclosed by a bilayer membrane generatedfrom the self-assembly of amphiphilic block copolymers [451]. In comparison with liposomes, polymer vesicles possessing better stability, higher toughness, tailorable membrane properties are attractive candidates for applications including drug delivery, encapsulation and nanoreactors. Polymeric vesicles have also been embedded in hydrogel to prepare functional nanocomposite hydrogels. For example, Litvinchuk *et al.* [442] reported a long-term stability of poly(2-methyl oxazoline-block-polydimethylsiloxane-block poly(2-methyl oxazoline) (PMOXA-PDMS-PMOXA) vesicles loaded in PVA hydrogels. The study found that the vesicles in the hydrogel are stable and the drugs could diffuse quickly throughout the PVA hydrogel, providing a promising controlled release system for water soluble drugs. Hao*et al.* [441] developed a thermal-responsive self-healing nanocomposite hydrogel by mixing hydrophobically modified chitosan with vesicle solution. This vesicle-based nanocomposite hydrogel exhibited an interesting self-healing feature in a matter of seconds through the autonomic reconstruction without the use of healing agent. Mart *et al.* [452] reported a new type of tissue-mimetic material, magnetically-sensitive vesicle-embedded composite ALG hydrogels. These vesicle gels contain vesicles crosslinked by Fe_3O_4 magnetic nanoparticles, as shown in Figure 14B. And these magnetic nanoparticle–vesicle assemblies render the prepared hydrogels the ability of hosting membrane-bound enzymes/glycolipids and responding to alternating magnetic fields to reach a magnetic trigger-controlled release of stored drugs or bio(macro)molecules, leading to potential applications as smart scaffolds for stem cells or remotely triggered *in vivo* drug delivery systems.

Typically, nanogel, a nanosized crosslinked polymer network, is an ideal nanocarrier for biomacromolecular drugs, as it can be effective in minimizing protein-denaturation by trapping the protein drug in a hydrated microenvironment. In one study, protein-loaded nanogels were physically encapsulated into a hyaluronan (HYA) hydrogel system for sustained delivery of protein [440]. The proteins were spontaneously trapped into the nanocomposite hydrogel without any obvious denaturation by the physical entrapment of nanogels in the HYA gel. The long-term release of protein both *in vitro* and *in vivo* was achieved by controlling the density of the HYA gel. This study offered a biocompatible, safe and sustained release formulation for proteins without the problem of denaturation of proteins in the formulation and during the formulation process. Park *et al.* [453] reported a hybrid hydrogel system composed of an *in situ* cross-linkable gelatin-poly(ethylene glycol)-tyramine (GPT) hydrogel and a self-assembled heparin-pluronic (HP) nanogel. As shown in Figure 14C, the hybrid hydrogels were prepared via enzymatic reaction in the presence of horseradish peroxidase and hydrogen peroxide. The injectable GPT hydrogels can play a role as a physical bulking reagent and the growth factor (GF)-loaded HP nanogels allow sustained release of GFs from the matrix to improve urethral tissue regeneration. The results indicated that the hybrid gel matrix can release GF continuously up to 28 days and stimulate the regeneration of the urethral muscle tissue around the urethral wall, which may offer an advanced therapeutic method for treating the urinary incontinence as well as an application for regenerative medicine.

Among the various types of polymeric NPs, dendrimer, a synthetic polymeric macromolecule of nanometer dimensions, are particularly fascinating, due to its unparalleled molecular uniformity, narrow molecular weight distribution, defined size and shape characteristics, highly branched structure, available internal cavity as well as a multifunctional terminal surface. These physicochemical characteristics of dendrimers, especially the multitude of functional groups on their periphery and abundant cavities within their interior, lead to higher reactivity and loading efficiency compared to the linear polymers, which make these macromolecules appropriate candidates for drug delivery [417,454,455]. Dendritic nanoparticles can also be incorporated into the hydrogel network via covalent or non-covalent interactions between dendrimers and the polymeric chains to obtain reinforced nanocomposite hydrogels with specific performance. For example, Desai *et al.* [420] developed a photocurable PEG hydrogel crosslinked with polyamidoamine (PAMAM) dendrimers, and the cytocompatibility, swelling and degradation of the resulting nanocomposite hydrogels can be controlled by modification of the chain length and charge density of PAMAM dendrimers. In another study, PAMAM NPs were physically integrated within COL scaffolds to improve the mechanical stiffness and structural integrity [416]. Interestingly, the human conjunctival fibroblast proliferation in COL hydrogels was significantly improved due to that the addition of PAMAM NPs can increase the stiffness of nanocomposite network. The use of these dendrimers-based nanocomposite hydrogels are envisioned in biomedical and pharmaceutical fields that require porous structure with controlled drug release properties [3]. Söntjens *et al.* [456] reported a novel hydrogel scaffold for cartilage repair, based on a water-soluble and multivalent triblock copolymer consisting of methacrylated poly(glycerol succinic acid) dendrimer terminal blocks and a poly(ethylene glycol) core. The terminal methacrylates allow biocompatible and mild photo-crosslinking with a visible light, which will facilitate *in vivo* filling of irregularly shaped defects with the dendrimer-based scaffold. The multivalent dendrimer components allow high crosslink densities, which will inhibit the swelling after crosslinking and simultaneously introduce biodegradation sites. The degradation properties, mechanical properties and water content of the hydrogel can easily be tailored by changing the biodendrimer concentration. Zhang *et al.* [457] reported a photocrosslinkable hyperbranched polyesters (HEP)-based hydrogel with sustained drug release profiles for cellular therapies. These photocrosslinkable HPE not only can encapsulate hydrophobic drug molecules in the HPE cavities, but also can form hybrid hydrogels with a mechanically tough network and a highly porous interconnected structure upon UV exposure. The physical properties of the resulting hydrogels, such as the microstructure, compressive modulus and drug retainability, as well as cell adhesion and proliferation can be tunable by changing the crosslinking density. The nanocomposite hydrogel showed a spatial control of cell adhesion and controlled release of hydrophobic drug, which is difficult to achieve when using conventional hydrogels made from linear polymers. Wang *et al.* [418] reported a PAMAM dendrimer-based poly(lactic acid)-poly(ethylene glycol)-poly(lactic acid) (PLA-PEG-PLA) hydrogel with a highly porous structure. In comparison with hydrogel without incorporating dendrimer, the nanocomposite hydrogels exhibit reduced swelling, enhanced mechanical stiffness, and better cell adhesion, differentiation, and proliferation. These studies indicated that the hyperbranched dendrimers can be used to reinforce polymeric network and impart hydrogels a stable attachment and a high binding capacity to proteins and drugs. The nanocomposite hydrogels from dendrimers NPs with tailored physicochemical properties exhibited a promising potential to serve as a model for developing advanced materials for various tissue engineering and drug delivery applications.

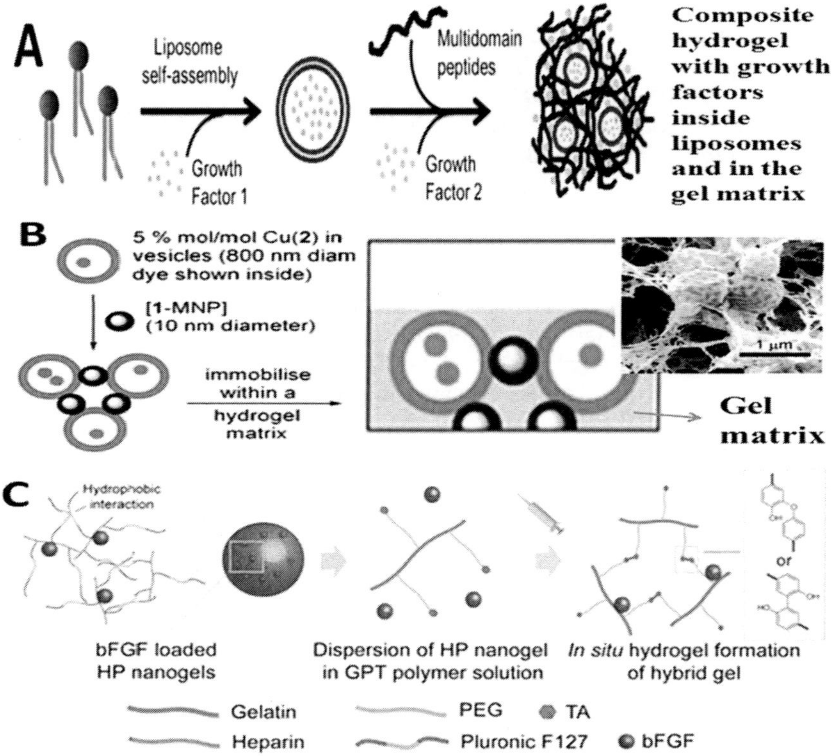

Figure 14. Schematic illustrations of nanocomposite hydrogels (**A**): from hydrogel and liposomes, and the formation of hybrid hydrogels by stepwise orthogonal self-assembly, reproduced with permission from [423]. Copyright American Chemical Society, 2014; (**B**): from hydrogel and vesicles, the formation of hybrid hydrogels by the sol–gel transition of vesicle-embedded composite hydrogels, reproduced with permission from [452]. Copyright Royal Society of Chemistry, 2009; (**C**): from hydrogel and polymer nanogels, and the formation of hybrid hydrogels via an enzymatic reaction, reproduced with permission from [453]. Copyright American Chemical Society, 2014

Overall, the fabrication of composite materials by integrating polymer NPs into hydrogels can not only surmount individual shortcomings, but also generate newborn characters. The hydrogels can preserve the structural integrity and the functionalities of the contained NPs and anchor the NPs at the target site. Meanwhile, NPs entrapped into hydrogel allow a more flexible loading and controlled release of drugs. Apparently, due to the synergistic benefits, mechanically stiff hydrogels with controlled drug delivery characteristics will provide an efficient route to improve the performances and expand application scopes of hydrogels and NPs. However, it was worth pointing out that additional animal experiments are required to examine their behavior *in vivo*, as most of the studies on polymer NPs-based hybrid hydrogels have been performed *in vitro*.

4. CONCLUSIONS AND FUTURE PERSPECTIVE

During the past several decades, hydrogels and nanoparticles have already exerted a dramatic impact in biological, biomedical, pharmaceutical and diagnostic fields. However, their some intrinsic shortcomings severely restrict their practical applications. Significant efforts have been paid to improve the performance of hydrogels and NPs. The fabrication of composite materials by combining two or more components in a single entity can surmount individual shortcomings and give rise to synergistic functions that are absent in the individual components. The incorporation of NPs in three-dimensional polymeric hydrogel matrix as an innovative means to obtain nanocomposite hydrogels with improved properties and multiple functionalities has gained enormous attention in many areas. On the basis of this review, we can found that various types of NPs, such as carbon-based NPs, silicon-based NPs, metal NPs and polymeric NPs are combined with the polymeric hydrogel network to create multicomponent systems. The porous structure and free space within the hydrogel networks can not only provide an ideal hydrated environment for the stabilization of NPs without aggregations or disintegration and the protection of NPs from degradation or denaturation, but also work as a reservoir to localize NPs at the target site. More importantly, the coatings of hydrogels around the NPs endow a hydration layer, which often is an essential prerequisite for the biomedical and biological applications of inorganic and metal NPs, as the hydration layer can significantly improve the biocompatibility and reduce the cytotoxicity. On the other hand, the incorporation of NPs into hydrogels can not only markedly improve their mechanical, elastic and adhesive properties as well as physicochemical and thermal stability, but also promote the cell attachment and proliferation and improve drug loading capacity and drug release profiles. More significantly, the encapsulation of carbon, metal colloidal particles or quantum dots into polymer hydrogel networks will impart them with exclusive thermal, sonic, optical, electrical or magnetic properties, which are not achieved by individual polymeric systems and are highly appropriate for various applications, especially for therapeutic and diagnostic applications. Apparently, the benefits of the combination of NPs and hydrogels have resulted in generation of a new class of advanced materials with unique properties that have a wide spectral range of biomedical applications, ranging from controlled drug delivery depots, cell and tissue adhesive matrices, wound dressing and tissue engineering scaffolds, stem cell engineering and regenerative medicines, biosensors, actuators and other biomedical devices.

However, despite of the vast potential applications of nanocomposite hydrogels in biomedical fields, there are still lots of challenges to be overcome before they can be applied in clinical use. For example, the improved performance of the nanocomposite hydrogels is mainly ascribed to the enhanced interactions between the NPs and polymer chains. Therefore, apart from particle parameters (size, shape, surface properties of NPs), the quantity and homogeneity of NPs integration are still a matter of concern. To obtain excellent performance, NPs should be abundantly and homogeneously dispersed within the hydrogel matrix. However, due to a large surface area, hydrophobic NPs physically embedded in the hydrophilic polymer matrix often tend to aggregate, leading to the failure of anticipative enhancement of properties. Therefore, there is still a need to elucidate the mechanisms and interactions between NPs and polymer chains inside the nanocomposite hydrogels, and the simple, cost-effective, scalable and reproducible preparations of nanocomposite hydrogels with desirable properties need to be investigated thoroughly. Apparently, the nanocomposite hydrogels combining the individual functions of NPs and hydrogels are ideal candidates for multimodal drug delivery platform with simultaneous capabilities in drug encapsulation, targeting delivery, photothermal therapy and *in vivo* imaging. This synergistic performance is usually not be achieved by an individual. However, the more detailed pharmacological studies need to demonstrate the therapeutic efficacy and biological responses. In addition, most of nanocomposite hydrogels have been performed only *in vitro*. Therefore, in future studies, the detailed performances (gelation time, swelling, elastic modulus, responsivity and functionality), long-term toxicity and biodegradability as well as the biological properties such as protein adsorption, cell adhesion, tissue compatibility and whole-body effect of such hydrogels under *in vivo* conditions should be clearly addressed. In summary, there should be a coordinated and comprehensive research to establish fundamental interactions among nanoparticulate materials, hydrogel matrices and biological systems, before hydrogels will become practical and useful in this exciting field. All these challenges will drive the effective collaboration of scientists from the fields of chemistry, materials, engineering, biology, medicine and nanoscience. We look forward to seeing many exciting research accomplishments in this burgeoning field of bio-related nanocomposite hydrogels.

ACKNOWLEDGMENTS

This project was supported by the National Natural Science Foundation of China (No. 31470925) and Tianjin Research Program of Application Foundation and Advanced Technology (No. 15JCQNJC03000).

AUTHOR CONTRIBUTIONS

J.Z and A.D. designed the subject of this review and wrote the manuscript. F.Z., D.Y., R.G. and L.D. conceived the project, discussed several related articles, and wrote the manuscript. All authors reviewed and approved the manuscript.

CONFLICTS OF INTEREST

The authors declare no conflict of interest.

REFERENCES

1. Hoare, T.R.; Kohane, D.S. Hydrogels in drug delivery: Progress and challenges. Polymer 2008, 49, 1993–2007.

2. Li, Y.; Rodrigues, J.; Tomas, H. Injectable and biodegradable hydrogels: Gelation, biodegradation and biomedical applications. Chem. Soc. Rev. 2012, 41, 2193–2221.

3. Annabi, N.; Tamayol, A.; Uquillas, J.A.; Akbari, M.; Bertassoni, L.E.; Cha, C.; Camci-Unal, G.; Dokmeci, M.R.; Peppas, N.A.; Khademhosseini, A. 25th Anniversary article: Rational design and applications of hydrogels in regenerative medicine. Adv. Mater. 2014, 26, 85–124.

4. Caló, E.; Khutoryanskiy, V.V. Biomedical applications of hydrogels: A review of patents and commercial products.Eur. Polym. J. 2015, 65, 252–267.

5. Seliktar, D. Designing cell-compatible hydrogels for biomedical applications. Science 2012, 336, 1124–1128.

6. Hoffman, A.S. Hydrogels for biomedical applications. Adv. Drug Deliv. Rev. 2012, 64, 18–23.

7. Peppas, N.A.; Hilt, J.Z.; Khademhosseini, A.; Langer, R. Hydrogels in biology and medicine: From molecular principles to bionanotechnology. Adv. Mater. 2006, 18, 1345–1360.

8. Slaughter, B.V.; Khurshid, S.S.; Fisher, O.Z.; Khademhosseini, A.; Peppas, N.A. Hydrogels in regenerative medicine.Adv. Mater. 2009, 21, 3307–3329.

9. Miyata, T.; Uragami, T.; Nakamae, K. Biomolecule-sensitive hydrogels. Adv. Drug Deliv. Rev. 2002, 54, 79–98.

10. Qiu, Y.; Park, K. Environment-sensitive hydrogels for drug delivery. Adv. Drug Deliv. Rev. 2012, 64, 49–60.

11. Hoffman, A.S. Stimuli-responsive polymers: Biomedical applications and challenges for clinical translation. Adv. Drug Deliv. Rev. 2013, 65, 10–16.

12. Ward, M.A.; Georgiou, T.K. Thermoresponsive polymers for biomedical applications. Polymers 2011, 3, 1215–1242.

13. Wichterle, O.; Lim, D. Hydrophilic gels for biological use. Nature 1960, 185, 117–118.

14. Zhu, J.M. Bioactive modification of poly(ethylene glycol) hydrogels for tissue engineering. Biomaterials 2010, 31, 4639–4656.

15. Lin, C.-C. Recent advances in crosslinking chemistry of biomimetic poly(ethylene glycol) hydrogels. RSC Adv. 2015, 5, 39844–39853.

16. Maitra, J.; Shukla, V.K. Cross-linking in Hydrogels—A Review. Am. J. Polym. Sci. 2014, 4, 25–31.

17. Ebara, M.; Kotsuchibashi, Y.; Narain, R.; Idota, N.; Kim, Y.-J.; Hoffman, J.M.; Uto, K.; Aoyagi, T. Smart Hydrogels. InSmart Biomaterials; Ebara, M., Kotsuchibashi, Y., Narain, R., Idota, N., Kim, Y.J., Hoffman, J.M., Uto, K., Aoyagi, T., Eds.; Springer: Tokyo, Japan, 2014; pp. 9–65.

18. Huang, P.; Wang, W.; Zhou, J.; Zhao, F.; Zhang, Y.; Liu, J.; Liu, J.; Dong, A.; Kong, D.; Zhang, J. Amphiphilic polyelectrolyte/prodrug nanoparticles constructed by synergetic electrostatic and hydrophobic interactions with cooperative pH-sensitivity for controlled doxorubicin delivery. ACS Appl. Mater. Interfaces 2015, 7, 6340–6350.

19. Zhang, J.; Lin, X.; Liu, J.; Zhao, J.; Dong, H.; Deng, L.; Liu, J.; Dong, A. Sequential thermo-induced self-gelation and acid-triggered self-release process of drug-conjugated nanoparticles: A strategy for sustained and controlled drug delivery to tumor. J. Mater. Chem. B 2013, 1, 4667–4677.

20. Thomas, V.; Namdeo, M.; Mohan, Y.M.; Bajpai, S.K.; Bajpai, M. Review on polymer, hydrogel and microgel metal nanocomposites: A facile nanotechnological approach. J. Macromol. Sci. Pure Appl. Chem. 2008, 45, 107–119.

21. Hu, Z.; Chen, G. Novel nanocomposite hydrogels consisting of layered double hydroxide with ultrahigh tensibility and hierarchical porous structure at low inorganic content. Adv. Mater. 2014, 26, 5950–5956.

22. Merino, S.; Martin, C.; Kostarelos, K.; Prato, M.; Vazquez, E. Nanocomposite Hydrogels: 3D Polymer-Nanoparticle Synergies for On-Demand Drug Delivery. ACS Nano 2015, 9, 4686–4697.

23. Bonanno, L.M.; Segal, E. Nanostructured porous silicon-polymer-based hybrids: From biosensing to drug delivery.Nanomedicine 2011, 6, 1755–1770.

24. Ye, E.; Loh, X.J. Polymeric Hydrogels and Nanoparticles: A Merging and Emerging Field. Aust. J. Chem. 2013, 66, 997–1007.

25. Gaharwar, A.K.; Peppas, N.A.; Khademhosseini, A. Nanocomposite hydrogels for biomedical applications. Biotechnol. Bioeng. 2014, 111, 441–453.

26. Xu, S.; Deng, L.; Zhang, J.; Yin, L.; Dong, A. Composites of electrospun-fibers and hydrogels: A potential solution to current challenges in biological and biomedical field. J. Biomed. Mater. Res. B 2015.

27. Thoniyot, P.; Tan, M.J.; Karim, A.A.; Young, D.J.; Loh, X.J. Nanoparticle-hydrogel composites: Concept, design,and applications of these promising, multi-functional materials. Adv. Sci. 2015, 1–2.

28. De, M.; Ghosh, P.S.; Rotello, V.M. Applications of Nanoparticles in Biology. Adv. Mater. 2008, 20, 4225–4241.

29. Thanh, N.T.K.; Green, L.A.W. Functionalisation of nanoparticles for biomedical applications. Nano Today 2010, 5, 213–230.

30. Kim, T.; Hyeon, T. Applications of inorganic nanoparticles as therapeutic agents. Nanotechnology. 2014, 25.

31. Doane, T.L.; Burda, C. The unique role of nanoparticles in nanomedicine: Imaging, drug delivery and therapy. Chem. Soc. Rev. 2012, 41, 2885–2911.

32. Nicolas, J.; Mura, S.; Brambilla, D.; Mackiewicz, N.; Couvreur, P. Design, functionalization strategies and biomedical applications of targeted biodegradable/biocompatible polymer-based nanocarriers for drug delivery. Chem. Soc. Rev.2013, 42, 1147–1235.

33. Arvizo, R.R.; Bhattacharyya, S.; Kudgus, R.A.; Giri, K.; Bhattacharya, R.; Mukherjee, P. Intrinsic therapeutic applications of noble metal nanoparticles: Past, present and future. Chem. Soc. Rev. 2012, 41, 2943–2970.

34. Elsabahy, M.; Wooley, K.L. Design of polymeric nanoparticles for biomedical delivery applications. Chem. Soc. Rev.2012, 41, 2545–2561.

35. Hong, H.; Chen, F.; Cai, W. Pharmacokinetic issues of imaging with nanoparticles: Focusing on carbon nanotubes and quantum dots. Mol. Imaging Biol. 2013, 15, 507–520.

36. Chung, C.; Kim, Y.K.; Shin, D.; Ryoo, S.R.; Hong, B.H.; Min, D.H. Biomedical applications of graphene and graphene oxide. Acc. Chem. Res. 2013, 46, 2211–2224.

37. Li, Z.; Barnes, J.C.; Bosoy, A.; Stoddart, J.F.; Zink, J.I. Mesoporous silica nanoparticles in biomedical applications.Chem. Soc. Rev. 2012, 41, 2590–2605.

38. Bitar, A.; Ahmad, N.M.; Fessi, H.; Elaissari, A. Silica-based nanoparticles for biomedical applications. Drug Discov. Today 2012, 17, 1147–1154.

39. Soenen, S.J.; Parak, W.J.; Rejman, J.; Manshian, B. (Intra)cellular stability of inorganic nanoparticles: Effects on cytotoxicity, particle functionality, and biomedical applications. Chem. Rev. 2015, 115, 2109–2135.

40. Dou, Q.Q.; Guo, H.C.; Ye, E. Near-infrared upconversion nanoparticles for bio-applications. Mater. Sci. Eng. C 2014, 45, 635–643.

41. Min, Y.; Li, J.; Liu, F.; Padmanabhan, P.; Yeow, E.; Xing, B. Recent advance of biological molecular imaging based on lanthanide-doped upconversion-luminescent nanomaterials. Nanomaterials 2014, 4, 129–154.

42. Bhattacharya, R.; Mukherjee, P. Biological properties of "naked" metal nanoparticles. Adv. Drug Deliv. Rev. 2008, 60, 1289–1306.

43. Dykman, L.; Khlebtsov, N. Gold nanoparticles in biomedical applications: Recent advances and perspectives. Chem. Soc. Rev. 2012, 41, 2256–2282.

44. Dreaden, E.C.; Alkilany, A.M.; Huang, X.; Murphy, C.J.; El-Sayed, M.A. The golden age: Gold nanoparticles for biomedicine. Chem. Soc. Rev. 2012, 41, 2740–2779.

45. Tiwari, P.; Vig, K.; Dennis, V.; Singh, S. Functionalized gold nanoparticles and their biomedical applications.Nanomaterials 2011, 1, 31–63.

46. Wei, L.; Lu, J.; Xu, H.; Patel, A.; Chen, Z.S.; Chen, G. Silver nanoparticles: Synthesis, properties, and therapeutic applications. Drug Discov. Today 2015, 20, 595–601.

47. Gupta, A.K.; Gupta, M. Synthesis and surface engineering of iron oxide nanoparticles for biomedical applications.Biomaterials 2005, 26, 3995–4021.

48. Ling, D.; Hyeon, T. Chemical design of biocompatible iron oxide nanoparticles for medical applications. Small 2013, 9, 1450–1466.

49. Gregoriadis, G. Engineering liposomes for drug delivery: Progress and problems. Trends Biotechnol. 1995, 13, 527–537.

50. Lasic, D.D. Novel applications of liposomes. Trends Biotechnol. 1998, 16, 307–321.

51. Torchilin, V.P. Recent advances with liposomes as pharmaceutical carriers. Nat. Rev. Drug Discov. 2005, 4, 145–160.

52. Shang, L.; Nienhaus, K.; Nienhaus, G.U. Engineered nanoparticles interacting with cells: Size matters. J. Nanobiotechnol. 2014, 12, 5–15.

53. Verma, A.; Stellacci, F. Effect of surface properties on nanoparticle-cell interactions. Small 2010, 6, 12–21.

54. Duan, X.; Li, Y. Physicochemical characteristics of nanoparticles affect circulation, biodistribution, cellular internalization, and trafficking. Small 2013, 9, 1521–1532.

55. Jo, D.H.; Kim, J.H.; Lee, T.G.; Kim, J.H. Size, surface charge, and shape determine therapeutic effects of nanoparticles on brain and retinal diseases. Nanomedicine 2015, 11, 1603–1611.

56. Zhang, L.; Li, Y.; Yu, J.C. Chemical modification of inorganic nanostructures for targeted and controlled drug delivery in cancer treatment. J. Mater. Chem. B 2014, 2, 452–470.

57. Kobayashi, K.; Wei, J.; Iida, R.; Ijiro, K.; Niikura, K. Surface engineering of nanoparticles for therapeutic applications.Polym. J. 2014, 46, 460–468.

58. Albanese, A.; Tang, P.S.; Chan, W.C. The effect of nanoparticle size, shape, and surface chemistry on biological systems. Annu. Rev. Biomed. Eng. 2012, 14, 1–16.

59. Erathodiyil, N.; Ying, J.Y. Functionalization of inorganic nanoparticles for bioimaging applications. Acc. Chem. Res.2011, 44, 925–935.

60. Deng, H.; Liu, J.; Zhao, X.; Zhang, Y.; Liu, J.; Xu, S.; Deng, L.; Dong, A.; Zhang, J. PEG-b-PCL copolymer micelles with the ability of pH-controlled negative-to-positive charge reversal for intracellular delivery of doxorubicin.Biomacromolecules 2014, 15, 4281–4292.

61. Shin, S.; Song, I.; Um, S. Role of physicochemical properties in nanoparticle toxicity. Nanomaterials 2015, 5, 1351–1365.

62. Decuzzi, P.; Lee, S.; Bhushan, B.; Ferrari, M. A theoretical model for the margination of particles within blood vessels.Ann. Biomed. Eng. 2005, 33, 179–190.

63. Zhao, J.; Wang, H.; Liu, J.; Deng, L.; Liu, J.; Dong, A.; Zhang, J. Comb-like amphiphilic copolymers bearing acetal-functionalized backbones with the ability of acid-triggered hydrophobic-to-hydrophilic transition as effective nanocarriers for intracellular release of curcumin. Biomacromolecules 2013, 14, 3973–3984.

64. Song, F.; Li, X.; Wang, Q.; Liao, L.; Zhang, C. Nanocomposite hydrogels and their applications in drug delivery and tissue engineering. J. Biomed. Nanotechnol. 2015, 11, 40–52.

65. Usuki, A.; Kojima, Y.; Kawasumi, M.; Okada, A.; Fukushima, Y.; Kurauchi, T.; Kamigaito, O. Mechanical properties of nylon 6-clay hybrid. J. Mater. Res. 1993, 8, 1179–1184.

66. Yano, K.; Usuki, A.; Okada, A.; Kurauch, T.; Kamigaito, O. Synthesis and properties of polyimide-clay hybrid. J. Polym. Sci. A Polym. Chem. 1993, 31, 2493–2498.

67. Kokabi, M.; Sirousazar, M.; Hassan, Z.M. PVA-clay nanocomposite hydrogels for wound dressing. Eur. Polym. J. 2007,43, 773–781.

68. Liu, Y.; Meng, H.; Konst, S.; Sarmiento, R.; Rajachar, R.; Lee, B.P. Injectable dopamine-modified poly(ethylene glycol) nanocomposite hydrogel with enhanced adhesive property and bioactivity. ACS Appl. Mater. Interfaces 2014, 6, 16982–16992.

69. Guo, R.; Du, X.; Zhang, R.; Deng, L.; Dong, A.; Zhang, J. Bioadhesive film formed from a novel organic-inorganic hybrid gel for

transdermal drug delivery system. Eur. J. Pharm. Biopharm. 2011, 79, 574–583.

70. Mohan, Y.M.; Lee, K.; Premkumar, T.; Geckeler, K.E. Hydrogel networks as nanoreactors: A novel approach to silver nanoparticles for antibacterial applications. Polymer 2007, 48, 158–164.

71. Suvorova, E.I.; Klechkovskaya, V.V.; Kopeikin, V.V.; Buffat, P.A. Stability of Ag nanoparticles dispersed in amphiphilic organic matrix. J. Cryst. Growth 2005, 275, 2351–2356.

72. Xu, S.; Zhang, J.; Paquet, C.; Lin, Y.; Kumacheva, E. From hybrid microgels to photonic crystals. Adv. Funct. Mater.2003, 13, 468–472.

73. Sershen, S.R.; Westcott, S.L.; Halas, N.J.; West, J.L. Independent optically addressable nanoparticle-polymer optomechanical composites. Appl. Phys. Lett. 2002, 80, 4609–4611.

74. Holtz, J.H.; Asher, S.A. Polymerized colloidal crystal hydrogel films as intelligent chemical sensing materials. Nature1997, 389, 829–832.

75. Xia, Y.; Yang, P.; Sun, Y.; Wu, Y.; Mayers, B.; Gates, B.; Yin, Y.; Kim, F.; Yan, H. One-dimensional nanostructures: Synthesis, characterization, and applications. Adv. Mater. 2003, 15, 353–389.

76. Pardo-Yissar, V.; Gabai, R.; Shipway, A.N.; Bourenko, T.; Willner, I. Gold nanoparticle/hydrogel composites with solvent-switchable electronic properties. Adv. Mater. 2001, 13, 1320–1323.

77. Sheeney-Haj-Ichia, L.; Sharabi, G.; Willner, I. Control of the electronic properties of thermosensitive poly(N-isopropylacrylamide) and Au-nano-particle/poly(N-isopropylacrylamide) composite hydrogels upon phase transition.Adv. Funct. Mater. 2002, 12, 27–31.

78. Jones, C.D.; Lyon, L.A. Photothermal patterning of microgel/gold nanoparticle composite colloidal crystals. J. Am. Chem. Soc. 2003, 125, 460–465.

79. Wang, C.; Flynn, N.T.; Langer, R. Controlled structure and properties of thermoresponsive nanoparticle-hydrogel composites. Adv. Mater. 2004, 16, 1074–1079.

80. Castaneda, L.; Valle, J.; Yang, N.; Pluskat, S.; Slowinska, K. Collagen cross-linking with Au nanoparticles.Biomacromolecules 2008, 9, 3383–3388.

81. Rose, S.; Prevoteau, A.; Elziere, P.; Hourdet, D.; Marcellan, A.; Leibler, L. Nanoparticle solutions as adhesives for gels and biological tissues. Nature 2014, 505, 382–385.

82. Wu, H.; Yu, G.; Pan, L.; Liu, N.; McDowell, M.T.; Bao, Z.; Cui, Y. Stable Li-ion battery anodes by in-situ polymerization of conducting hydrogel to conformally coat silicon nanoparticles. Nat. Commun. 2013, 4.

83. Cha, C.; Shin, S.R.; Annabi, N.; Dokmeci, M.R.; Khademhosseini, A. Carbon-Based Nanomaterials: Multifunctional Materials for Biomedical Engineering. ACS Nano 2013, 7, 2891–2897.

84. Goenka, S.; Sant, V.; Sant, S. Graphene-based nanomaterials for drug delivery and tissue engineering. J. Control. Release 2014, 173, 75–88.

85. Adhikari, B.; Banerjee, A. Short peptide based hydrogels: Incorporation of graphene into the hydrogel. Soft Matter2011, 7, 9259–9266.

86. Shin, S.R.; Hojae, B.; Cha, J.M.; Mun, J.Y.; Chen, Y.C.; Tekin, H.; Shin, H.; Farshch, S.; Dokmeci, M.R.; Tang, S.; et al. Carbon nanotube reinforced hybrid microgels as scaffold materials for cell encapsulation. ACS Nano 2012, 6, 362–372.

87. Arslantunali, D.; Budak, G.; Hasirci, V. Multiwalled CNT-pHEMA composite conduit for peripheral nerve repair. J. Biomed. Mater. Res. A 2014, 102, 828–841.

88. Davaran, S.; Alimirzalu, S.; Nejati-Koshki, K.; Nasrabadi, H.T.; Akbarzadeh, A.; Khandaghi, A.A.; Abbasian, M.; Alimohammadi, S. Physicochemical characteristics of Fe3O4 magnetic nanocomposites based on poly(N-isopropylacrylamide) for anti-cancer drug delivery. Asian Pac. J. Cancer Prev. 2014, 15, 49–54.

89. Evingur, G.A.; Pekcan, O. Effect of multiwalled carbon nanotube (MWNT) on the behavior of swelling of polyacrylamide-MWNT composites. J. Reinf. Plast. Comp. 2014, 33, 1199–1206.

90. Curcio, M.; Spizzirri, U.G.; Cirillo, G.; Vittorio, O.; Picci, N.; Nicoletta, F.P.; Iemma, F.; Hampel, S. On demand delivery of ionic drugs from electro-responsive CNT hybrid films. RSC Adv. 2015, 5, 44902–44911.

91. Toumia, Y.; Orlanducci, S.; Basoli, F.; Licoccia, S.; Paradossi, G. "Soft" confinement of graphene in hydrogel matrixes.J. Phys. Chem. B 2015, 119, 2051–2061.

92. Baniasadi, H.; Ramazani, S.A.A.; Mashayekhan, S. Fabrication and characterization of conductive chitosan/gelatin-based scaffolds for nerve tissue engineering. Int. J. Biol. Macromol. 2015, 74, 360–366.

93. Tian, L.; Pei, X.; Zeng, Y.; He, R.; Li, Z.; Wang, J.; Wan, Q.; Li, X. Functionalized nanoscale graphene oxide for high efficient drug delivery of cisplatin. J. Nanopart. Res. 2014, 16, 1–4.

94. Guo, Q.F.; Cao, H.; Li, X.H.; Liu, S.W. Thermosensitive hydrogel drug delivery system containing doxorubicin loaded CS–GO nanocarriers for controlled release drug in situ. Mater. Technol. 2015, 30, 294–300.

95. Byrne, M.T.; Guńko, Y.K. Recent advances in research on carbonnanotube-polymer composites. Adv. Mater. 2010, 22, 1672–1688.

96. Li, W.; Liu, M.; Chen, H.; Xu, J.; Gao, Y.; Li, H. Phenylboronate-diol crosslinked polymer/SWCNT hybrid gels with reversible sol-gel transition. Polym. Adv. Technol. 2014, 25, 233–239.

97. Huang, X.; Qi, X.; Boey, F.; Zhang, H. Graphene-based composites. Chem. Soc. Rev. 2012, 41, 666–686.

98. Novoselov, K.S.; Falko, V.I.; Colombo, L.; Gellert, P.R.; Schwab, M.G.; Kim, K. A roadmap for graphene. Nature 2012,490, 192–200.

99. Wu, J.; Chen, A.; Qin, M.; Huang, R.; Zhang, G.; Xue, B.; Wei, J.; Li, Y.; Cao, Y.; Wang, W. Hierarchical construction of a mechanically stable peptide-graphene oxide hybrid hydrogel for drug delivery and pulsatile triggered release in vivo. Nanoscale 2015, 7, 1655–1660.

100. Li, Y.; Yuan, H.; Vondem-Bussche, A.; Creighton, M.; Hurt, R.H.; Kane, A.B.; Gaoa, H. Graphene microsheets enter cells through spontaneous membrane penetration at edge asperities and corner sites. Proc. Natl. Acad. Sci. USA 2013,110, 12295–12300.

101. Mac Kenna, N.; Calvert, P.; Morrin, A.; Wallacec, G.G.; Moulton, S.E. Electro-stimulated release from a reduced graphene oxide composite hydrogel. J. Mater. Chem. B 2015, 3, 2530–2537.

102. Shen, J.; Xu, G.; Xin, X.; Wang, L.; Song, Z.; Zhang, H.; Tong, L.; Yang, Z. Supramolecular hydrogels of α-cyclodextrin/reverse poloxamines/carbon-based nanomaterials and its multi-functional application. RSC Adv. 2015, 5, 40173–40182.

103. Luo, H.; Xiong, G.; Yang, Z.; Raman, S.R.; Si, H.; Wan, Y. A novel three-dimensional graphene/bacterial cellulose nanocomposite prepared by in situ biosynthesis. RSC Adv. 2014, 4, 14369–14372.

104. Kun, J.; Ye, C.; Zhang, P.; Wang, X.; Zhao, Y. One-pot controlled synthesis of homopolymers and diblock copolymers grafted graphene oxide using couplable raft agents. Macromolecules 2012, 45, 1346–1355.

105. Mohammad Mahdi Dadfar, S.; Kavoosi, G.; Mohammad Ali Dadfar, S. Investigation of mechanical properties, antibacterial features, and water vapor permeability of polyvinyl alcohol thin films reinforced by glutaraldehyde and multiwalled carbon nanotube. Polym. Compos. 2014, 35, 1736–1743.

106. Choi, E.J.; Shin, J.; Khaleel, Z.H.; Cha, I.; Yun, S.-H.; Cho, S.W.; Song, C. Synthesis of electroconductive hydrogel films by an electro-controlled click reaction and their application to drug delivery systems. Polym. Chem. 2015, 6, 4473–4478.

107. Wang, P.; Wang, S.; Lian, L.; Liu, Y.; Zhang, M. Preparation and properties of novel microporous hydrogels with poly(ethylene glycol) dimethacrylate and carboxylated carbon nanotubes. J. Control. Release 2015, 213, e86.

108. Mu, S.; Liang, Y.; Chen, S.; Zhang, L.; Liu, T. MWNT-hybrided supramolecular hydrogel for hydrophobic camptothecin delivery. Mater. Sci. Eng. C 2015, 50, 294–299.

109. Giri, A.; Bhunia, T.; Goswami, L.; Panda, A.B.; Bandyopadhyay, A. Fabrication of acrylic acid grafted guar gum-multiwalled carbon nanotube hydrophobic membranes for transdermal drug delivery. RSC Adv. 2015, 5, 41736–41744.

110. Shi, X.; Zheng, Y.; Wang, C.; Yue, L.; Qiao, K.; Wang, G.; Wang, L.; Quan, H. Dual stimulus responsive drug release under the interaction of pH value and pulsatile electric field for a bacterial cellulose/sodium alginate/multi-walled carbon nanotube hybrid hydrogel. RSC Adv. 2015, 5, 41820–41829.

111. Zinchenko, A.; Taki, Y.; Sergeyev, V.; Murata, S. DNA-assisted solubilization of carbon nanotubes and construction of DNA-MWCNT cross-linked hybrid hydrogels. Nanomaterials 2015, 5, 270–283.

112. Sun, Y.X.; Ren, K.F.; Chang, G.X.; Zhao, Y.X.; Liu, X.S.; Ji, J. Enhanced electrochemical stimuli multilayers based on a ferrocene-containing polymer. Sci. Bull. 2015, 60, 936–942.

113. Zhou, J.; Chen, J.; Sun, H.; Qiu, X.; Mou, Y.; Liu, Z.; Zhao, Y.; Li, X.; Han, Y.; Duan, C.; et al. Engineering the heart: Evaluation of conductive nanomaterials for improving implant integration and cardiac function. Sci. Rep. 2014, 4, 313–317.

114. Venkatesan, J.; Jayakumar, R.; Mohandas, A.; Bhatnagar, I.; Kim, S.K. Antimicrobial activity of chitosan-carbon nanotube hydrogels. Materials 2014, 7, 3946–3955.

115. Lee, J.H.; Lee, J.Y.; Yang, S.H.; Lee, E.J.; Kim, H.W. Carbon nanotube-collagen three-dimensional culture of mesenchymal stem cells promotes expression of neural phenotypes and secretion of neurotrophic factors. Acta Biomater. 2014, 10, 4425–4436.

116. Mandal, D.; Kar, T.; Das, P.K. Pyrene-based fluorescent ambidextrous gelators: Scaffolds for mechanically robust SWNT-gel nanocomposites. Chem. Eur. J. 2014, 20, 1349–1358.

117. Sheikholeslam, M.; Pritzker, M.; Chen, P. Hybrid peptide-carbon nanotube dispersions and hydrogels. Carbon 2014,71, 284–293.

118. Tardani, F.; La Mesa, C. Effects of single-walled carbon nanotubes on lysozyme gelation. Colloid Surface B 2014, 121, 165–170.

119. Liu, J.; Cui, L.; Kong, N.; Barrow, C.J.; Yang, W. RAFT controlled synthesis of graphene/polymer hydrogel with enhanced mechanical property for pH-controlled drug release. Eur. Polym. J. 2014, 50, 9–17.

120. Servant, A.; Leon, V.; Jasim, D.; Methven, L.; Limousin, P.; Fernandez-Pacheco, E.V.; Prato, M.; Kostarelos, K. Graphene-based electroresponsive scaffolds as polymeric implants for on-demand drug delivery. Adv. Healthc. Mater.2014, 3, 1334–1343.

121. Chen, P.; Wang, X.; Wang, G.Y.; Duo, Y.R.; Zhang, X.Y.; Hu, X.H.; Zhang, X.J. Double network self-healing graphene hydrogel by two

step method for anticancer drug delivery. Mater. Technol. 2014, 29, 210–213.

122. Faghihi, S.; Karimi, A.; Jamadi, M.; Imani, R.; Salarian, R. Graphene oxide/poly(acrylic acid)/gelatin nanocomposite hydrogel: Experimental and numerical validation of hyperelastic model. Mater. Sci. Eng. C 2014, 38, 299–305.

123. Piao, Y.; Chen, B. Self-assembled graphene oxide-gelatin nanocomposite hydrogels: Characterization, formation mechanisms, and pH-sensitive drug release behavior. J. Polym. Sci. Polym. Phys. 2015, 53, 356–367.

124. Shen, J.; Xin, X.; Zhang, Y.; Song, L.; Wang, L.; Tang, W.; Ren, Y. Manipulation the behavior of supramolecular hydrogels of alpha-cyclodextrin/star-like block copolymer/carbon-based nanomaterials. Carbohydr. Polym. 2015, 117, 592–599.

125. Wang, J.; Liu, C.; Shuai, Y.; Cui, X.; Nie, L. Controlled release of anticancer drug using graphene oxide as a drug-binding effector in konjac glucomannan/sodium alginate hydrogels. Colloid. Surface. B 2014, 113, 223–229.

126. Wei, H.; Han, L.; Tang, Y.; Ren, J.; Zhao, Z.; Jia, L. Highly flexible heparin-modified chitosan/graphene oxide hybrid hydrogel as a super bilirubin adsorbent with excellent hemocompatibility. J. Mater. Chem. B 2015, 3, 1646–1654.

127. Zhang, E.; Wang, T.; Zhao, L.; Sun, W.; Liu, X.; Tong, Z. Fast self-healing of graphene oxide-hectorite clay-poly(N,N-dimethylacrylamide) hybrid hydrogels realized by near-infrared irradiation. ACS Appl. Mater. Interfaces 2014, 6, 22855–22861.

128. Katiyar, R.; Bag, D.S.; Nigam, I. Fullerene (C60) Containing Poly(N-isopropylacrylamide) Thermo-responsive Smart Hydrogels and their Swelling Behavior. J. Polym. Mater. 2013, 30, 15–26.

129. Katiyar, R.; Bag, D.S.; Nigam, I. Synthesis and evaluation of swelling characteristics of fullerene (C60) containing cross-linked poly(2-hydroxyethyl methacrylate) hydrogels. Adv. Mater. Lett. 2014, 5, 214–222.

130. Kim, S.; Lee, D.J.; Kwag, D.S.; Lee, U.Y.; Youn, Y.S.; Lee, E.S. Acid pH-activated glycol chitosan/fullerene nanogels for efficient tumor therapy. Carbohydr. Polym. 2014, 101, 692–698.

131. Hoppe, A.; Güldal, N.S.; Boccaccini, A.R. A review of the biological response to ionic dissolution products from bioactive glasses and glassceramics. Biomaterials 2011, 32, 2757–2774.

132. Song, W.; Markel, D.C.; Jin, X.; Shi, T.; Ren, W. Poly(vinyl alcohol)/collagen/hydroxyapatite hydrogel: Properties andin vitro cellular response. J. Biomed. Mater. Res. A 2012, 100, 3071–3079.

133. Kokubo, T.; Takadama, H. How useful is SBF in predicting in vivo bone bioactivity? Biomaterials 2006, 27, 2907–2915.

134. Gaharwar, A.K.; Mihaila, S.M.; Swami, A.; Patel, A.; Sant, S.; Reis, R.L.; Marques, A.P.; Gomes, M.E.; Khademhosseini, A. Bioactive silicate nanoplatelets for osteogenic diffrentiation of human human mesenchymal stem cells. Adv. Mater.2013, 25, 3329–3336.

135. Zhao, L.; Weir, M.D.; Xu, H.H. An injectable calcium phosphate-alginate hydrogelumbilical cord mesenchymal stem cell paste for bone tissue engineering. Biomaterials 2010, 31, 6502–6510.

136. Lee, J.H.; El-Fiqi, A.; Han, C.M.; Kim, H.W. Physically-strengthened collagen bioactive nanocomposite gels for bone: A feasibility study. Tissue Eng. Regen. Med. 2015, 12, 90–97.

137. Aguzzi, C.; Capra, P.; Bonferoni, C.; Cerezo, P.; Salcedo, I.; Sánchez, R.; Caramella, C.; Viseras, C. Chitosan-silicate biocomposites to be used in modified drug release of 5-aminosalicylic acid (5-ASA). Appl. Clay Sci. 2010, 50, 106–111.

138. Kawachi, Y.; Kugimiya, S.; Kato, K. Preparation of silica hydrogels using a synthetic peptide for application as carriers for controlled drug release and mesoporous oxides. J. Ceram. Soc. Jpn. 2014, 122, 134–140.

139. Karnik, S.; Hines, K.; Mills, D.K. Nanoenhanced hydrogel system with sustained release capabilities. J. Biomed. Mater. Res. A 2015, 103, 2416–2426.

140. Ayyub, O.B.; Kofinas, P. Enzyme Induced stiffening of nanoparticle-hydrogel composites with structural color. ACS Nano 2015, 9, 8004–8011.

141. Liu, J.; Yang, X.; Wang, K.; Wang, Q.; Ji, H.; Wu, C.; Li, J.; He, X.; Tang, J.; Huang, J. Combining physical embedding and covalent bonding for stable encapsulation of quantum dots into agarose hydrogels. J. Mater. Chem. 2012, 22, 495–501.

142. Yuan, J.; Wen, D.; Gaponik, N.; Eychmuller, A. Enzyme-encapsulating quantum dot hydrogels and xerogels as biosensors: Multifunctional platforms for both biocatalysis and fluorescent probing. Angew. Chem. Int. Ed. Engl. 2013,52, 976–979.

143. Wang, Z.; Zhang, Y.; Zhang, J.; Huang, L.; Liu, J.; Li, Y.; Zhang, G.; Kundu, S.C.; Wang, L. Exploring natural silk protein sericin for regenerative medicine: An injectable, photoluminescent, cell-adhesive 3D hydrogel. Sci. Rep. 2014,4.

144. Li, J.; Hong, X.; Liu, Y.; Li, D.; Wang, Y.W.; Li, J.H.; Bai, Y.B.; Li, T.J. Highly photoluminescent cdte/poly(n-isopropylacrylamide) temperature-sensitive gels. Adv. Mater. 2005, 17, 163–166.

145. Amdursky, N.; Gazit, E.; Rosenman, G. Quantum confinement in self-assembled bioinspired peptide hydrogels. Adv. Mater. 2010, 22, 2311–2315.

146. Zhang, D.; Yang, J.; Bao, S.; Wu, Q.; Wang, Q. Semiconductor nanoparticle-based hydrogels prepared via self-initiated polymerization under sunlight, even visible light. Sci. Rep. 2013, 3.

147. Guo, M.; Jiang, M. Supramolecular hydrogels with CdS quantum dots incorporated by host-guest interactions.Macromol. Rapid Commun. 2010, 31, 1736–1739.

148. Lu, Z.S.; Li, C.M. Quantum dot-based nanocomposites for biomedical applications. Curr. Med. Chem. 2011, 18, 3516–3528.

149. Biju, V.; Itoh, T.; Ishikawa, M. Delivering quantum dots to cells: Bioconjugated quantum dots for targeted and nonspecific extracellular and intracellular imaging. Chem. Soc. Rev. 2010, 39, 3031–3056.

150. Zhou, L.; Zhang, F. Thermo-sensitive and photoluminescent hydrogels: Synthesis, characterization, and their drug-release property. Mater. Sci. Eng. C Mater. Biol. Appl. 2011, 31, 1429–1435.

151. Kim, J.H.; Lim, S.Y.; Nam, D.H.; Ryu, J.; Ku, S.H.; Park, C.B. Self-assembled, photoluminescent peptide hydrogel as a versatile platform for enzyme-based optical biosensors. Biosens. Bioelectron. 2011, 26, 1860–1865.

152. Zhang, W.; He, X.W.; Li, W.Y.; Zhang, Y.K. Thermo-sensitive imprinted polymer coating CdTe quantum dots for target protein specific recognition. Chem. Commun. 2012, 48, 1757–1759.

153. Li, P.; Huang, L.; Lin, Y.; Shen, L.; Chen, Q.; Shi, W. Printable temperature-responsive hybrid hydrogels with photoluminescent carbon nanodots. Nanotechnology 2014, 25, 055603.

154. Fang, Y.; Chen, L.; Wang, C.-F.; Chen, S. Facile synthesis of fluorescent quantum dot-polymer nanocomposites via frontal polymerization. J. Polym. Sci. A Polym. Chem. 2010, 48, 2170–2177.

155. Gattas-Asfura, K.M.; Zheng, Y.; Micic, M.; Snedaker, M.J.; Ji, X.; Sui, G.; Orbulescu, J.; Andreopoulos, F.M.; Pham, S.M.; Wang, C.; et al. Immobilization of quantum dots in the photo-cross-linked poly(ethylene glycol)-based hydrogel.J. Phys. Chem. B 2003, 107, 10464–10469.

156. Xie, X.; Ma, D.; Zhang, L.-M. Fabrication and properties of a supramolecular hybrid hydrogel doped with CdTe quantum dots. RSC Adv. 2015, 5, 58746–58754.

157. Gong, Y.; Gao, M.; Wang, D.; Möhwald, H. Incorporating fluorescent CdTe nanocrystals into a hydrogel via hydrogen bonding: Toward fluorescent microspheres with temperature-responsive properties. Chem. Mater. 2005, 17, 2648–2653.

158. Bai, S.; Nguyen, T.L.; Mulvaney, P.; Wang, D. Using hydrogels to accommodate hydrophobic nanoparticles in aqueous media via solvent exchange. Adv. Mater. 2010, 22, 3247–3250.

159. Li, X.; Chen, R.; Xu, S.; Liu, H.; Hu, Y. Thermoresponsive behavior and rheology of SiO2-hyaluronic acid/poly(N-isopropylacrylamide) (NaHA/PNIPAm) core-shell structured microparticles. J. Chem. Technol. Biotechnol. 2015, 90, 407–414.

160. Dogru, O.; Abdurrahmanoglu, S.; Kayaman-Apohan, N. Preparation and characterization of modified nanosilica/PNIPAm hybrid cryogels. Polym. Bull. 2015, 72, 993–1005.

161. Garnweitner, G.; Smarsly, B.; Assink, R.; Dunphy, D.R.; Scullin, C.; Brinker, C.J. Characterization of self-assembled lamellar thermoresponsive silica-hydrogel nanocomposite films. Langmuir 2004, 20, 9811–9820.

162. Jin, S.W.; Kim, H.I. Preparation of temperature-responsive hydrogel nanocapsules with interfacial modification of colloidal silica. Compos. Interfaces 2006, 13, 277–284.

163. Kurumada, K.I.; Nakamura, T.; Suzuki, A.; Umeda, N.; Kishimoto, N.; Hiro, M. Nanoscopic replication of cross-linked hydrogel in high-porosity nanoporous silica. J. Non-Cryst. Solids 2007, 353, 4839–4844.

164. Van den Brom, C.R.; Anac, I.; Roskamp, R.F.; Retsch, M.; Jonas, U.; Menges, B.; Preece, J.A. The swelling behaviour of thermoresponsive hydrogel/silica nanoparticle composites. J. Mater. Chem. 2010, 20, 4827–4839.

165. Alam, M.A.; Takafuji, M.; Ihara, H. Thermosensitive hybrid hydrogels with silica nanoparticle-cross-linked polymer networks. J. Colloid Interface Sci. 2013, 405, 109–117.

166. Ham, M.J.; Kim, Y.H. Property Modulation of poly(N-isopropylacrylamide) hydrogels by incorporation of modified silica. Polym. Eng. Sci. 2008, 48, 2439–2445.

167. Van Durme, K.; van Mele, B.; Loos, W.; Du Prez, F.E. Introduction of silica into thermo-responsive poly(N-isopropyl acrylamide) hydrogels: A novel approach to improve response rates. Polymer 2005, 46, 9851–9862.

168. Zaragoza, J.; Babhadiashar, N.; O'Brien, V.; Chang, A.; Blanco, M.; Zabalegui, A.; Lee, H.; Asuri, P. Experimental investigation of mechanical and thermal properties of silica nanoparticle-reinforced poly(acrylamide) nanocomposite hydrogels. PLoS ONE 2015, 10, e0136293.

169. Wu, L.; Zeng, L.; Chen, H.; Zhang, C. Effects of silica sol content on the properties of poly(acrylamide)/silica composite hydrogel. Polym. Bull. 2012, 68, 309–316.

170. Wang, Q.; Hou, R.; Cheng, Y.; Fu, J. Super-tough double-network hydrogels reinforced by covalently compositing with silica-nanoparticles. Soft Matter 2012, 8, 6048–6056.

171. Yang, J.; Deng, L.-H.; Han, C.-R.; Duan, J.-F.; Ma, M.-G.; Zhang, X.-M.; Xu, F.; Sun, R.-C. Synthetic and viscoelastic behaviors of silica nanoparticle reinforced poly(acrylamide) core-shell nanocomposite hydrogels. Soft Matter 2013, 9, 1220–1230.

172. Yang, J.; Han, C.R. Dynamics of Silica-nanoparticle-filled hybrid hydrogels: Nonlinear viscoelastic behavior and chain entanglement network. J. Phys. Chem. C 2013, 117, 20236–20243.

173. Yang, J.; Zhao, J.J.; Han, C.R.; Duan, J.F. Keys to enhancing mechanical properties of silica nanoparticle composites hydrogels: The role of network structure and interfacial interactions. Compos. Sci. Technol. 2014, 95, 1–7.

174. Yin, P.; Huang, G.B.; Tse, W.H.; Bao, Y.G.; Denstedt, J.; Zhang, J. Nanocomposited silicone hydrogels with a laser-assisted surface modification for inhibiting the growth of bacterial biofilm. J. Mater. Chem. B 2015, 3, 3234–3241.

175. Lin, M.; Xu, P.; Zhong, W. Preparation, characterization, and release behavior of aspirin-loaded poly(2-hydroxyethyl acrylate)/silica hydrogels. J. Biomed. Mater. Res. B 2012, 100, 1114–1120.

176. Zhang, Y.; Ye, L. Structure and property of polyvinyl alcohol/precipitated silica composite hydrogels for microorganism immobilization. Compos. B 2014, 56, 749–755.

177. Lewandowska-Lancucka, J.; Fiejdasz, S.; Rodzik, L.; Koziel, M.; Nowakowska, M. Bioactive hydrogel-nanosilica hybrid materials: A potential injectable scaffold for bone tissue engineering. Biomed. Mater. 2015, 10.

178. Zhu, M.; Zhu, Y.; Zhang, L.; Shi, J. Preparation of chitosan/mesoporous silica nanoparticle composite hydrogels for sustained co-delivery of biomacromolecules and small chemical drugs. Sci. Technol. Adv. Mater. 2013, 14, 045005.

179. Lou, Y.R.; Kanninen, L.; Kaehr, B.; Townson, J.L.; Niklander, J.; Harjumaki, R.; Jeffrey Brinker, C.; Yliperttula, M. Silica bioreplication preserves three-dimensional spheroid structures of human pluripotent stem cells and HepG2 cells. Sci. Rep. 2015, 5.

180. Panic, V.V.; Spasojevic, P.M.; Radoman, T.S.; Dzunuzovic, E.S.; Popovic, I.G.; Velickovic, S.J. Methacrylic acid based polymer networks with a high content of unfunctionalized nanosilica: Particle distribution, swelling, and rheological properties. J. Phys. Chem. C 2015, 119, 610–622.

181. Yang, J.; Zhao, J. Preparation and mechanical properties of silica nanoparticles reinforced composite hydrogels. Mater. Lett. 2014, 120, 36–38.

182. Schlossmacher, U.; Schroeder, H.C.; Wang, X.; Feng, Q.; Diehl-Seifert, B.; Neumann, S.; Trautwein, A.; Mueller, W.E.G. Alginate/silica composite hydrogel as a potential morphogenetically active scaffold for three-dimensional tissue engineering. RSC Adv. 2013, 3, 11185–11194.

183. Lucia Foglia, M.; Edhit Camporotondi, D.; Solange Alvarez, G.; Heinemann, S.; Hanke, T.; Javier Perez, C.; Eduardo Diaz, L.; Federico Desimone, M. A new method for the preparation of biocompatible silica coated-collagen hydrogels.J. Mater. Chem. B 2013, 1, 6283–6290.

184. Solange Alvarez, G.; Helary, C.; Mathilde Mebert, A.; Wang, X.; Coradin, T.; Federico Desimone, M. Antibiotic-loaded silica nanoparticle-collagen composite hydrogels with prolonged antimicrobial activity for wound infection prevention. J. Mater. Chem. B 2014, 2, 4660–4670.

185. Lu, C.; Zahedi, P.; Forman, A.; Allen, C. Multi-arm PEG/silica hydrogel for sustained ocular drug delivery. J. Pharm. Sci. 2014, 103, 216–226.

186. Giray, S.; Bal, T.; Kartal, A.M.; Kizilel, S.; Erkey, C. Controlled drug delivery through a novel PEG hydrogel encapsulated silica aerogel system. J. Biomed. Mater. Res. A 2012, 100, 1307–1315.

187. Zhao, P.; Liu, H.; Deng, H.; Xiao, L.; Qin, C.; Du, Y.; Shi, X. A study of chitosan hydrogel with embedded mesoporous silica nanoparticles loaded by ibuprofen as a dual stimuli-responsive drug release system for surface coating of titanium implants. Colloid. Surface. B 2014, 123, 657–663.

188. Torres, C.C.; Urbano, B.F.; Campos, C.H.; Rivas, B.L.; Reyes, P. Composite hydrogel based on surface modified mesoporous silica and poly[(2-acryloyloxy)ethyl trimethylammonium chloride]. Mater. Chem. Phys. 2015, 152, 69–76.

189. Liu, Y.; Zhou, Y.; Nie, W.; Song, L.; Chen, P. Formation and surface properties of raspberry-like silica particles: Effect of molecular weight of the coating poly(methacrylic acid) brushes. J. Sol-Gel Sci. Technol. 2014, 72, 122–129.

190. Miyamoto, N.; Shimasaki, K.; Yamamoto, K.; Shintate, M.; Kamachi, Y.; Bastakoti, B.P.; Suzuki, N.; Motokawa, R.; Yamauchi, Y. Mesoporous silica particles as topologically crosslinking fillers for poly(N-isopropylacrylamide) hydrogels. Chemistry 2014, 20, 14955–14958.

191. Sadat-Shojai, M.; Khorasani, M.T.; Jamshidi, A. 3-Dimensional cell-laden nano-hydroxyapatite/protein hydrogels for bone regeneration applications. Mater. Sci. Eng. C 2015, 49, 835–843.

192. Zuo, Y.; Liu, X.; Wei, D.; Sun, J.; Xiao, W.; Zhao, H.; Guo, L.; Wei, Q.; Fan, H.; Zhang, X. Photo-crosslinkable methacrylated gelatin and hydroxyapatite hybrid hydrogel for modularly engineering biomimetic osteon. ACS Appl. Mater. Interfaces 2015, 7, 10386–10394.

193. Mozafari, M.; Moztarzadeh, F.; Rabiee, M.; Azami, M.; Maleknia, S.; Tahriri, M.; Moztarzadeh, Z.; Nezafati, N. Development of macroporous nanocomposite scaffolds of gelatin/bioactive glass prepared through layer solvent casting combined with lamination technique for bone tisue engineering. Ceram. Int. 2010, 36, 2431–2439.

194. Jiang, D.; Wang, G.; Zheng, F.; Han, J.; Wu, X. Novel thermo-sensitive hydrogels containing polythioether dendrons: Facile tuning of LCSTs, strong absorption of Ag ions, and embedment of smaller Ag nanocrystals. Polym. Chem. 2015,6, 625–632.

195. Ribeiro, M.; de Moraes, M.A.; Beppu, M.M.; Garcia, M.P.; Fernandes, M.H.; Monteiro, F.J.; Ferraz, M.P. Development of silk fibroin/nanohydroxyapatite composite hydrogels for bone tissue engineering. Eur. Polym. J. 2015, 67, 66–77.

196. Yang, M.; He, W.; Shuai, Y.; Min, S.; Zhu, L. Nucleation of hydroxyapatite crystalsby self-assembled Bombyx mori silk fibroin. J. Polym. Sci. B 2013, 51, 742–748.

197. Shi, P.; Teh, T.K.; Toh, S.L.; Goh, J.C. Variation of the effect of calcium phosphateenhancement of implanted silk fibroin ligament bone integration. Biomaterials 2013, 34, 5947–5957.

198. Yokoi, T.; Kawashit, M.; Kikuta, K.; Ohtsuki, C. Biomimetic mineralization of calcium phosphate crystals in polyacrylamide hydrogel: Effect of concentrations of calcium and phosphate ions on

crystalline phases and morphology. Mater. Sci. Eng. C 2010, 30, 154–159.

199. Dhivya, S.; Saravanan, S.; Sastry, T.P.; Selvamurugan, N. Nanohydroxyapatite-reinforced chitosan composite hydrogel for bone tissue repair in vitro and in vivo. J. Nanobiotechnol. 2015, 13.

200. Caridade, S.G.; Merino, E.G.; Alves, N.M.; Bermudez Vde, Z.; Boccaccini, A.R.; Mano, J.F. Chitosan membranes containing micro or nano-size bioactive glass particles: Evolution of biomineralization followed by in situ dynamic mechanical analysis. J. Mech. Behav. Biomed. Mater. 2013, 20, 173–183.

201. Couto, D.S.; Hong, Z.; Mano, J.F. Development of bioactive and biodegradable chitosan-based injectable systems containing bioactive glass nanoparticles. Acta Biomater. 2009, 5, 115–123.

202. Couto, D.S.; lves, N.M.; Mano, J.F. Nanostructured multilayer coatings combining chitosan with bioactive glass nanoparticles. J. Nanosci. Nanotechnol. 2009, 9, 1741–1748.

203. Iwatsubo, T.; Kishi, R.; Miura, T.; Ohzono, T.; Yamaguchi, T. Formation of Hydroxyapatite Skeletal Materials from Hydrogel Matrices via Artificial Biomineralization. J. Phys. Chem. B 2015, 119, 8793–8799.

204. Taguchi, T.; Kishida, A.; Akashi, M. Hydroxyapatite formation on/in poly(vinyl alcohol) hydrogel matrices using a novel alternate soaking process. Chem. Lett. 1998, 8, 711–712.

205. Arora, A.; Sharma, P.; Katti, D.S. Pullulan-based composite scaffolds for bone tissue engineering: Improved osteoconductivity by pore wall mineralization. Carbohydr. Polym. 2015, 123, 180–189.

206. Fu, S.Z.; Guo, G.; Gong, C.Y.; Zeng, S.; Liang, H.; Luo, F.; Zhang, X.N.; Zhao, X.; Wei, Y.Q.; Qian, Z.Y. Injectable biodegradable thermosensitive hydrogel composite for orthopedic tissue engineering 1: Preparation and characterization of nanohydroxyapatite/poly(ethylene glycol)-poly(ε-caprolactone)-poly(ethylene glycol) hydrogel nanocomposites. J. Phys. Chem. B 2009, 113, 16518–16525.

207. Ito, T.; Sasaki, M.; Taguchi, T. Enhanced ALP activity of MG63 cells cultured on hydroxyapatite-poly(ethylene glycol) hydrogel

composites prepared using EDTA-OH. Biomed. Mater. 2015, 10, 015025.

208. Gaharwar, A.K.; Dammu, S.A.; Canter, J.M.; Wu, C.J.; Schmidt, G. Highly extensible, tough, and elastomeric nanocomposite hydrogels from poly(ethylene glycol) and hydroxyapatite nanoparticles. Biomacromolecules 2011, 12, 1641–1650.

209. Rottensteiner, U.; Sarker, B.; Heusinger, D.; Dafinova, D.; Rath, S.; Beier, J.; Kneser, U.; Horch, R.; Detsch, R.; Boccaccini, A.; et al. In vitro and in vivo biocompatibility of alginate dialdehyde/gelatin hydrogels with and without nanoscaled bioactive glass for bone tissue engineering applications. Materials 2014, 7, 1957–1974.

210. Wang, X.; Tolba, E.; Schroder, H.C.; Neufurth, M.; Feng, Q.; Diehl-Seifert, B.; Muller, W.E. Effect of bioglass on growth and biomineralization of SaOS-2 cells in hydrogel after 3D cell bioprinting. PLoS ONE 2014, 9.

211. Valenzuela, F.; Covarrubias, C.; Martinez, C.; Smith, P.; Diaz-Dosque, M.; Yazdani-Pedram, M. Preparation and bioactive properties of novel bone-repair bionanocomposites based on hydroxyapatite and bioactive glass nanoparticles. J. Biomed. Mater. Res. B 2012, 100, 1672–1682.

212. Yang, S.; Guo, Q.; Shores, L.S.; Aly, A.; Ramakrishnan, M.; Kim, G.H.; Lu, Q.; Su, L.; Elisseeff, J.H. Use of a chondroitin sulfate bioadhesive to enhance integration of bioglass particles for repairing critical-size bone defects. J. Biomed. Mater. Res. A 2015, 103, 235–242.

213. Eglin, D.; Maalheem, S.; Livage, J.; Coradin, T. In vitro apatite forming ability of type I collagen hydrogels containing bioactive glass and silica sol-gel particles. J. Mater. Sci. Mater. Med. 2006, 17, 161–167.

214. Gleeson, J.P.; Plunkett, N.A.; O'Brien, F.J. Addition of hydroxyapatite improves stiffness, interconnectivity and osteogenic potential of a highly porous collagen-based scaffold for bone tissue regeneration. Eur. Cell Mater. 2010, 20, 218–230.

215. Marelli, B.; Ghezzi, C.E.; Mohn, D.; Stark, W.J.; Barralet, J.E.; Boccaccini, A.R.; Nazhat, S.N. Accelerated mineralization of dense collagen-nano bioactive glass hybrid gels increases scaffold stiffness

and regulates osteoblastic function.Biomaterials 2011, 32, 8915–8926.

216. Li, C.; Mu, C.; Lin, W.; Ngai, T. Gelatin Effects on the Physicochemical and hemocompatible properties of gelatin/pAAm/laponite nanocomposite hydrogels. ACS Appl. Mater. Interfaces 2015, 7, 18732–18741.

217. Patel, R.G.; Purwada, A.; Cerchietti, L.; Inghirami, G.; Melnick, A.; Gaharwar, A.K.; Singh, A. Microscale bioadhesive hydrogel arrays for cell engineering applications. Cell. Mol. Bioeng. 2014, 7, 394–408.

218. Li, J.; Jiang, J.; Li, C.; Lin, M.Y.; Schwarz, S.A.; Rafailovich, M.H.; Sokojov, J. Effect of temperature on shear-induced anisotropic structure in polymer clay hydrogels. Macromol. Rapid Commun. 2006, 27, 1787–1791.

219. Chang, C.W.; Spreeuwel, A.V.; Zhang, C.; Varghese, S. PEG/clay nanocomposite hydrogel: A mechanically robust tissue engineering scaffold. Soft Matter 2010, 6, 5157–5164.

220. Ding, X.; Vegesna, G.K.; Meng, H.; Winter, A.; Lee, B.P. Nitro-group functionalization of dopamine and its contribution to the viscoelastic properties of catechol-containing nanocomposite hydrogels. Macromol. Chem. Phys.2015, 216, 1109–1119.

221. Li, P.; Kim, N.H.; Lee, J.H. Swelling behavior of polyacrylamide/laponite clay nanocomposite hydrogels: pH-sensitive property. Compos. B 2009, 40, 275–283.

222. Okay, O.; Oppermann, W. Polyacrylamide-clay nanocomposite hydrogels: Rheological and light scattering characterization. Macromolecules 2007, 40, 3378–3387.

223. Xia, M.; Wu, W.; Liu, F.; Theato, P.; Zhu, M. Swelling behavior of thermosensitive nanocomposite hydrogels composed of oligo(ethylene glycol) methacrylates and clay. Eur. Polym. J. 2015, 69, 472–482.

224. Tan, Y.; Wu, R.; Li, H.; Ren, W.; Du, J.; Xu, S.; Wang, J. Electric field-induced gradient strength in nanocomposite hydrogel through gradient crosslinking of clay. J. Mater. Chem. B 2015, 3, 4426–4430.

225. Wang, J.; Lin, L.; Cheng, Q.F.; Jiang, L. A strong bio-inspired layered PNIPAM-clay nanocomposite hydrogel. Angew. Chem. Int. Ed. 2012, 51, 4676–4678.

226. Haraguchi, K.; Li, H.J.; Matsuda, K.; Takehisa, T.; Elliott, E. Mechanism of forming organic/inorganic network structures during in-situ free-radical polymerization in PNIPA-clay nanocomposite hydrogels. Macromolecules 2005,38, 3482–3490.

227. Haraguchi, K.; Li, H.J.; Wu, C.J.; Gaharwar, A.K.; Chan, B.K.; Schmidt, G. Mechanically tough pluronic F127/laponite nanocomposite hydrogels from covalently and physically cross-linked networks. Macromolecules 2011, 44, 8215–8224.

228. Wu, L.; Ohtani, M.; Tamesue, S.; Ishida, Y.; Aida, T. High water content clay-nanocomposite hydrogels incorporating guanidinium-pendant methacrylamide: Tuning of mechanical and swelling properties by supramolecular approach. J. Polym. Sci. A 2014, 52, 839–847.

229. Neufeld, L.; Bianco-Peled, H. Designing a biocompatible hydrogel for the delivery of mesalamine. Int. J. Pharm. 2015,491, 170–179.

230. Ferfera Harrar, H.; Aiouaz, N.; Dairi, N.; Hadj-Hamou, A.S. Preparation of chitosan-g-poly (acrylamide)/montmorillonite superabsorbent polymer composites: Studies on swelling, thermal, and antibacterial properties. J. Appl. Polym. Sci. 2014, 131, 39747–39761.

231. Dadkhah, D.; Navarchian, A.H.; Aref, L.; Tavakoli, N. Application of taguchi method to investigate the drug release behavior of poly(acrylamide-co-maleic acid)/montmorillonite nanocomposite hydrogels. Adv. Polym. Technol. 2014, 33.

232. Ibraeva, Z.E.; Zhumaly, A.A.; Blagih, E.; Kudaibergenov, S.E. Preparation and characterization of organic-inorganic composite materials based on poly(acrylamide) hydrogels and clay minerals. Macromol. Symp. 2015, 351, 97–111.

233. Salimi, F.; Sefti, M.V.; Jarrahian, K.; Rafipoor, M.; Ghorashi, S.S. Preparation and investigation of the physical and chemical properties of clay-based polyacrylamide/Cr (III) hydrogels as a water shut-off agent in oil reservoirs. Korean J. Chem. Eng. 2014, 31, 986–993.

234. Li, T.; Xiang, S.; Ma, P.; Bai, H.; Dong, W.; Chen, M. Nanocomposite hydrogel consisting of Na-montmorillonite with enhanced mechanical properties. J. Polym. Sci. B Polym. Phys. 2015, 53, 1020–1026.

235. Kevadiya, B.D.; Rajkumar, S.; Bajaj, H.C.; Chettiar, S.S.; Gosai, K.; Brahmbhatt, H.; Bhatt, A.S.; Barvaliya, Y.K.; Dave, G.S.; Kothari, R.K. Biodegradable gelatin-ciprofloxacin-montmorillonite composite hydrogels for controlled drug release and wound dressing application. Colloid. Surface. B 2014, 122, 175–183.

236. Pour-Esmaeil, S.; Taheri Qazvini, N.; Mahdavi, H. Interpenetrating polymer networks (IPN) based on gelatin/poly(ethylene glycol) dimethacrylate/clay nanocomposites: Structure-properties relationship. Mater. Chem. Phys. 2014, 143, 1396–1403.

237. Pavlukhina, S.; Zhuk, I.; Mentbayeva, A.; Rautenberg, E.; Chang, W.; Yu, X.; van de Belt-Gritter, B.; Busscher, H.J.; van der Mei, H.C.; Sukhishvili, S.A. Small-molecule-hosting nanocomposite films with multiple bacteria-triggered responses. NPG Asia Mater. 2014, 6, e121.

238. Zhang, S.; Guan, Y.; Fu, G.-Q.; Chen, B.-Y.; Peng, F.; Yao, C.-L.; Sun, R.-C. Organic/inorganic superabsorbent hydrogels based on xylan and montmorillonite. J. Nanomater. 2014, 2014, 2.

239. Nistor, M.T.; Vasile, C.; Chiriac, A.P. Hybrid collagen-based hydrogels with embedded montmorillonite nanoparticles.Mater. Sci. Eng. C 2015, 53, 212–221.

240. Haraguchi, K.; Takehisa, T.; Fan, S. Effects of clay content on the properties of nanocomposite hydrogels composed of poly(N-isopropylacrylamide) and clay. Macromolecules 2002, 35, 10162–10171.

241. Ma, R.-F.; Yuan, J.; Li, C.-C.; Peng, J.; Dung, Z.; Xu, L.; Li, J.-Q.; Zhai, M.-L. Radiation synthesis of polymer/clay nanocomposite hydrogels with high mechanical strength. Nucl. Sci. Tech. 2014, 25, 060301.

242. Stempfle, B.; Grosse, A.; Ferse, B.; Arndt, K.-F.; Woell, D. Anomalous diffusion in thermoresponsive polymer-clay composite hydrogels probed by wide-field fluorescence microscopy. Langmuir 2014, 30, 14056–14061.

243. Yao, C.; Liu, Z.; Yang, C.; Wang, W.; Ju, X.-J.; Xie, R.; Chu, L.-Y. Poly(N-isopropylacrylamide)-clay nanocomposite hydrogels with responsive bending property as temperature-controlled manipulators. Adv. Funct. Mater. 2015, 25, 2980–2991.

244. Wan, T.; Cheng, W.; Zhou, Z.; Xu, M.; Zou, C.; Li, R. Influence of crosslinker amount on swelling and gel properties of hectorite/poly(acrylamide/itaconic acid) nanocomposite hydrogels. Korean J. Chem. Eng. 2015, 32, 1434–1439.

245. Wan, T.; Zou, C.; Wang, L.; Wu, D.; Cheng, W.; Li, R.; Xu, M. Hectorite effects on swelling and gel properties of hectorite/poly(AM/IA) nanocomposite hydrogels. Polym. Bull. 2015, 72, 1113–1125.

246. Wang, L.A.N.; Cheng, W.; Wan, T.A.O.; Hu, Z.; Xu, M.I.N.; Li, R.; Zou, C. Synthesis and properties of hectorite/poly(AM/IA) nanocomposite hydrogels with high gel strength. J. Chem. Sci. 2015, 127, 19–23.

247. Ganguly, S.; Das, N.C. Synthesis of a novel pH responsive phyllosilicate loaded polymeric hydrogel based on poly(acrylic acid-co-N-vinylpyrrolidone) and polyethylene glycol for drug delivery: Modelling and kinetics study for the sustained release of an antibiotic drug. RSC Adv. 2015, 5, 18312–18327.

248. Niu, Y.; Dong, W.; Guo, H.; Deng, Y.; Guo, L.; An, X.; He, D.; Wei, J.; Li, M. Mesoporous magnesium silicate-incorporated poly(epsilon-caprolactone)-poly(ethylene glycol)-poly(epsilon-caprolactone) bioactive composite beneficial to osteoblast behaviors. Int. J. Nanomed. 2014, 9, 2665–2675.

249. Gao, G.; Du, G.; Cheng, Y.; Fu, J. Tough nanocomposite double network hydrogels reinforced with clay nanorods through covalent bonding and reversible chain adsorption. J. Mater. Chem. B 2014, 2, 1539–1548.

250. Shikinaka, K.; Yokoi, T.; Koizumi-Fujii, Y.; Shimotsuya, M.; Shigehara, K. Robust imogolite hydrogels with tunable physical properties. RSC Adv. 2015, 5, 46493–46500.

251. Schexnailder, P.; Schmidt, G. Nanocomposite polymer hydrogels. Colloid Polym. Sci. 2009, 287, 1–11.

252. Skardal, A.; Zhang, J.; McCoard, L.; Oottamasathien, S.; Prestwich, G.D. Dynamically crosslinked gold nanoparticle—Hyaluronan hydrogels. Adv. Mater. 2010, 22, 4736–4740.

253. Balazs, A.C.; Emrick, T.; Russell, T.P. Nanoparticle polymer composites: Where two small worlds meet. Science 2006,314, 1107–1110.

254. Vimala, K.; Sivudu, K.S.; Mohan, Y.M.; Sreedhar, B.; Raju, K.M. Controlled silver nanoparticles synthesis in semi-hydrogel networks of poly(acrylamide) and carbohydrates: A rational methodology for antibacterial application.Carbohydr. Polym. 2009, 75, 463–471.

255. Thomas, V.; Yallapu, M.M.; Sreedhar, B.; Bajpai, S.K. A versatile strategy to fabricate hydrogel-silver nanocomposites and investigation of their antimicrobial activity. J. Colloid Interface Sci. 2007, 315, 389–395.

256. Endo, T.; Ikeda, R.; Yanagida, Y.; Hatsuzawa, T. Stimuli-responsive hydrogel-silver nanoparticles composite for development of localized surface plasmon resonance-based optical biosensor. Anal. Chim. Acta 2008, 611, 205–211.

257. Juby, K.A.; Dwivedi, C.; Kumar, M.; Kota, S.; Misra, H.S.; Bajaj, P.N. Silver nanoparticle-loaded PVA/gum acacia hydrogel: Synthesis, characterization and antibacterial study. Carbohydr. Polym. 2012, 89, 906–913.

258. Zhang, F.; Wu, J.; Kang, D.; Zhang, H. Development of a complex hydrogel of hyaluronan and PVA embedded with silver nanoparticles and its facile studies on Escherichia coli. J. Biomater. Sci. Polym. Ed. 2013, 24, 1410–1425.

259. Mohan, Y.M.; Premkumar, T.; Lee, K.; Geckeler, K.E. Fabrication of silver nanoparticles in hydrogel networks.Macromol. Rapid Commun. 2006, 27, 1346–1354.

260. Mbhele, Z.H.; Salemane, M.G.; van Sittert, C.G.C.E.; Nedeljkovic, J.M.; Djokovic, V.; Luyt, A.S. Fabrication and characterization of silver-polyvinyl alcohol nanocomposites. Chem. Mater. 2003, 15, 5019–5024.

261. Yadollahi, M.; Namazi, H.; Aghazadeh, M. Antibacterial carboxymethyl cellulose/Ag nanocomposite hydrogels cross-linked

with layered double hydroxides. Int. J. Biol. Macromol. 2015, 79, 269–277.

262. Xiang, Y.; Chen, D. Preparation of a novel pH-responsive silver nanoparticle/poly (HEMA-PEGMA-MAA) composite hydrogel. Eur. Polym. J. 2007, 43, 4178–4187.

263. Boonkaew, B.; Barber, P.M.; Rengpipat, S.; Supaphol, P.; Kempf, M.; He, J.; John, V.T.; Cuttle, L. Development and characterization of a novel, antimicrobial, sterile hydrogel dressing for burn wounds: Single-step production with gamma irradiation creates silver nanoparticles and radical polymerization. J. Pharm. Sci. 2014, 103, 3244–3253.

264. Fan, Z.; Liu, B.; Wang, J.; Zhang, S.; Lin, Q.; Gong, P.; Ma, L.; Yang, S. A novel wound dressing based on Ag/graphene polymer hydrogel: Effectively kill bacteria and accelerate wound healing. Adv. Func. Mater. 2014, 24, 3933–3943.

265. Ahearn, D.G.; Grace, D.T.; Jennings, M.J.; Borazjani, R.N.; Boles, K.J.; Rose, L.J.; Simmons, R.B.; Ahanotu, E.N. Effects of hydrogel/silver coatings on in vitro adhesion to catheters of bacteria associated with urinary tract infections. Cur. Microbiol. 2000, 41, 120–125.

266. Rao, K.M.; Rao, K.S.; Ramanjaneyulu, G.; Rao, K.C.; Subha, M.C.; Ha, C.S. Biodegradable sodium alginate-based semi-interpenetrating polymer network hydrogels for antibacterial application. J. Biomed. Mater. Res. A 2014, 102, 3196–3206.

267. Reithofer, M.R.; Lakshmanan, A.; Ping, A.T.; Chin, J.M.; Hauser, C.A. In situ synthesis of size-controlled, stable silver nanoparticles within ultrashort peptide hydrogels and their anti-bacterial properties. Biomaterials 2014, 35, 7535–7542.

268. Manjula, B.; Varaprasad, K.; Sadiku, R.; Ramam, K.; Reddy, G.V.; Raju, K.M. Development of microbial resistant thermosensitive Ag nanocomposite (gelatin) hydrogels via green process. J. Biomed. Mater. Res. A 2014, 102, 928–934.

269. Abdel-Halim, E.S.; Al-Deyab, S.S. Electrically conducting silver/guar gum/poly(acrylic acid) nanocomposite. Int. J. Biol. Macromol. 2014, 69, 456–463.

270. Qu, Y.; Chu, B.Y.; Peng, J.R.; Liao, J.F.; Qi, T.T.; Shi, K.; Zhang, X.N.; Wei, Y.Q.; Qian, Z.Y. A biodegradable thermo-responsive hybrid

hydrogel: Therapeutic applications in preventing the post-operative recurrence of breast cancer.NPG Asia Mater. 2015, 7, e207.

271. Yu, J.; Ha, W.; Sun, J.; Shi, Y. Supramolecular Hybrid Hydrogel Based on Host-Guest Interaction and Its Application in Drug Delivery. ACS Appl. Mater. Interfaces 2014, 6, 19544–19551.

272. Chen, R.; Chen, Q.; Huo, D.; Ding, Y.; Hu, Y.; Jiang, X. In situ formation of chitosan-gold hybrid hydrogel and its application for drug delivery. Colloid. Surface. B 2012, 97, 132–137.

273. Song, J.; Hwang, S.; Im, K.; Hur, J.; Nam, J.; Hwang, S.; Ahn, G.O.; Kim, S.; Park, N. Light-responsible DNA hydrogel-gold nanoparticle assembly for synergistic cancer therapy. J. Mater. Chem. B 2015, 3, 1537–1543.

274. Wu, X.; Gao, Y.; Dong, C.M. Polymer/gold hybrid nanoparticles: From synthesis to cancer theranostic applications.RSC Adv. 2015, 5, 13787–13796.

275. Wang, Y.; Zhang, B.; Zhu, L.; Li, Y.; Huang, F.; Li, S.; Shen, Y.; Xie, A. Preparation and multiple antitumor properties of AuNRs/Spinach Extract/PEGDA composite hydrogel. ACS Appl. Mater. Interfaces 2014, 6, 15000–15006.

276. Jang, B.; Park, J.Y.; Tung, C.H.; Kim, I.H.; Choi, Y. Gold nanorod-photosensitizer complex for near-infrared fluorescence imaging and photodynamic/photothermal therapy in vivo. ACS Nano 2011, 5, 1086–1094.

277. Sershen, S.R.; Westcott, S.L.; Halas, N.J.; West, J.L. Temperature-sensitive polymer-nanoshell composites for photothermally modulated drug delivery. J. Biomed. Mater. Res. 2000, 5, 293–298.

278. Kowalczyk, A.; Wagner, B.; Karbarz, M.; Nowicka, A.M. A dual DNA biosensor based on two redox couples with a hydrogel sensing platform functionalized with carboxyl groups and gold nanoparticles. Sensor. Actuat. B 2015, 208, 220–227.

279. Wang, X.; Li, S.; Yan, C.; Liu, P.; Ding, J. Fabrication of RGD micro/nanopattern and corresponding study of stem cell differentiation. Nano Lett. 2015, 15, 1457–1467.

280. Heo, D.N.; Ko, W.K.; Bae, M.S.; Lee, J.B.; Lee, D.W.; Byun, W.; Lee, C.H.; Kim, E.C.; Jung, B.-Y.; Kwon, I.K. Enhanced bone regeneration

with a gold nanoparticle-hydrogel complex. J. Mater. Chem. B 2014, 2, 1584–1593.

281. Pérez, R.A.; Won, J.E.; Knowles, J.C.; Kim, H.W. Naturally and synthetic smart composite biomaterials for tissue regeneration. Adv. Drug Deliv. Rev. 2013, 65, 471–496.

282. Samal, S.K.; Dash, M.; Shelyakova, T.; Declercq, H.A.; Uhlarz, M.; Banobre-Lopez, M.; Dubruel, P.; Cornelissen, M.; Herrmannsdorfer, T.; Rivas, J.; et al. Biomimetic magnetic silk scaffolds. ACS Appl. Mater. Interfaces 2015, 7, 6282–6292.

283. Liu, T.Y.; Hu, S.H.; Liu, T.Y.; Liu, D.M.; Chen, S.Y. Magnetic-sensitive behavior of intelligent ferrogels for controlled release of drug. Langmuir 2006, 22, 5974–5978.

284. Satarkar, N.S.; Zach Hilt, J. Hydrogel nanocomposites as remote-controlled biomaterials. Acta Biomater. 2008, 4, 11–16.

285. Ghaemy, M.; Naseri, M. Synthesis of chitosan networks: Swelling, drug release, and magnetically assisted BSA separation using Fe3O4 nanoparticles. Carbohydr. Polym. 2012, 90, 1265–1272.

286. Zhao, Q.; Wang, L.; Cheng, R.; Mao, L.; Arnold, R.D.; Howerth, E.W.; Chen, Z.G.; Platt, S. Magnetic nanoparticle-based hyperthermia for head & neck cancer in mouse models. Theranostics 2012, 2, 113–121.

287. Lee, N.; Cho, H.R.; Oh, M.H.; Lee, S.H.; Kim, K.; Kim, B.H.; Shin, K.; Ahn, T.-Y.; Choi, J.W.; Kim, Y.-W. Multifunctional Fe3O4/TaOx Core/Shell Nanoparticles for Simultaneous Magnetic Resonance Imaging and X-ray Computed Tomography. J. Am. Chem. Soc. 2012, 134, 10309–10312.

288. Zhang, N.; Lock, J.; Sallee, A.; Liu, H. Magnetic Nanocomposite Hydrogel for Potential Cartilage Tissue Engineering: Synthesis, Characterization, and Cytocompatibility with Bone Marrow Derived Mesenchymal Stem Cells. ACS Appl. Mater. Interfaces 2015, 7, 20987–20998.

289. Thiruvengadam, V.; Vitta, S. Ni–bacterial cellulose nanocomposite: A magnetically active inorganic–organic hybrid gel. RSC Adv. 2013, 3, 12765–12773.

290. Fuhrer, R.; Athanassiou, E.K.; Luechinger, N.A.; Stark, W.J. Crosslinking metal nanoparticles into the polymer backbone of

hydrogels enables preparation of soft, magnetic field-driven actuators with muscle-like flexibility. Small2009, 5, 383–388.

291. Price, R.L.; Gutwein, L.G.; Kaledin, L.; Tepper, F.; Webster, T.J. Osteoblast function on nanophase alumina materials: Influence ofchemistry, phase, and topography. J. Biomed. Mater. Res. A 2003, 67, 1284–1293.

292. Webster, T.J.; Siegel, R.W.; Bizios, R. Nanoceramic surface roughness enhances osteoblast and osteoclast functions for improved orthopaedic/dental implant efficacy. Scr. Mater. 2001, 44, 1639–1642.

293. Vimala, K.; Kanny, K.; Varaprasad, K.; Kumar, N.M.; Reddy, G.S. Novel-porous-Ag0 nanocomposite hydrogels via green process for advanced antibacterial applications. J. Biomed. Mater. Res. A 2014, 102, 4616–4624.

294. Vimala, K.; Varaprasad, K.; Sadiku, R.; Ramam, K.; Kanny, K. Development of novel protein-Ag nanocomposite for drug delivery and inactivation of bacterial applications. Int. J. Biol. Macromol. 2014, 63, 75–82.

295. Varaprasad, K.; Mohan, Y.M.; Vimala, K.; Raju, K.M. Synthesis and characterization of hydrogel-silver nanoparticle-curcumin composites for wound dressing and antibacterial application. J. Appl. Polym. Sci. 2011, 121, 784–796.

296. Murthy, P.S.K.; Mohan, Y.M.; Varaprasad, K.; Sreedhar, B.; Raju, K.M. First successful design of semi-IPN hydrogel-silver nanocomposites: A facile approach for antibacterial application. J. Colloid Interface Sci. 2008, 318, 217–224.

297. Varaprasad, K.; Mohan, Y.M.; Ravindra, S.; Reddy, N.N.; Vimala, K.; Monika, K.; Sreedhar, B.; Raju, K.M. Hydrogel-silver nanoparticle composites: A new generation of antimicrobials. J. Appl. Polym. Sci. 2010, 115, 1199–1207.

298. Jovanović, Ž.; Radosavljević, A.; Stojkovska, J.; Nikolić, B.; Obradovic, B.; Kačarević-Popović, Z.; Mišković-Stanković, V. Silver/poly(N-vinyl-2-pyrrolidone) hydrogel nanocomposites obtained by electrochemical synthesis of silver nanoparticles inside the polymer hydrogel aimed for biomedical applications. Polym. Compos. 2014, 35, 217–226.

299. Oliveira, R.N.; Rouze, R.; Quilty, B.; Alves, G.G.; Soares, G.D.A.; Thire, R.M.; McGuinness, G.B. Mechanical properties and in vitro characterization of polyvinyl alcohol-nano-silver hydrogel wound dressings. Interface Focus 2014, 4.

300. Ghasemzadeh, H.; Ghanaat, F. Antimicrobial alginate/PVA silver nanocomposite hydrogel, synthesis and characterization. J. Polym. Res. 2014, 21, 1–14.

301. Eghbalifam, N.; Frounchi, M.; Dadbin, S. Antibacterial silver nanoparticles in polyvinyl alcohol/sodium alginate blend produced by gamma irradiation. Int. J. Biol. Macromol. 2015, 80, 170–176.

302. Zan, X.; Kozlov, M.; McCarthy, T.J.; Su, Z. Covalently attached, silver-doped poly(vinyl alcohol) hydrogel films on poly(L-lactic acid). Biomacromolecules 2010, 11, 1082–1088.

303. Deen, G.; Chua, V. Synthesis and properties of new "stimuli" responsive nanocomposite hydrogels containing silver nanoparticles. Gels 2015, 1, 117–134.

304. Spasojević, J.; Radosavljević, A.; Krstić, J.; Jovanović, D.; Spasojević, V.; Kalagasidis-Krušić, M.; Kačarević-Popović, Z. Dual responsive antibacterial Ag-poly(N-isopropylacrylamide/itaconic acid) hydrogel nanocomposites synthesized by gamma irradiation. Eur. Polym. J. 2015, 69, 168–185.

305. Bajpai, S.K.; Bajpai, M.; Sharma, L. In situ formation of silver nanoparticles in poly(N-isopropyl acrylamide) hydrogel for antibacterial applications. Des. Monomers Polym. 2011, 14, 383–394.

306. Bazzaz, B.S.F.; Khameneh, B.; Jalili-Behabadi, M.M.; Malaekeh-Nikouei, B.; Mohajeri, S.A. Preparation, characterization and antimicrobial study of a hydrogel (soft contact lens) material impregnated with silver nanoparticles. Cont. Lens. Anterio Eye 2014, 37, 149–152.

307. Guadalupe Rodriguez-Delgado, M.; Yanez-Flores, I.G.; Sanchez-Valdes, S.; Rodriguez-Fernandez, O.S.; Betancourt-Galindo, R.; Lozano-Ramirez, T.; Ramirez-Vargas, E.; Ortega-Ortiz, H. Preparation and characterization of acrylic acid/itaconic acid hydrogel coatings containing silver nanoparticles. J. Appl. Polym. Sci. 2013, 130, 2713–2721.

308. Wei, Q.B.; Fu, F.; Zhang, Y.Q.; Tang, L. Preparation, characterization, and antibacterial properties of pH-responsive P(MMA-co-MAA)/silver nanocomposite hydrogels. J. Polym. Res. 2014, 21, 1–9.

309. Gils, P.S.; Ray, D.; Sahoo, P.K. Designing of silver nanoparticles in gum arabic based semi-IPN hydrogel. Int. J. Biol. Mmacromol. 2010, 46, 237–244.

310. Cho, S.-A.; Sung, A.Y. Change of physical properties of hydrogel lens polymer containing isocyanate group with Ag nanoparticle. J. Chosun Nat. Sci. 2014, 7, 5–10.

311. Boonkaew, B.; Kempf, M.; Kimble, R.; Supaphol, P.; Cuttle, L. Antimicrobial efficacy of a novel silver hydrogel dressing compared to two common silver burn wound dressings: Acticoat™ and PolyMem Silver®. Burns 2014, 40, 89–96.

312. Young Yook, J.; Choi, G.H.; Hack Suh, D. A novel method for preparing silver nanoparticle-hydrogel nanocomposites via pH-induced self-assembly. Chem. Commun. 2012, 48, 5001–5003.

313. Mandal, S.K.; Brahmachari, S.; Das, P.K. In situ synthesised silver nanoparticle-infusedl-lysine-based injectable hydrogel: Development of a biocompatible, antibacterial, soft nanocomposite. ChemPlusChem 2014, 79, 1733–1746.

314. Adhikari, B.; Banerjee, A. Short-peptide-based hydrogel: A template for the in situ synthesis of fluorescent silver nanoclusters by using sunlight. Chem. Eur. J. 2010, 16, 13698–13705.

315. Bardajee, G.R.; Hooshyar, Z.; Rezanezhad, H. A novel and green biomaterial based silver nanocomposite hydrogel: Synthesis, characterization and antibacterial effect. J. Inorg. Biochem. 2012, 117, 367–373.

316. Das, A.; Kumar, A.; Patil, N.B.; Viswanathan, C.; Ghosh, D. Preparation and characterization of silver nanoparticle loaded amorphous hydrogel of carboxymethylcellulose for infected wounds. Carbohydr. Polym. 2015, 130, 254–261.

317. Hebeish, A.; Hashem, M.; El-Hady, M.M.A.; Sharaf, S. Development of CMC hydrogels loaded with silver nano-particles for medical applications. Carbohydr. Polym. 2013, 92, 407–413.

318. Rattanaruengsrikul, V.; Pimpha, N.; Supaphol, P. In vitro efficacy and toxicology evaluation of silver nanoparticle-loaded gelatin hydrogel pads as antibacterial wound dressings. J. Appl. Polym. Sci. 2012, 124, 1668–1682.

319. Sahiner, M.; Alpaslan, D.; Bitlisli, B.O. Collagen-based hydrogel films as drug-delivery devices with antimicrobial properties. Polym. Bull. 2014, 71, 3017–3033.

320. Garcia-Astrain, C.; Chen, C.; Buron, M.; Palomares, T.; Eceiza, A.; Fruk, L.; Corcuera, M.A.; Gabilondo, N. Biocompatible hydrogel nanocomposite with covalently embedded silver nanoparticles. Biomacromolecules 2015, 16, 1301–1310.

321. Trivedi, T.J.; Rao, K.S.; Kumar, A. Facile preparation of agarose–chitosan hybrid materials and nanocomposite ionogels using an ionic liquid via dissolution, regeneration and sol–gel transition. Green Chem. 2014, 16, 320–330.

322. Martins, A.F.; Monteiro, J.P.; Bonafe, E.G.; Gerola, A.P.; Silva, C.T.P.; Girotto, E.M.; Rubira, A.F.; Muniz, E.C. Bactericidal activity of hydrogel beads based on N,N,N-trimethyl chitosan/alginate complexes loaded with silver nanoparticles. Chin. Chem. Lett. 2015, 26, 1129–1132.

323. Grade, S.; Eberhard, J.; Neumeister, A.; Wagener, P.; Winkel, A.; Stiesch, M.; Barcikowski, S. Serum albumin reduces the antibacterial and cytotoxic effects of hydrogel-embedded colloidal silver nanoparticles. RSC Adv. 2012, 2, 7190–7196.

324. Hauser, A.W.; Evans, A.A.; Na, J.-H.; Hayward, R.C. Photothermally reprogrammable buckling of nanocomposite gel sheets. Angew. Chem. Int. Ed. 2015, 54, 5434–5437.

325. Kim, J.H.; Lee, T.R. Hydrogel-templated growth of large gold nanoparticles: Synthesis of thermally responsive hydrogel-nanoparticle composites. Langmuir 2007, 23, 6504–6509.

326. Kim, J.H.; Lee, T.R. Discrete thermally responsive hydrogel-coated gold nanoparticles for use as drug-delivery vehicles. Drug Dev. Res. 2006, 67, 61–69.

327. Rong, Q.; Han, H.; Feng, F.; Ma, Z. Network nanostructured polypyrrole hydrogel/Au composites as enhanced electrochemical biosensing platform. Sci. Rep. 2015, 5.

328. Takahashi, D.; Sasaki, M.; Nagamine, K.; Torimitsu, K.; Nishizawa, M. Electroporation of adherent cells by direct lamination of hydrogel-based microelectrode substrates. Chem. Lett. 2014, 43, 444–446.

329. Cheng, C.; Tang, M.C.; Wu, C.S.; Simon, T.; Ko, F.-H. New synthesis route of hydrogel through a bioinspired supramolecular approach: Gelation, binding interaction, and in vitro dressing. ACS Appl. Mater. Interfaces 2015, 7, 19306–19315.

330. Huang, Y.; Ma, Y.; Chen, Y.; Wu, X.; Fang, L.; Zhu, Z.; Yang, C.J. Target-responsive DNAzyme cross-linked hydrogel for visual quantitative detection of lead. Anal. Chem. 2014, 86, 11434–11439.

331. Li, J.; Xie, J.; Gao, L.; Li, C.M. Au nanoparticles-3D graphene hydrogel nanocomposite to boost synergistically in situdetection sensitivity toward cell-released nitric oxide. ACS Appl. Mater. Interfaces 2015, 7, 2726–2734.

332. Mahdavinia, G.R.; Etemadi, H.; Soleymani, F. Magnetic/pH-responsive beads based on caboxymethyl chitosan and kappa-carrageenan and controlled drug release. Carbohydr. Polym. 2015, 128, 112–121.

333. Mahdavinia, G.R.; Etemadi, H. In situ synthesis of magnetic CaraPVA IPN nanocomposite hydrogels and controlled drug release. Mater. Sci. Eng. C 2014, 45, 250–260.

334. Bardajee, G.R.; Hooshyar, Z.; Rastgo, F. Kappa carrageenan-g-poly (acrylic acid)/SPION nanocomposite as a novel stimuli-sensitive drug delivery system. Colloid Polym. Sci. 2013, 291, 2791–2803.

335. Mahdavinia, G.R.; Rahmani, Z.; Karami, S.; Pourjavadi, A. Magnetic/pH-sensitive kappa-carrageenan/sodium alginate hydrogel nanocomposite beads: Preparation, swelling behavior, and drug delivery. J. Biomater. Sci. Polym. Ed. 2014,25, 1891–1906.

336. Bardajee, G.R.; Hooshyar, Z. One-pot synthesis of biocompatible superparamagnetic iron oxide nanoparticles/hydrogel based on salep: Characterization and drug delivery. Carbohydr. Polym. 2014, 101, 741–751.

337. Bardajee, G.R.; Hooshyar, Z.; Asli, M.J.; Shahidi, F.E.; Dianatnejad, N. Synthesis of a novel supermagnetic iron oxide nanocomposite

hydrogel based on graft copolymerization of poly((2-dimethylamino)ethyl methacrylate) onto salep for controlled release of drug. Mater. Sci. Eng. C 2014, 36, 277–286.

338. Bardajee, G.R.; Hooshyar, Z. A novel biocompatible magnetic iron oxide nanoparticles/hydrogel based on poly (acrylic acid) grafted onto starch for controlled drug release. J. Polym. Res. 2013, 20, 1–13.

339. Uva, M.; Mencuccini, L.; Atrei, A.; Innocenti, C.; Fantechi, E.; Sangregorio, C.; Maglio, M.; Fini, M.; Barbucci, R. On the Mechanism of drug release from polysaccharide hydrogels cross-linked with magnetite nanoparticles by applying alternating magnetic fields: The case of Dox delivery. Gels 2015, 1, 24–43.

340. Giani, G.; Fedi, S.; Barbucci, R. Hybrid magnetic hydrogel: A potential system for controlled drug delivery by means of alternating magnetic fields. Polymers 2012, 4, 1157–1169.

341. Ziv-Polat, O.; Skaat, H.; Shahar, A.; Margel, S. Novel magnetic fibrin hydrogel scaffolds containing thrombin and growth factors conjugated iron oxide nanoparticles for tissue engineering. Int. J. Nanomed. 2012, 7, 1259–1274.

342. Zhang, Y.; Sun, Y.; Yang, X.; Hilborn, J.; Heerschap, A.; Ossipov, D.A. Injectable in situ forming hybrid iron oxide-hyaluronic acid hydrogel for magnetic resonance imaging and drug delivery. Macromol. Biosci. 2014, 14, 1249–1259.

343. Zhang, D.; Sun, P.; Li, P.; Xue, A.; Zhang, X.; Zhang, H.; Jin, X. A magnetic chitosan hydrogel for sustained and prolonged delivery of Bacillus Calmette-Guerin in the treatment of bladder cancer. Biomaterials 2013, 34, 10258–10266.

344. Demarchi, C.A.; Debrassi, A.; Buzzi Fde, C.; Correa, R.; Filho, V.C.; Rodrigues, C.A.; Nedelko, N.; Demchenko, P.; Slawska-Waniewska, A.; Dluzewski, P.; et al. A magnetic nanogel based on O-carboxymethylchitosan for antitumor drug delivery: Synthesis, characterization and in vitro drug release. Soft Matter 2014, 10, 3441–3450.

345. Zhang, L.L.; Li, P.; Li, Y.M.; Wang, A.Q. Preparation and characterization of magnetic alginate-chitosan hydrogel beads loaded matrine. Drug Dev. Ind. Pharm. 2012, 38, 872–882.

346. Zhao, L.B.; Pan, L.; Zhang, K.; Guo, S.S.; Liu, W.; Wang, Y.; Chen, Y.; Zhao, X.Z.; Chan, H.L.W. Generation of Janus alginate hydrogel particles with magnetic anisotropy for cell encapsulation. Lab Chip 2009, 9, 2981–2986.

347. Murali, R.; Vidhya, P.; Thanikaivelan, P. Thermoresponsive magnetic nanoparticle-aminated guar gum hydrogel system for sustained release of doxorubicin hydrochloride. Carbohydr. Polym. 2014, 110, 440–445.

348. Kim, J.I.; Lee, B.S.; Chun, C.; Cho, J.-K.; Kim, S.-Y.; Song, S.-C. Long-term theranostic hydrogel system for solid tumors. Biomaterials 2012, 33, 2251–2259.

349. Koc, K.; Alveroglu, E. Adsorption and desorption studies of lysozyme by Fe3O4-polymer nanocomposite via fluorescence spectroscopy. J. Mol. Struc. 2015, 1089, 66–72.

350. Reddy, N.N.; Mohan, Y.M.; Varaprasad, K.; Ravindra, S.; Joy, P.A.; Raju, K.M. Magnetic and electric responsive hydrogel-magnetic nanocomposites for drug-delivery application. J. Appl. Polym. Sci. 2011, 122, 1364–1375.

351. Reddy, N.N.; Varaprasad, K.; Ravindra, S.; Reddy, G.V.S.; Reddy, K.M.S.; Reddy, K.M.M.; Raju, K.M. Evaluation of blood compatibility and drug release studies of gelatin based magnetic hydrogel nanocomposites. Colloid. Surface. A2011, 385, 20–27.

352. Hu, K.; Sun, J.; Guo, Z.; Wang, P.; Chen, Q.; Ma, M.; Gu, N. A novel magnetic hydrogel with aligned magnetic colloidal assemblies showing controllable enhancement of magnetothermal effect in the presence of alternating magnetic field. Adv. Mater. 2015, 27, 2507–2514.

353. Satarkar, N.S.; Hilt, J.Z. Magnetic hydrogel nanocomposites for remote controlled pulsatile drug release. J. Control. Release 2008, 130, 246–251.

354. Nazli, C.; Demirer, G.S.; Yar, Y.; Acar, H.Y.; Kizilel, S. Targeted delivery of doxorubicin into tumor cells via MMP-sensitive PEG hydrogel-coated magnetic iron oxide nanoparticles (MIONPs). Colloid. Surface. B 2014, 122, 674–683.

355. Nazli, C.; Ergenc, T.I.; Yar, Y.; Acar, H.Y.; Kizilel, S. RGDS-functionalized polyethylene glycol hydrogel-coated magnetic iron

oxide nanoparticles enhance specific intracellular uptake by HeLa cells. Int. J. Nanomed. 2012, 7, 1903–1920.

356. Adriane, K.; Huang, J.; Ding, G.W.; Chen, J.; Liu, Y.F. Self assembled magnetic PVP/PVA hydrogel microspheres; magnetic drug targeting of VX2 auricular tumours using pingyangmycin. J. Drug Target. 2006, 14, 243–253.

357. Fu, C.Y.; Lin, C.Y.; Chu, W.C.; Chang, H.Y. A simple cell patterning method using magnetic particle-containing photosensitive poly (ethylene glycol) hydrogel blocks: A Technical Note. Tissue Eng. C 2011, 17, 871–877.

358. Mahdavinia, G.R.; Ettehadi, S.; Amini, M.; Sabzi, M. Synthesis and characterization of hydroxypropyl methylcellulose-g-poly (acrylamide)/LAPONITE® nanocomposites as novel magnetic-and pH-sensitive carriers for controlled drug release. RSC Adv. 2015, 5, 44516–44523.

359. Chou, F.Y.; Shih, C.M.; Tsai, M.C.; Chiu, W.Y.; Lue, S.J. Functional acrylic acid as stabilizer for synthesis of smart hydrogel particles containing a magnetic Fe3O4 core. Polymer 2012, 53, 2839–2846.

360. Fan, T.; Li, M.; Wu, X.; Li, M.; Wu, Y. Preparation of thermoresponsive and pH-sensitivity polymer magnetic hydrogel nanospheres as anticancer drug carriers. Colloid. Surface. B 2011, 88, 593–600.

361. Anirudhan, T.S.; Divya, P.L.; Nima, J. Synthesis and characterization of silane coated magnetic nanoparticles/glycidylmethacrylate-grafted-maleated cyclodextrin composite hydrogel as a drug carrier for the controlled delivery of 5-fluorouracil. Mater. Sci. Eng. C 2015, 55, 471–481.

362. Bayramoglu, G.; Altintas, B.; Arica, M.Y. Immobilization of glucoamylase onto polyaniline-grafted magnetic hydrogel via adsorption and adsorption/cross-linking. Appl. Microbiol. Biotechnol. 2013, 97, 1149–1159.

363. Ghosh, R.; Goswami, U.; Ghosh, S.S.; Paul, A.; Chattopadhyay, A. Synergistic anticancer activity of fluorescent copper nanoclusters and cisplatin delivered through a hydrogel nanocarrier. ACS Appl. Mater. Interfaces 2015, 7, 209–222.

364. Cometa, S.; Iatta, R.; Ricci, M.A.; Ferretti, C.; de Giglio, E. Analytical characterization and antimicrobial properties of novel

copper nanoparticle-loaded electrosynthesized hydrogel coatings. J. Bioact. Compat. Polym. 2013, 28, 508–522.

365. Yadollahi, M.; Gholamali, I.; Namazi, H.; Aghazadeh, M. Synthesis and characterization of antibacterial carboxymethylcellulose/CuO bio-nanocomposite hydrogels. Int. J. Biol. Macromol. 2015, 73, 109–114.

366. Staneva, D.; Atanasova, D.; Vasileva-Tonkova, E.; Lukanova, V.; Grabchev, I. A cotton fabric modified with a hydrogel containing ZnO nanoparticles: Preparation and properties study. Appl. Surf. Sci. 2015, 345, 72–80.

367. Yadollahi, M.; Gholamali, I.; Namazi, H.; Aghazadeh, M. Synthesis and characterization of antibacterial carboxymethyl cellulose/ZnO nanocomposite hydrogels. Int. J. Biol. Macromol. 2015, 74, 136–141.

368. Hashem, M.; Sharaf, S.; Abd El-Hady, M.M.; Hebeish, A. Synthesis and characterization of novel carboxymethylcellulose hydrogels and carboxymethylcellulolse hydrogel ZnO nanocomposites. Carbohydr. Polym.2013, 95, 421–427.

369. Kumar, S.P.T.; Lakshmanan, V.K.; Raj, M.; Biswas, R.; Hiroshi, T.; Nair, S.V.; Jayakumar, R. Evaluation of wound healing potential of beta-chitin hydrogel/nano Zinc oxide composite bandage. Pharm. Res. 2013, 30, 523–537.

370. Zhang, Q.; Fang, Z.; Cao, Y.; Du, H.; Wu, H.; Beuerman, R.; Chan-Park, M.B.; Duan, H.; Xu, R. High refractive index inorganic–organic interpenetrating polymer network (IPN) hydrogel nanocomposite toward artificial cornea implants. ACS Macro Lett. 2012, 1, 876–881.

371. Weltin, A.; Kieninger, J.; Enderle, B.; Gellner, A.K.; Fritsch, B.; Urban, G.A. Polymer-based, flexible glutamate and lactate microsensors for in vivo applications. Biosens. Bioelectron. 2014, 61, 192–199.

372. Bai, Y.H.; Zhang, H.; Xu, J.J.; Chen, H.Y. Relationship between nanostructure and electrochemical/biosensing properties of MnO2 nanomaterials for H2O2/choline. J. Phys. Chem. C 2008, 112, 18984–18990.

373. Xu, B.; Jiang, H.; Li, H.; Zhang, G.; Zhang, Q. High strength nanocomposite hydrogel bilayer with bidirectional bending and

shape switching behaviors for soft actuators. RSC Adv. 2015, 5, 13167–13170.

374. Zhang, H.; Liu, Y.; Yao, D.; Yang, B. Hybridization of inorganic nanoparticles and polymers to create regular and reversible self-assembly architectures. Chem. Soc. Rev. 2012, 41, 6066–6088.

375. Kastellorizios, M.; Papadimitrakopoulos, F.; Burgess, D.J. Multiple tissue response modifiers to promote angiogenesis and prevent the foreign body reaction around subcutaneous implants. J. Control. Release 2015, 214, 103–111.

376. Wang, Y.; Papadimitrakopoulos, F.; Burgess, D.J. Polymeric "smart" coatings to prevent foreign body response to implantable biosensors. J. Control. Release 2013, 169, 341–347.

377. Kastellorizios, M.; Papadimitrakopoulos, F.; Burgess, D.J. Prevention of foreign body reaction in a pre-clinical large animal model. J. Control. Release 2015, 202, 101–107.

378. Gu, B.; Burgess, D.J. Prediction of dexamethasone release from PLGA microspheres prepared with polymer blends using a design of experiment approach. Int. J. Pharm. 2015, 495, 393–403.

379. Nie, L.; Zhang, G.H.; Hou, R.X.; Xu, H.P.; Li, Y.P.; Fu, J. Controllable promotion of chondrocyte adhesion and growth on PVA hydrogels by controlled release of TGF-beta 1 from porous PLGA microspheres. Colloid. Surface. B 2015, 125, 51–57.

380. Zhang, G.H.; Hou, R.X.; Zhan, D.X.; Cong, Y.; Cheng, Y.J.; Fu, J. Fabrication of hollow porous PLGA microspheres for controlled protein release and promotion of cell compatibility. Chin. Chem. Lett. 2013, 24, 710–714.

381. Rahmani-Neishaboor, E.; Hartwell, R.; Jalili, R.; Jackson, J.; Brown, E.; Ghahary, A. Localized controlled release of stratifin reduces implantation-induced dermal fibrosis. Acta Biomater. 2012, 8, 3660–3668.

382. Jo, J.H.; Choi, S.W.; Choi, J.W.; Paik, D.H.; Kang, S.S.; Kim, S.E.; Jeon, Y.-C.; Huh, J.B. Effects of different rhBMP-2 release profiles in defect areas around dental implants on bone regeneration. Biomed. Mater. 2015, 10, 045007.

383. Yoon, Y.M.; Lewis, J.S.; Carstens, M.R.; Campbell-Thompson, M.; Wasserfall, C.H.; Atkinson, M.A.; Keselowsky, B.G. A combination

hydrogel microparticle-based vaccine prevents type 1 diabetes in non-obese diabetic mice. Sci. Rep.2015, 5, 13155.

384. Sellers, D.L.; Kim, T.H.; Mount, C.W.; Pun, S.H.; Horner, P.J. Poly(lactic-co-glycolic) acid microspheres encapsulated in Pluronic F-127 prolong hirudin delivery and improve functional recovery from a demyelination lesion. Biomaterials2014, 35, 8895–8902.

385. Geng, H.Q.; Song, H.; Qi, J.; Cui, D.X. Sustained release of VEGF from PLGA nanoparticles embedded thermo-sensitive hydrogel in full-thickness porcine bladder acellular matrix. Nanoscale Res. Lett. 2011, 6, 1–8.

386. Karam, J.P.; Muscari, C.; Sindji, L.; Bastiat, G.; Bonafè, F.; Venier-Julienne, M.-C.; Montero-Menei, N.C. Pharmacologically active microcarriers associated with thermosensitive hydrogel as a growth factor releasing biomimetic 3D scaffold for cardiac tissue-engineering. J. Control. Release 2014, 192, 82–94.

387. Acarregui, A.; Herrán, E.; Igartua, M.; Blanco, F.J.; Pedraz, J.L.; Orive, G.; Hernandez, R.M. Multifunctional hydrogel-based scaffold for improving the functionality of encapsulated therapeutic cells and reducing inflammatory response.Acta Biomater. 2014, 10, 4206–4216.

388. Yonet-Tanyeri, N.; Rich, M.H.; Lee, M.; Lai, M.H.; Jeong, J.H.; DeVolder, R.J.; Kong, H. The spatiotemporal control of erosion and molecular release from micropatterned poly(ethylene glycol)-based hydrogel. Biomaterials 2013, 34, 8416–8423.

389. Elliott Donaghue, I.; Tator, C.H.; Shoichet, M.S. Sustained delivery of bioactive neurotrophin-3 to the injured spinal cord. Biomater. Sci. 2015, 3, 65–72.

390. Elliott Donaghue, I.; Shoichet, M.S. Controlled release of bioactive PDGF-AA from a hydrogel/nanoparticle composite.Acta Biomater. 2015, 25, 35–42.

391. Wang, Y.F.; Cooke, M.J.; Sachewsky, N.; Morshead, C.M.; Shoichet, M.S. Bioengineered sequential growth factor delivery stimulates brain tissue regeneration after stroke. J. Control. Release 2013, 172, 1–11.

392. Kang, C.E.; Baumann, M.D.; Tator, C.H.; Shoichet, M.S. Localized and sustained delivery of fibroblast growth factor-2 from a

nanoparticle-hydrogel composite for treatment of spinal cord injury. Cells Tissues Organs 2013, 197, 55–63.

393. Stanwick, J.C.; Baumann, M.D.; Shoichet, M.S. Enhanced neurotrophin-3 bioactivity and release from a nanoparticle-loaded composite hydrogel. J. Control. Release 2012, 160, 666–675.

394. Stanwick, J.C.; Baumann, M.D.; Shoichet, M.S. In vitro sustained release of bioactive anti-NogoA, a molecule in clinical development for treatment of spinal cord injury. Int. J. Pharm. 2012, 426, 284–290.

395. Baumann, M.D.; Kang, C.E.; Tator, C.H.; Shoichet, M.S. Intrathecal delivery of a polymeric nanocomposite hydrogel after spinal cord injury. Biomaterials 2010, 31, 7631–7639.

396. Wang, F.; Gao, W.W.; Thamphiwatana, S.; Luk, B.T.; Angsantikul, P.; Zhang, Q.Z.; Hu, C.M.J.; Fang, R.H.; Copp, J.A.; Pornpattananangkul, D.; et al. Hydrogel retaining toxin-absorbing nanosponges for local treatment of methicillin-resistant staphylococcus aureus infection. Adv. Mater. 2015, 27, 3437–3443.

397. Pakulska, M.M.; Vulic, K.; Tam, R.Y.; Shoichet, M.S. Hybrid crosslinked methylcellulose hydrogel: A predictable and tunable platform for local drug delivery. Adv. Mater. 2015, 27, 5002–5008.

398. Posadowska, U.; Parizek, M.; Filova, E.; Wlodarczyk-Biegun, M.; Kamperman, M.; Bacakova, L.; Pamula, E. Injectable nanoparticle-loaded hydrogel system for local delivery of sodium alendronate. Int. J. Pharm. 2015, 485, 31–40.

399. Baghaei, B.; Jafari, S.H.; Khonakdar, H.A.; Wagenknecht, U.; Heinrich, G. Novel thermosensitive hydrogel composites based on poly(d,l-lactide-co-glycolide) nanoparticles embedded in poly(N-isopropyl acrylamide) with sustained drug-release behavior. J. Appl. Polym. Sci. 2014, 131.

400. Yang, H.; Tyagi, P.; Kadam, R.S.; Holden, C.A.; Kompella, U.B. Hybrid dendrimer hydrogel/PLGA nanoparticle platform sustains drug delivery for one week and antiglaucoma effects for four days following one-time topical administration. ACS Nano 2012, 6, 7595–7606.

401. Chronopoulou, L.; Sennato, S.; Bordi, F.; Giannella, D.; di Nitto, A.; Barbetta, A.; Dentini, M.; Togna, A.R.; Togna, G.I.; Moschini, S.; et al.

Designing unconventional Fmoc-peptide-based biomaterials: Structure and related properties. Soft Matter 2014, 10, 1944–1952.

402. Gause, S.; Chauhan, A. Nanoparticle-loaded UV-blocking contact lenses. J. Appl. Polym. Sci. 2015, 132.

403. Rossi, F.; Ferrari, R.; Castiglione, F.; Mele, A.; Perale, G.; Moscatelli, D. Polymer hydrogel functionalized with biodegradable nanoparticles as composite system for controlled drug delivery. Nanotechnology 2015, 26, 015602.

404. Murab, S.; Samal, J.; Shrivastava, A.; Ray, A.R.; Pandit, A.; Ghosh, S. Glucosamine loaded injectable silk-in-silk integrated system modulate mechanical properties in bovine ex vivo degenerated intervertebral disc model.Biomaterials 2015, 55, 64–83.

405. Pacheco, D.P.; Amaral, M.H.; Reis, R.L.; Marques, A.P.; Correlo, V.M. Development of an injectable PHBV microparticles-GG hydrogel hybrid system for regenerative medicine. Int. J. Pharm. 2015, 478, 398–408.

406. Appel, E.A.; Tibbitt, M.W.; Webber, M.J.; Mattix, B.A.; Veiseh, O.; Langer, R. Self-assembled hydrogels utilizing polymer-nanoparticle interactions. Nat. Commun. 2015, 6.

407. Samanta, D.; Meiser, J.L.; Zare, R.N. Polypyrrole nanoparticles for tunable, pH-sensitive and sustained drug release.Nanoscale 2015, 7, 9497–9504.

408. Luo, R.C.; Lim, Z.H.; Li, W.; Shi, P.; Chen, C.H. Near-infrared light triggerable deformation-free polysaccharide double network hydrogels. Chem. Commun. 2014, 50, 7052–7055.

409. Peng, Q.; Sun, X.; Gong, T.; Wu, C.Y.; Zhang, T.; Tan, J.; Zhang, Z.R. Injectable and biodegradable thermosensitive hydrogels loaded with PHBHHx nanoparticles for the sustained and controlled release of insulin. Acta Biomater. 2013,9, 5063–5069.

410. Jung, H.J.; Abou-Jaoude, M.; Carbia, B.E.; Plummer, C.; Chauhan, A. Glaucoma therapy by extended release of timolol from nanoparticle loaded silicone-hydrogel contact lenses. J. Control. Release 2013, 165, 82–89.

411. Dyondi, D.; Webster, T.J.; Banerjee, R. A nanoparticulate injectable hydrogel as a tissue engineering scaffold for multiple

growth factor delivery for bone regeneration. Int. J. Nanomed. 2013, 8, 47–59.

412. Rossi, F.; Ferrari, R.; Papa, S.; Moscatelli, D.; Casalini, T.; Forloni, G.; Perale, G.; Veglianese, P. Tunable hydrogel-Nanoparticles release system for sustained combination therapies in the spinal cord. Colloid. Surface. B 2013, 108, 169–177.

413. Segovia, N.; Pont, M.; Oliva, N.; Ramos, V.; Borrós, S.; Artzi, N. Hydrogel doped with nanoparticles for local sustained release of siRNA in breast cancer. Adv. Healthc. Mater. 2015, 4, 271–280.

414. Fan, M.; Guo, Q.F.; Luo, J.C.; Luo, F.; Xie, P.; Tang, X.H.; Qian, Z.Y. Preparation and in vitro characterization of dexamethasone-loaded poly(D,L-lactic acid) microspheres embedded in poly(ethylene glycol)-poly(-caprolactone)-poly(ethylene glycol) hydrogel for orthopedic tissue engineering. J. Biomater. Appl. 2013, 28, 288–297.

415. Hu, J.J.; Liu, S.; Zheng, W.L.; Xiu, Z.L.; Zhang, Q. Dendrimer complexed hydrogel for local delivery of PTX to inhibit recurrence of breast cancer. J. Control. Release 2015, 213, e54.

416. Zhong, S.; Yung, L.Y. Enhanced biological stability of collagen with incorporation of PAMAM dendrimer. J. Biomed. Mater. Res. A 2009, 91, 114–122.

417. Bi, X.D.; Luckanagul, J.A.; Allen, A.; Ramaboli, M.; Campbell, E.; West, D.; Maturavongsadit, P.; Brummett, K.; Wang, Q. Synthesis of PAMAM dendrimer-based fast cross-linking hydrogel for biofabrication. J. Biomater. Sci. Polym. Ed.2015, 26, 669–682.

418. Wang, Y.; Zhao, Q.; Zhang, H.; Yang, S.; Jia, X.R. A Novel poly(amido amine)-dendrimer-based hydrogel as a mimic for the extracellular matrix. Adv. Mater. 2014, 26, 4163–4167.

419. Winnicka, K.; Wroblewska, M.; Wieczorek, P.; Sacha, P.T.; Tryniszewska, E. Hydrogel of ketoconazole and pamam dendrimers: Formulation and antifungal activity. Molecules 2012, 17, 4612–4624.

420. Desai, P.N.; Yuan, Q.; Yang, H. Synthesis and characterization of photocurable polyamidoamine dendrimer hydrogels as a versatile platform for tissue engineering and drug delivery. Biomacromolecules 2010, 11, 666–673.

421. Din, F.U.; Mustapha, O.; Kim, D.W.; Rashid, R.; Park, J.H.; Choi, J.Y.; Ku, S.K.; Yong, C.S.; Kim, J.O.; Choi, H.G. Novel dual-reverse thermosensitive solid lipid nanoparticle-loaded hydrogel for rectal administration of flurbiprofen with improved bioavailability and reduced initial burst effect. Eur. J. Pharm. Biopharm. 2015, 94, 64–72.

422. Gao, W.W.; Vecchio, D.; Li, J.M.; Zhu, J.Y.; Zhang, Q.Z.; Fu, V.; Li, J.Y.; Thamphiwatana, S.; Lu, D.N.; Zhang, L.F. Hydrogel containing nanoparticle-stabilized liposomes for topical antimicrobial delivery. ACS Nano 2014, 8, 2900–2907.

423. Wickremasinghe, N.C.; Kumar, V.A.; Hartgerink, J.D. Two-step self-assembly of liposome-multidomain peptide nanofiber hydrogel for time-controlled release. Biomacromolecules 2014, 15, 3587–3595.

424. El Kechai, N.; Bochot, A.; Huang, N.; Nguyen, Y.; Ferrary, E.; Agnely, F. Effect of liposomes on rheological and syringeability properties of hyaluronic acid hydrogels intended for local injection of drugs. Int. J. Pharm. 2015, 487, 187–196.

425. Widjaja, L.K.; Bora, M.; Chan, P.N.P.H.; Lipik, V.; Wong, T.T.L.; Venkatraman, S.S. Hyaluronic acid-based nanocomposite hydrogels for ocular drug delivery applications. J. Biomed. Mater. Res. A 2014, 102, 3056–3065.

426. Dong, J.; Jiang, D.H.; Wang, Z.; Wu, G.Z.; Miao, L.Y.; Huang, L.X. Intra-articular delivery of liposomal celecoxib-hyaluronate combination for the treatment of osteoarthritis in rabbit model. Int. J. Pharm. 2013, 441, 285–290.

427. Lajavardi, L.; Camelo, S.; Agnely, F.; Luo, W.; Goldenberg, B.; Naud, M.C.; Behar-Cohen, F.; de Kozak, Y.; Bochot, A. New formulation of vasoactive intestinal peptide using liposomes in hyaluronic acid gel for uveitis. J. Control. Release2009, 139, 22–30.

428. Zhou, X.; Shao, H.; Liu, H. Preparation and characterization of film-forming raspberry-like polymer/silica nanocomposites via soap-free emulsion polymerization and the sol–gel process. Colloid Polym. Sci. 2012, 291, 1181–1190.

429. Suri, A.; Campos, R.; Rackus, D.G.; Spiller, N.J.S.; Richardson, C.; Palsson, L.O.; Kataky, R. Liposome-doped hydrogel for implantable tissue. Soft Matter 2011, 7, 7071–7077.

430. Popescu, M.T.; Mourtas, S.; Pampalakis, G.; Antimisiaris, S.G.; Tsitsilianis, C. pH-responsive hydrogel/liposome soft nanocomposites for tuning drug release. Biomacromolecules 2011, 12, 3023–3030.

431. Wang, Y.J.; Chen, L.J.; Tan, L.W.; Zhao, Q.; Luo, F.; Wei, Y.Q.; Qian, Z.Y. PEG-PCL based micelle hydrogels as oral docetaxel delivery systems for breast cancer therapy. Biomaterials 2014, 35, 6972–6985.

432. Gong, C.Y.; Wu, Q.J.; Wang, Y.J.; Zhang, D.D.; Luo, F.; Zhao, X.; Wei, Y.Q.; Qian, Z.Y. A biodegradable hydrogel system containing curcumin encapsulated in micelles for cutaneous wound healing. Biomaterials 2013, 34, 6377–6387.

433. Jin, N.X.; Morin, E.A.; Henn, D.M.; Cao, Y.; Woodcock, J.W.; Tang, S.C.; He, W.; Zhao, B. Agarose hydrogels embedded with pH-responsive diblock copolymer micelles for triggered release of substances. Biomacromolecules 2013,14, 2713–2723.

434. Li, Y.; Yang, C.; Khan, M.; Liu, S.Q.; Hedrick, J.L.; Yang, Y.Y.; Ee, P.L.R. Nanostructured PEG-based hydrogels with tunable physical properties for gene delivery to human mesenchymal stem cells. Biomaterials 2012, 33, 6533–6541.

435. Hashimoto, Y.; Mukai, S.A.; Sawada, S.I.; Sasaki, Y.; Akiyoshi, K. Nanogel tectonic porous gel loading biologics, nanocarriers, and cells for advanced scaffold. Biomaterials 2015, 37, 107–115.

436. Meid, J.; Friedrich, T.; Tieke, B.; Lindner, P.; Richtering, W. Composite hydrogels with temperature sensitive heterogeneities: Influence of gel matrix on the volume phase transition of embedded poly-(N-isopropylacrylamide) microgels. Phys. Chem. Chem. Phys. 2011, 13, 3039–3047.

437. Sekine, Y.; Moritani, Y.; Ikeda-Fukazawa, T.; Sasaki, Y.; Akiyoshi, K. A hybrid hydrogel biomaterial by nanogel engineering: Bottom-up design with nanogel and liposome building blocks to develop a multidrug delivery system.Adv. Healthc. Mater. 2012, 1, 722–728.

438. Molinos, M.; Carvalho, V.; Silva, D.M.; Gama, F.M. Development of a hybrid dextrin hydrogel encapsulating dextrin nanogel as protein delivery system. Biomacromolecules 2012, 13, 517–527.

439. Sivakumaran, D.; Maitland, D.; Hoare, T. Injectable microgel-hydrogel composites for prolonged small-molecule drug delivery. Biomacromolecules 2011, 12, 4112–4120.

440. Hirakura, T.; Yasugi, K.; Nemoto, T.; Sato, M.; Shimoboji, T.; Aso, Y.; Morimoto, N.; Akiyoshi, K. Hybrid hyaluronan hydrogel encapsulating nanogel as a protein nanocarrier: New system for sustained delivery of protein with a chaperone-like function. J. Control. Release 2010, 142, 483–489.

441. Hao, X.; Liu, H.; Xie, Y.J.; Fang, C.; Yang, H.Y. Thermal-responsive self-healing hydrogel based on hydrophobically modified chitosan and vesicle. Colloid Polym. Sci. 2013, 291, 1749–1758.

442. Litvinchuk, S.; Lu, Z.; Rigler, P.; Hirt, T.D.; Meier, W. Calcein release from polymeric vesicles in blood plasma and PVA hydrogel. Pharm. Res. 2009, 26, 1711–1717.

443. Marianecci, C.; Carafa, M.; di Marzio, L.; Rinaldi, F.; di Meo, C.; Alhaique, F.; Matricardi, P.; Coviello, T. A new vesicle-loaded hydrogel system suitable for topical applications: Preparation and characterization. J. Pharm. Pharm. Sci. 2011, 14, 336–346.

444. Jeon, O.; Wolfson, D.W.; Alsberg, E. In situ formation of growth-factor-loaded coacervate microparticle-embedded hydrogels for directing encapsulated stem cell fate. Adv. Mater. 2015, 27, 2216–2223.

445. Luckanagul, J.; Lee, L.A.; Nguyen, Q.L.; Sitasuwan, P.; Yang, X.M.; Shazly, T.; Wang, Q. Porous alginate hydrogel functionalized with virus as three-dimensional scaffolds for bone differentiation. Biomacromolecules 2012, 13, 3949–3958.

446. Lampe, K.J.; Kern, D.S.; Mahoney, M.J.; Bjugstad, K.B. The administration of BDNF and GDNF to the brain via PLGA microparticles patterned within a degradable PEG-based hydrogel: Protein distribution and the glial response. J. Biomed. Mater. Res. A 2011, 96, 595–607.

447. Xiao, L.; Liu, C.; Zhu, J.; Pochan, D.J.; Jia, X. Hybrid, elastomeric hydrogels crosslinked by multifunctional block copolymer micelles. Soft Matter 2010, 6, 5293–5297.

448. Weiner, A.L.; Carpentergreen, S.S.; Soehngen, E.C.; Lenk, R.P.; Popescu, M.C. Liposome–collagen gel matrix: A novel sustained drug delivery system. J. Pharm. Sci. 1985, 74, 922–925.

449. Carvalho, I.M.; Marques, C.S.; Oliveira, R.S.; Coelho, P.B.; Costa, P.C.; Ferreira, D.C. Sustained drug release by contact lenses for glaucoma treatment: A review. J. Control. Release 2015, 202, 76–82.

450. Gulsen, D.; Chauhan, A. Dispersion ofmicroemulsion drops in HEMA hydrogel: A potential ophthalmic drug delivery vehicle. Int. J. Pharm. 2005, 292, 95–117.

451. Li, M.-H.; Keller, P. Stimuli-responsive polymer vesicles. Soft Matter 2009, 5, 927–937.

452. Mart, R.J.; Liem, K.P.; Webb, S.J. Magnetically-controlled release from hydrogel-supported vesicle assemblies. Chem. Commun. 2009, 2009, 2287–2289.

453. Park, K.M.; Son, J.Y.; Choi, J.H.; Kim, I.G.; Lee, Y.; Lee, J.Y.; Park, K.D. Macro/nano-gel composite as an injectable and bioactive bulking material for the treatment of urinary incontinence. Biomacromolecules 2014, 15, 1979–1984.

454. Akhtar, S.; Al-Zaid, B.; El-Hashim, A.Z.; Chandrasekhar, B.; Attur, S.; Yousif, M.H.M.; Benter, I.F. Cationic polyamidoamine dendrimers as modulators of EGFR signaling in vitro and in vivo. PLoS ONE 2015, 10, e0132215.

455. Gillies, E.R.; Frechet, J.M. Dendrimers and dendritic polymers in drug delivery. Drug Discov. Today 2005, 10, 35–43.

456. Söntjens, S.H.; Nettles, D.L.; Carnahan, M.A.; Setton, L.A.; Grinstaff, M.W. Biodendrimer-based hydrogel scaffolds for cartilage tissue repair. Biomacromolecules 2006, 7, 310–316.

457. Zhang, H.; Patel, A.; Gaharwar, A.K.; Mihaila, S.M.; Iviglia, G.; Mukundan, S.; Bae, H.; Yang, H.; Khademhosseini, A. Hyperbranched polyester hydrogels with controlled drug release and cell adhesion properties. Biomacromolecules2013, 14, 1299–1310.

5

MICROMECHANISMS CONTROLLING THE STRUCTURAL EVOLUTION OF TRIBOSYSTEMS

Dmitry Lubimov [1], Kirill Dolgopolov[1] and [1]

[1] "LIC" Ltd, Russia

1. INTRODUCTION

Development of scientific programs on the wear resistance is determined by economical significance of this issue for the development of productive potential of the world's countries. Deterioration is the main reason for removal of machinery and equipment from service; that is why controlling of the process of wear is the central core for such national scale issues as saving of non-renewable energy resources like hydrocarbon fuel and reducing the consumption of lubricants and structural materials. The condition of selective transfer in friction discovered by D.N. Garkunov and I.V. Kragelsky is the only effect theoretically admitting wearlessness (Garkunov, 2001). However, insufficiently developed physical and chemical basis for the theory of selective transfer hinders its wide deployment in engineering practices. The most controversial issue in the theory of selective transfer is considered to be the nature of products of chemical modification of friction surface. Traditional notions that are associated with the formation on a surface of either relatively simple inorganic compounds or tribopolymers are not sufficient to explain the unique tribotechnical characteristics in self-organizing friction systems. Considering the electronic structure of the rubbing metals, chemical properties of molecules of the active lubricant components, as well as the conditions of a frictional contact, one can expect that during friction of metals in the condition of selective transfer, in addition to normally expected products of the friction surface chemical modification, coordination compounds are generated. Those compounds, being more stable than tribopolymers and less stable than inorganic products of chemical modification can ensure a mass transfer process under the conditions of almost no-wear friction (Garkunov, 2001). Therefore, development of the fundamentals of the selective transfer theory and development of lubricants to put it into practice require considering the dynamics of physical and chemical processes and factors catalyzing the formation of complex compounds during friction, and their subsequent coordination on the surface. It is incontrovertible that with the help of additives it is possible to control almost all properties of lubricants. This being the case, oil acts as a carrier of the chemical reaction components, and a friction assembly act as a reactor, where processes are controlled not only by the composition of the lubricating medium and the nature of the rubbing surfaces, but also by the presence of external force action. In this connection, the analysis of the development trends for lubrication systems and existing advanced developments in chemmotology allows to highlight the following research directions in this research area:

a) decreasing the range of mineral lubricants replacing them with the highly compounded synthetic lubricants;

b) development of lubricants applied to the surface of a tribological conjunction once in the entire lifetime of a friction assembly;

c) development and creation of adaptive lubricating devices that represent an oil feeder controlled by a sensor responding to changes in key parameters of a friction pair;

d) application of centralized combined lubrication systems controlled by a microprocessor that will automatically change a lubrication mode either according to a preset program or depending on the operational performance of the friction assemblies controlled by sensors.

Active interference in electrochemical processes of a frictional contact is a very promising direction to prevent corrosion, to reduce oxidizing, mechanochemical and hydrogen wear, for instance, in metal-polymeric friction assemblies. It is known that the rate of electrochemical corrosion can be greatly reduced if a metal part is subjected to polarization. Polarization of a friction system pursues different objectives: to improve the wettability of the solid surface and adsorption of environment components. The friction system polarization method is that the potential of the system, with the help of an external source of polarization, is shifted to the area that is optimal for the friction process behavior, and then it is maintained at the required level during the work of a friction pair. The main difficulty in dealing with this kind of research is that the value of an electrode potential during friction may differ significantly from the electrode potential of the metal measured in static conditions (Goldade et al., 1993). In the first place, this is caused by mechanical cleaning of the surface from oxide films. In the second place, during friction secondary structures are generated shifting the potential to the positive or negative side. And, finally, the potential shift is affected by the plastic yield that leads to emergence of dislocated atoms in the crystal lattice of the metal. Being a fundamental characteristic of the electrical state of a "metal – electrolyte" interface, the electrode potential of the system can make a significant impact on the friction and wear of metals. The effect of polarization on the friction coefficient and surface damage of metals was discovered by Bowden and Young and further studied by other authors. Decreased friction factor and increased surface charge during polarization are associated with the increased wedge effect of the double electrical layer between the rubbing surfaces. Polymeric materials used in friction assemblies add some specific features to the electrochemical processes in the friction zone. During friction of metal-polymeric pairs, the polymeric components acquire the properties of surface-active substances, which substantially change the electrochemical activity of the metals (Pinchuk et al., 2004). The results of these studies are presented in details in the papers (Garkunov, 2001; Goldade et al., 1993; Kostetsky, 1980), where the influence of a friction system electrode potential on the adsorption, diffusion, oxidation, friction and wear of materials is noted. Unlike processes normally considered in electrochemistry, electrochemical processes during friction occur under the conditions of moving deformable discrete contacts of separate microroughnesses. This leads to generation of currents between the rubbing surfaces during friction (Ryzhikov & Dolgopolov, 2005). Roughnesses of two sliding bodies get elastic or plastic impacts; and consequently the equilibrium state corresponding to the minimal energy of the strained area is disturbed. Under the impacts, the microroughnesses oscillate with the frequency which is close to that of external action. Dynamic loads, causing a variable strain of the material in the mating parts contact, lead to emergence of an alternating magnetic flux in the strained layer. Changed magnetic flux induces an induction emf in the circuit generated by the mating parts. The electrical resistance between these parts conditioned by the properties of the oxide films and lubricant leads to a alternating potential difference in the contact area. If the body and counterbody are electrically isolated from each other with a dielectric layer (e.g., lubricant), the friction couple can be considered as a capacitor. When moving one body against the other, a point contact occurs due to existing surface roughness or other surface imperfections. When the point contacts diverge, the hydrodynamic effects occur like squeezing lubricant out of the friction area, generation of current-conducting seizure bridges from the products of erosion. Consequently, a kind of capacitor electric breakdown occurs, as a result of which a large amount of electrical energy accumulated by the friction surfaces is released. The lubricant and material of the surface layers are decomposed into the components of ionic, radical or ion-radical nature, transferring to the lubricating medium. In this specific short-circuited galvanic microelement, the potential probability is created of having on the friction surfaces the redox reactions of sedimentation of the lubricant medium's active components. Thus, during friction pair operation, an automatically functioning electrochemical mechanism of wear control appears. And this mechanism depends on such factors as the electromotive force of the galvanic cell, the shift of potentials during the circuit closing, the polarizability of the electrochemical circuit, and the rates of the electrode processes. Therefore, creation of alternative adaptive sources of polarization using design and technological features of a friction assembly itself, is of current importance. The main principle of operation of such adaptive lubricating device can be control of friction and wear processes via "external agents" - the force fields of different nature, which was named a "field effect" (Belyi, 1985; Shvedkov & Rovinsky, 1979). The "field effect" means a change of a friction coefficient and wear rate under the influence of electric, magnetic, thermal and radiation fields. Force fields are capable to change values of friction parameters by several times, and in some cases reduce them to zero ("the effect of abnormally low friction" (Shvedkov & Rovinsky, 1979)). Electromagnetic fields have a special place among the force fields (Garkunov, 2001; Goldade et al., 1993; Lubimov et al., 1992); the study of their influence on the performance of materials has occupied the minds of scientists since long ago. One of the problematic issues in reducing wear and friction of metals is the development of reliable and compact sources of electric polarization. The point is that the methods of polarization of a friction system by external sources have not found wide application in friction assemblies due to the bulky dimensions of the electrical sources, which significantly complicate the design of machines. However, today, in the century of rapidly developing microelectronics engineering, this problem can be perfectly solved. Based on the above said, we shall dwell on such aspect of the study of tribosystems as the capability to control their friction parameters directly in the conditions of dynamic interaction. This kind of work does not require interfering in the design features of a machine, which significantly reduces the cost and simplifies the potential possibility of its introduction into the production cycle of any engineering company.

2. STRUCTURAL CHANGES OF TRIBOUNITS

2.1. Third Bodies Of Friction Assemblies

According to present-day thinking, all triboeffects become evident on a macro scale, but the phenomena that cause and accompany them occur at a micro level. Friction is a combination of processes covering the thin surface layers of the contacting bodies. The work of the friction force equals to the energy of destruction of the interfacial bonds; about 99% of it turns into the heat that is spent for structural and thermal activation of the friction surfaces, and the remaining 1% of the energy is stored by a very small portion of surface and promotes activation of the surface bonds (Lubimov & Ryzhikov, 2001; Lubimov & Ryzhikov, 2006). Formation, growth and destruction of the interfacial bonds are determined by the nature of the contacting surfaces, the chemical processes occurring on them, and the stressed state of the surface layers caused by the loading conditions. I.V. Kragelsky coined the term "the third body" combining the interfacial bonds, the products of their destruction and the surface layers where the strains are localized. Later on, M. Godet developed the concept of the third body having expanded its meaning (Godet, 1984) that adds up to significant differences between the surface and bulk properties of the tribounit matter. The third bodies often have the form of film structures arising on the friction surfaces, which significantly affect the frictional characteristics of the friction assemblies. This phenomenon is especially evident when the effect of wearlessness or selective transfer takes place. In accordance with the model of Godet, the copper of the third body during selective transfer has a loose structure, different from a blocked one. It has a weak resistance to shear deformation and a low friction coefficient. These third bodies were named “servovite” films. Along with the servovite films, during a selective transfer the following structures that respectively belong to the third bodies are seen:

- “Surfing” film - a molecular film structure consisting of associated coordination compounds;
- Metal-cladding film – a protective film where a vacansion-dislocation deformation mechanism is partially realized;

- “Divical” film – a metal protective film formed as a result of discharge of metal ions in the contact area during friction;
- “Nubial” film – a tribopolymer protective film consisting of tribopolymers chemisorbed on the active surface of the servovite film, which arise during destruction of lubricants (Polyakov, 1990).

Another type of third bodies appears on a friction surface during interaction of polymer or metal-polymer friction units. Such third bodies were named the friction transfer films; development of the theory of the friction transfer films is in many respects associated with the papers by V.A. Belyi and his school (Pogosyan & Oganesyan, 1986). In general case, during polymer-to-metal friction, as a rule, the transfer of the polymer onto the metal surface is observed. During the frictional transfer of the polymer, as a result of the molecular interaction, fine surface particles are separated and transferred to the active areas of the friction surfaces. During the transition to a steady friction state, in metal-polymer tribounits, actually the polymer rubs against polymer. When friction transfer films are generated, there are the following main types of adhesive interaction: adhesion of separate particles and a layer of particles, melt adhesion, wetting of the counterbody’s surface, and film adhesion. Generation of the friction transfer films has been especially well studied during frictional interaction between polytetrafluorethylene (PTFE) and a steel counterbody. As it has been established (Nikolsky et al., 1988), during the interaction of PTFE with a metal surface, the PTFE molecules are decomposed, and during this process the fluorine atoms are coming off and are being replaced by the hydrogen atoms. Furthermore, generation of a chemical bond between the fluorine and the metal is observed, which determines the adhesion of the PTFE to the metal counterbody (Buckley, D. (1986). The friction transfer films are formed not only during the friction of metal-polymer tribounits, but also during the interaction of metal with graphite. The emergence of third bodies in the frictional contact area in most cases leads to reduced friction forces and wear levels of the tribounit materials (Sysoev et al., 1990). The main reason for such an effect is that the atoms and molecules of a third body passivate the surfaces of friction, thus reducing the adhesive component of the friction force. Intensity of the formation of third bodies depends on a host of aspects related to the fact that a friction assembly is a single tribosystem, the evolution of which is described via the laws of synergetics and thermodynamics of open systems (Garkunov, 2001).

2.2. Higher Energy States Of A Tribosystem Matter

As stated above, the energy released during friction in the form of heat and stored by the surface layer of the material, leads to structural and thermal activation of the surface. The structural-thermal activation means the origination and emergence at the surface of dislocations and point defects. Therefore, energy in tribosystems is degraded, its mechanical form is continuously transformed into the thermal one, and the entropy is constantly reproduced. The discovery of the wearlessness effect has made it possible to single out the general principle of evolution inherent to all friction systems, which consists in their thermodynamic "openness", ability to exchange energy with the external environment, and possibility to reduce entropy. On this criterion, a friction assembly by the level of self-organization approaches to biological systems (Kadolich et al., 2001; Pinchuk et al., 2007). The decreased entropy growth in an open thermodynamic system is determined by the interaction of two energy flows. The thermodynamic forces emerging in them have an influence not only within their own energy flow but also on the adjacent flow. In tribology, such flows are the mass transfer (diffusion) and the heat transfer, and their interaction determines entropy change and the friction assembly structure. American tribologist Buckley has associated a mass transfer with the processes of self-regeneration of friction assembly materials, which prevents the destruction provoked by the heat flow. In the first approximation, the influence of the flows on the tribosystem entropy looks as follows: the diffusion decreases the entropy of the friction assembly; the heat transfer, on the contrary, increases it. Based on the above said, the conclusion can be made that the decrease of tribosystem's entropy relates to the system's structural complication - structural adaptability, and consequently, is identical to the optimization of tribosystem's properties which, in thermodynamic terms, are expressed in minimized entropy growth, and in tribotechnical terms - in reduced wear and friction forces. The structural adaptability is preceded by the changes that have been recently named "tribomutation" (Voinov, 2010). Tribomutation is recrystallization of the thinnest surface layers, decreasing of the free surface energy, elimination of the boundaries of surface and near-surface crystallites, and annihilations of dislocations as they emerge at the surface. The well-known Soviet physicist Zeldovich has demonstrated that in the near-surface layers, the defects with a minimum elastic energy occur; resulting in the loss of elastic properties by the solid bodies, their becoming plastic and fluid, and their transition to a "liquid-like" state. During friction, such physical constants of the material as Young modulus, Poisson ratio, dynamic strength, and acoustic stiffness lose physical meaning. The described structural changes are explained by the action of the processes accompanying the structural and thermal activation of the surfaces matter of a tribounit. Though the percentage of the frictional energy reserved by the surface is small in an absolute value, but, if we take into account that it is accumulated by thin surface layers with a thickness of fractions of micrometer and calculate the density of this energy, then the received numbers can reach critical values for this aggregate state of the material. The surface is supersaturated with dislocations and point defects; its behavior obeys the hydrodynamic laws, and in some extreme cases - the gas dynamics laws. In this extreme case, the movement of dislocations, according to Kuhlmann – Wilsdorf, is described by the equation (Cahn, 1968):

$$\frac{\partial^2 W_z}{\partial x'^2} - \frac{\partial^2 W_z}{\partial y^2} = 0 \qquad (1)$$

$x' = \dfrac{(x - \upsilon t)}{\beta}$; $\beta = (1 - \dfrac{\upsilon^2}{c^2})^{\frac{1}{2}}$ – – dimensionless parameter; c - velocity of elastic waves propagation in the matter; ϑ - velocity of dislocations propagation.

The energy of dislocation W ' moving with the velocity ϑ equals:

$$W' = \frac{W}{\beta} \qquad (2)$$

W – energy of immobile dislocation.

When the dislocation velocity ϑ tends to "c", the parameter $\beta \to 0$ and, consequently, W' indefinably grows. Kuhlmann-Wilsdorf, by analogy with the theory of relativity, recorded the ratio for the W 'through the speed "c ":

$$W' = c^2 \times m' \qquad (3)$$

$$m' = \frac{m}{\beta} \qquad (4)$$

m – mass of the dislocation at rest, equal to $m = \rho \times b^2$; ρ - density of dislocation; b – Burgers vector.

Thus, the movement of dislocations is described by the equations identical to the equations of the special relativity theory (SRT). Therefore, in accordance with the postulates of the SRT, the specific effects predicted by this theory should occur in such systems: change of the space metric, increased mass and energy carried by the defects out to the surface, and the local time slowed down. The effect of these processes is determined by the relativistic parameter β. The value β depends on the speed "c", which, in the conditions of transition of the surface layers to the "liquid-like" state and then to the gaseous state, significantly decreases. Under the conditions of the frictional contact, the parameter β varies from 0.99 to 0.7, which makes the contribution of the relativistic correction to the equations of movement of dislocations very important. Consequently, a dislocation moving during friction can be regarded as a relativistic object, then according to the opinion of Nobel laureate L.D. Landau stated in the section "Relativistic Hydrodynamics" in the book "Gidrodinamika" ("Hydrodynamics") (Landau & Lifshitz, 2001): "The need for relativistic effects consideration... may be associated not only with a high rate of macroscopic motion... the equations (describing this motion (author's note)) also change significantly when this rate is not high, but high are the rates of the microscopic motions of the particles that constitute it". In other words, if a large "classical body" consists of small parts obeying in their movement the laws of the relativity theory, the entire body will be partly "relativistic". Hence, if the constituent friction materials of a dislocation acquire relativistic properties due to structural and thermal activation, then the macro quantities of the tribounit matter start showing the special properties arising from the SRT equations. In this regard, it is very interesting to consider the time paradox that explains some anomalies in the physical and chemical processes of friction surfaces. The equations of the relativity theory do not make a distinction between the time directions; Einstein, as a matter of fact, stated that the past, the present and the future "are nothing but a human illusion devoid of physical sense..." At the same time, the equivalence of the future and the past leads to a number of paradoxes related to the violation of the causality principle, to obey which, the "time arrow" concept is introduced. The direction of the increasing entropy points which state in a couple of the adjacent events is the subsequent rather than the preceding. Many present-day physicists believe that the entropy is actually the arrow of time, which makes the time itself just a simple sequence of states. Time is a part of many physical laws, and it is quite possible that a change in the dimension of time can affect the behavior of various processes. In this connection, there are two interesting examples that illustrate the subsequent line of argument. The American theoretical physicist Feynman believed that an antielectron (positron) emerging in nuclear reactions is actually a normal electron moving backwards in time, and their positive and negative charges are nothing more than a direction indicator of time in which the particles make their way through the space-time at a given moment. The mathematical apparatus developed by Feynman has become a space-time interpretation of the quantum mechanics, for which he was awarded the Nobel Prize "for his fundamental contribution to the development of quantum electrodynamics, which had profound implications for particle physics" (Cholakov, 1986). Another leading theoretical physicist, the Englishman Hawking investigated changes in the entropy of a black hole system when a certain amount of substance gets to "beyond the event horizon". Since the latter is a measure of the entropy of a black hole, then, having become wider, it would increase its own entropy by exactly the same amount as was carried on itself by the substance falling into the black hole. But the substance has carried away into the black hole a part of the entropy which just equals to this increase, therefore the total entropy has not been changed, and its increment is zero. According to the "arrow of time" definition via entropy, Hawking has come to the conclusion about the stop of time in the black hole outskirts, relatively to an outside observer (Hawking & Penrose, 2007). This conclusion finds confirmation in astronomical observations of the movement of gas clouds in the areas of anticipated presence of black holes, where their speed is found to slow down. Based on this, we can assume that a tribosystem, in accordance with the definition given by Landau, can be classified as a relativistic object, representing an example of a relativistic solid state physics, which, according to Nobel laureate A. Heym, is "... a new scientific paradigm where quantum relativistic phenomena can be studied in normal laboratory conditions..." (Morozov, 2008). And the decreased entropy increment and the related decrease of friction coefficients and wear rates are explained by slowing of the tribosystem's local time created by fast dislocations. An indirect example in favor of such an unusual interpretation of changes in anti-friction characteristics is the effect of anomalously low friction, opened by the Russian tribologist Silin (Shvedkov & Rovinsky, 1979), where the friction factor was recorded to drop zero in the polyethylene-metal system when this friction pair was exposed to the radiation of α-particles flow and cooled down to the temperature of liquid nitrogen. Indeed, the exposure to low temperatures zeroes the entropy, and fast α-particles can be considered as "carriers of relativistic effects" into tribosystem. From a formal point of view, the zeroed entropy corresponds to the time stop within the tribosystem, which slows down the wear and decreases friction coefficients down to ultra low values. An extremely important factor influencing the change of tribosystems' surface states is the increased, due to relativistic effects, kinetic energy of dislocations which they pass to the matter as they emerge on the surface of the friction assembly. Given this adjustment to the energy reserved by the surface one can explain the transformations that happen with the aggregate state of the surface layer material. A discrete contact inherent to the rubbing surfaces leads to heavy stress fluctuations in the subsurface layers of the friction assembly at the initial moment of friction. Force and deformation processes in the spots of actual contact look like multiple short-time pulses the duration of which is limited by the time of the knock-on collision of the microroughness, ranging from 10^{-3} to 10^{-11}s. Under such loading conditions, the substance of the tribounit transforms to a high-energy state with significant concentrations of energy, nearing to the values leading to the phase transformations. A slightly increased energy flow associated with the relativistic increase of dislocation masses is capable to start the process of avalanche-like activation of the friction surfaces substance, the highest peak of which is the condition called triboplasma or magmaplasma (Garkunov, 2001; Heinicke, 1987). Triboplasma is an energetic bunch, into which the substance of the friction surface transfers. For a long time, triboplasma has been a hypothesis explaining many abnormal processes occurring during friction: mechanoemission by the friction surface of charged particles with energy of 5 eV; emission of electromagnetic radiation quantums, abnormal chemical activity, and many more. Since during friction the matter can not be fully ionized like it is observed to happen on stars or in a nuclear explosion, the triboplasma is classified as nonideal plasma.

In most studies, triboplasma is presented as a hypothetically postulated object, the properties of which are almost not described (Heinicke, 1987). In the works generalizing the previously published studies (Lubimov et al., 1998; Lubimov et al., 2007a; Lubimov et al., 2007b), reasonable assumptions are made about the properties of this aggregate state of substance. The listed papers are based on the study of the concentration profiles of the diffusion distribution of the matter in the surface layer of the friction materials, and the effect of the plasmatic states of the matter on this distribution. It was established that during friction the diffusion coefficients ($Д_э$) increase by tens compared with the diffusion coefficients typical for the solid state of matter. We have found the equation establishing the law of proportionality between the diffusion coefficient and the square root of the triboplasma temperature. This is typical for substances in a gaseous state, so it was assumed that triboplasma by its properties is close to gases (Lubimov et al., 2007b):

$$Д_э = \frac{2e}{h}\sqrt{\pi k} \times \frac{Q^*}{W}\sqrt{\frac{T}{n}} \tag{4}$$

T – temperature of triboplasma; n - average density of particles in the plasma, e - electron charge; Q * = kT + Q - value characterizing the kinetic energy of the thermal motion of the triboplasma particles; Q - heat released by friction; W – binding energy of the triboplasma components; h - Planck constant; k - Boltzmann constant.

Such a conclusion allows to simplify the discussion of the relaxation processes, applying the apparatus of the plasma kinetic theory and the wave physics. The value $\frac{Q^*}{W}$ is a characteristic for the degree of substance ionization as it shows by how many times the energy of the ionized plasma is greater than the energy of the same substance in a steady state. According to our evaluation, the ratio of $\frac{Q^*}{W}$ has an order of 10^{-7}. We have obtained the ratios allowing to evaluate the temperature, charge, mass and lifetime of triboplasma components (Lubimov et al., 2007a). Triboplasma lives for a very short time (less than 10^{-7} c) and has a very high local temperature (above 10^4 K) (Pinchuk, 2004; Lubimov et al., 1998). This is a higher excited state where the cohesive bonds in solids weaken considerably, and structural degradation processes occur associated with the emission of charged particles and photons of different wave lengths. In the gas plasma, in addition to the random thermal motion, particles are involved in the ordered processes of the so-called Langmuir waves having the plasma frequencies ω_p(Ginsburg, 1967). We have obtained the formula for the quantitative estimation of the ω_p value:

$$\omega_p = \frac{4\pi e^2}{h}\sqrt{\frac{k}{m}} \times \frac{Q^*}{W} \times \frac{\sqrt{T}}{Д_э} \tag{5}$$

m – electron mass.

The numerical solution of the equation (5) gives us the values of plasma frequencies in the gigahertz range of oscillations. Triboplasma as any high-energy metastable state is not stable and relaxes rapidly passing to the normal aggregate states of substance. This transition is a multistage process; in one of its intermediate stages, chemically very active postplasma states of substance are present, having a long lifetime. It is these long-lived postplasma states of the tribounit surface substance which cause the chemical processes to take place throughout the entire nominal contact area, not just in the spots of actual contact. Based on the plasma hypothesis, there exists a chemical model of "surfing" effect explaining the low values of the friction factor during a selective transfer. The latter process is associated with the time evolution of triboplasma's substance after it leaves the actual contact area and has step by step relaxation till the initial steady state. Complex compounds constituting the surfing film in the selective transfer mode and ensuring the minimal friction coefficient, are formed due to the tribochemical reactions associated with transformations occurring in triboplasma and postplasma phase (Garnovskyi, 1984). The modern day leading tribologists G. Heinicke, W. Ebeling, Belyi V.A., Garkunov D.N., Kostetsky, B.I., have pointed out the top-priority need to study tribochemical processes from the standpoint of plasma chemical substance transformations and the catalytic impact on them from physical fields.

2.3. Experimental Study Of Triboplasma Structural Elements

Short lifetime of triboplasma and the fact that the friction contact can not be accessed with the analytical equipment make it difficult to study this aggregate state of the friction surface substance. For the purpose of this study, an experiment design was proposed aiming at separation of triboplasma in its "pure form". This plan was implemented using the experimental equipment representing a friction machine working in "shaft- bushing" contact mode with polytetrafluoroethylene - steel coupling. Shaft material is PTFE, bushing material is steel. The friction assembly is placed under a vacuum bell jar connected to a fore vacuum pump. The potential range from 0.5 to 200V can be applied to the friction assembly, and also a magnetic field is generated with an intensity of 10^4 A m^{-1}. Electromagnetic oscillations of different frequencies within the range from 0 to 10 GHz are superimposed on the friction assembly, and the characteristic signals are read out. In the steel counterbody (bushing) there are plasmatic traps keeping the friction substance in an active state (Fig. 1). Signal from a plasmatic trap is sent to a spectrum analyzer that resolves it into the spectrum representing a set of harmonics with individual frequencies. The friction torque is recorded by a change of the current in the winding of the machine electrical motor (similar to the Timken friction machine (US)). Frictional parameters were controlled at a sliding velocity of 1 m/s and a load of 10N, test time is 5 hours.

Figure 1. Part of the laboratory system for the triboplasma study.

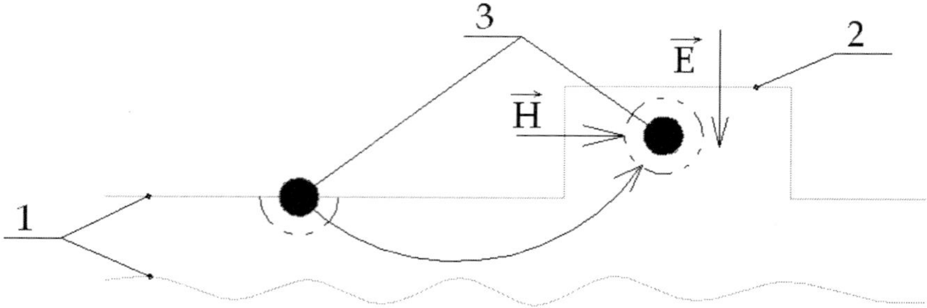

Figure 2. Circuit diagram of a plasmatic trap: 1) friction surfaces; 2) plasmatic trap; 3) triboplasma particle. and are intensities of the magnetic and electric field.

We assumed that triboplasma is an electrically active state of substance, one of the fundamental properties of which is shielding the electrical potential applied to it. That is why its exposure to an outer electric or magnetic field discloses those oscillation spectrum harmonics which are directly connected with the triboplasma. Exposure to an outer force field discovers a response from the spectrum harmonics in the megahertz and gigahertz frequency range. Depending on alteration of the electromagnetic spectrum profiles recorded by the measuring equipment when the friction assembly is exposed to the electric and magnetic fields of different directions, we have established that the oscillators constituting the triboplasma are electrons and positively and negatively charged ions. The frequency values of the free oscillations of these oscillators split up into the two ranges: 10÷100 MHz and 1÷1.5 GHz. These match the theoretically found values of the plasmatic frequencies (5). We have discovered that in each of the mentioned range of frequencies (megahertz and gigahertz) there are two types of oscillators. For the gigahertz range, one of the oscillator types is electrons. While plotting the graph of the relation between the friction factor, the harmonic intensities of the electromagnetic spectrum (electrons and oppositely charged ions), and the test time, we have discovered that the positively charged ions of triboplasma decrease the friction factor, and the negatively charged particles (ions and electrons), on the contrary, facilitate its increase (Fig. 3). The decreased friction factor is seen to coincide with the increased harmonics corresponding to the positively charged particles.

Based on the studies conducted, we can say that the exposure of the electrically active triboplasma components to an electromagnetic field of a certain direction and ripple frequency, leads to changes in the antifriction properties of a tribosystem. That is why, by changing the intensity of the electromagnetic field tuned to the resonance with the triboplasma components, it is possible to control the tribounit frictional parameters by intensifying the formation of lubricating structures on the contacting surfaces.

2.4. Impact Of Electromagnetic Control Of A Frictional Contact

It was shown that friction as a physical process generates electrically active responsive groups of triboplasma substance, which significantly influence the triboprocesses. Postplasmatic electrical states play a special role in that. They are complicated chemical structures carrying within them elements of free and bound charges, and also having extremely high chemical activity. They catalyze the formation of the "third bodies" driving a change in frictional characteristics of the friction assembly. In electrically conducting media, the influence of an electrical field upon the formation and growth of the thickness of the protective lubricating films on the friction surfaces, can be explained by the electrolytic transfer of the charges and substance. In dielectric media, which can be, for example, a lubricant or a friction transfer film of a metal-polymeric coupling, there are no free electrical charges in the form in which they are present in metals. Applying an electrical field of certain intensity to the frictional contact leads to redistribution of the charge densities in the constituting materials – a phenomenon of field polarization. The polarization is one of the fundamental properties of dielectrics, which consists in disequilibrium of charge distribution in the matter and appearance of a resultant moment different from zero and proportional to the electric field intensity value :

$$\vec{p} = \alpha \vec{E} \qquad (6)$$

α – substance polarizability, m^{-3}.

In the paper (Lubimov et al., 2009a) discussed the hydrocarbon lubricating media separating metal surfaces, it was proved that the interaction of an external electromagnetic field with a self electromagnetic field of the surface (induced by the frictional forces as a result of the dynamic contacting of the rubbing bodies), causes appearance of ponderomotive forces within the working layer of the friction assembly. These forces have a mechanical effect on the substance and cause to move the macrovolume of the friction substance. In the general case, the ponderomotive force value can be found from the ratio (7):

$$\vec{F} = \frac{m}{\mu} \alpha \omega \vec{S} \qquad (7)$$

– ponderomotive force; m – mass of the moving macrovolume; μ – molar mass of the moving substance; α – the substance polarizability; ω – frequency of the external, substance exciting factor; – the Umov-Poynting vector.Currents generated during friction, due to the contact discreteness, lead to appearance of electromagnetic fields in the gap between the rubbing bodies; that is why friction, as a physical process, per se generates forces of the electromagnetic nature. These forces, having orienting and structuring action on a very small amount of the lubricating medium near the surface of the solid body, facilitate emergence of ordered and stable states – multipoles, interfacing with the charged friction surface field. Under the electromagnetic field, the time limit of existence for the structurally activated substance of the gas triboplasma, increases due to non-dissipative electric drift of the plasma components in the electromagnetic field (Prokhorov, 1992). In other words, in the gas plasma, a weakly decaying electromagnetic wave is excited, which maintains the working layer substance in chemically active post-plasmatic state. Ponderomotive pressure exerted by the electromagnetic field upon the frictional substance depends on the properties of this substance (density, molecular mass, mass of polarizable volume), the field properties (field frequency and energy), and the contact geometry (gap size between the rubbing surfaces, dimensions of the metal surfaces). During the action of an electromagnetic field, favorable conditions are created for synthesis of the protective lubricating films covering the friction assembly surfaces. Our assumption is proved experimentally using the electromagnetic sanation method for the friction surface, during frictional interaction of the materials "epoxy-filled composite – zinc counterbody" (Lubimov et al., 2001). The amplitudes of the emergent high frequency harmonics of the electromagnetic oscillations spectrum depend upon the magnitude and direction of the electric field (Fig. 4). The recorded values of the harmonic amplitudes help to have a general idea of the quantity of the active particles that create oscillations in this frequency range, proved to be responsive to the external action.

From the plot one can see that with the applied potential (U, Volt) increased up to a certain value, the harmonic amplitudes increase correspondingly (ΔA). This is an evidence that more of the active centers (free charge states) response to the external action. The curve going parallel to the X-axis means that the mentioned charge states are present on the surface now not in the form of uncoordinated active centers, but in the form of multipoles constituting the "skeleton" of the protective lubricating structure – the "third body" – being formed on this surface. These multipoles passivate the solid body surface, and are electrically neutral formations; therefore, they do not react when exposed to an external field. Furthermore, the friction factor was recorded to drop during the transition from the negative to positive potential, which can serve as an indirect proof for the formation of the protective lubricating layer – the "third body".

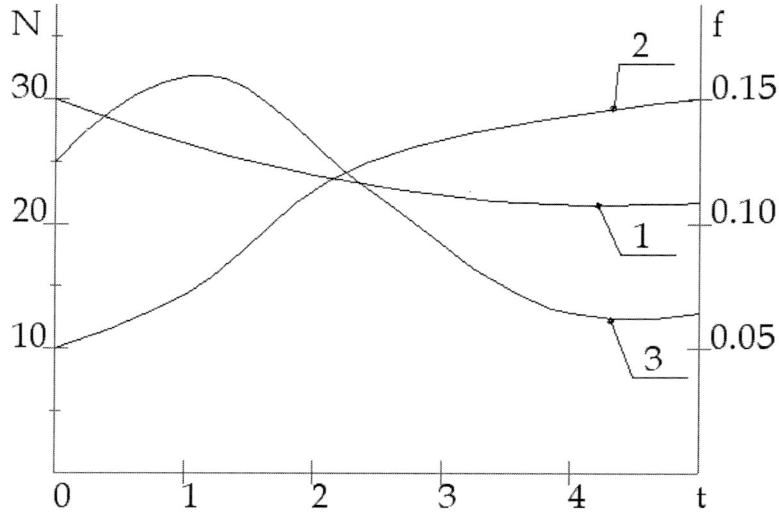

Figure 3. Relation between the harmonic intensities N (in conditional units) of electromagnetic emission of tribloplasma components and the friction factor (f) of the "PTFE-steel" coupling, and the friction time (t, hours). 1) plot f(t); 2) plot N(t) for positively charged articles; 3) plot N(t) for negatively charged articles

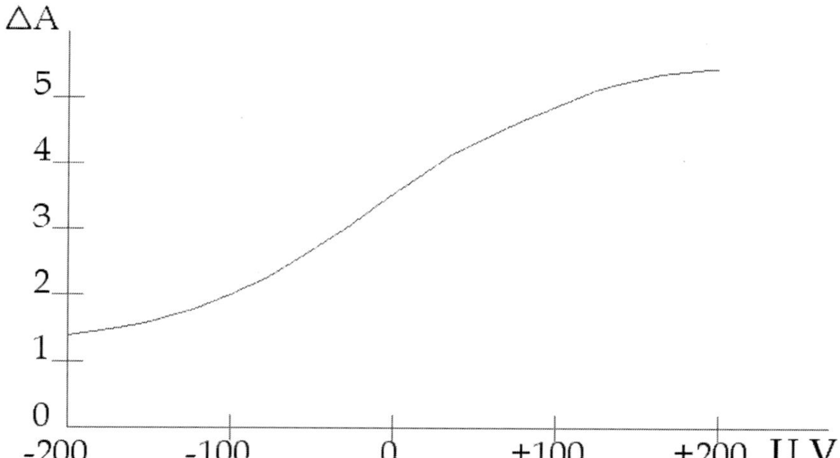

Figure 4. Impact of the electric field magnitude and direction on the harmonic amplitudes change (ΔA) in a frequency range of 40 MGz for the friction pair "epoxy-filled composite – zinc counterbody".

Table 1. Change of frictional properties of the materials exposed to the electric field

Item No.	Potential applied to the metal counterbody, V	Friction Factor
1	-200	0,36
2	0	0,25
3	+200	0,12

The applied electric field influences not only the kinetics of the tribochemical reactions within the working layer, significantly increasing their speed, but also the orientation of components constituting the third body. At the same time, the frictional body polarized by the external electrical field, forms the lubricating film much more actively in comparison with the similar materials in the absence of the field (Fig. 5).

As seen from the photos, when the field is present, the film structure is more homogeneous across the thickness, has no discontinuities and fully covers the zinc surface. On the right-hand photograph, the film has clearly seen new formations on the surface of the already existent layer. Besides that, the film, being developed under the action of the field, adheres to the metal surface more firmly. Such an intensive layer growth and visible increase of the third body volume filling the working layer, can be related to the better wetting of the metal surface by the polymer and higher adhesion of the polymer film. The electric potential in this example acts like the factor polarizing and maintaining the polarized state for the working volume matter. Molecule energy depends upon its position on the surface (Akhmatov, 1963). A polarized molecule (dipole and multipole) of a lubricating medium in the external force field of the surface, orients with regard to it so that its potential energy turns out to be minimal (8):

$$U(\alpha) = \frac{U_0\left(1 - \cos(n\alpha)\right)}{2} \tag{8}$$

$$p(\alpha)d\alpha = C \times \exp\left(-\frac{U(\alpha)}{kT}\right) \tag{9}$$

$U(\alpha)$ - dipole potential energy; U_0 – potential barrier; n – dipole order of symmetry; α – inclination angle of the dipole symmetry axis to the surface; p(α) – relative probability of the molecule orientation angle; C - constant.

(a) (b)

Figure 5. Photographs of friction paths in the coupling "PTFE – zinc counterbody" at the potentials applied to the metal counterbody: a) 0V; b) +100V.

In accordance with the above formulas (8) and (9), applying an external electric field results in the multipole molecules tending to take up a position perpendicular to the metal surface, orienting along the electric field force lines. Electrical action on a polarized molecule, ion, or radical fixed with their active center on the friction surface can lead to the molecule precession. This fact was recorded by S.Z. Zaitsev during the study of water molecule behavior with the help of a field ion microscope, and was discussed in the paper (Lubimov et al., 1994). As the observations showed, in an electric field, the water molecule, structurally being a dipole, starts to precess and nutate, i.e. make movements similar to the movement of a gyroscope with one fixed point. Schematic sketch of this process is shown inFig. 6.

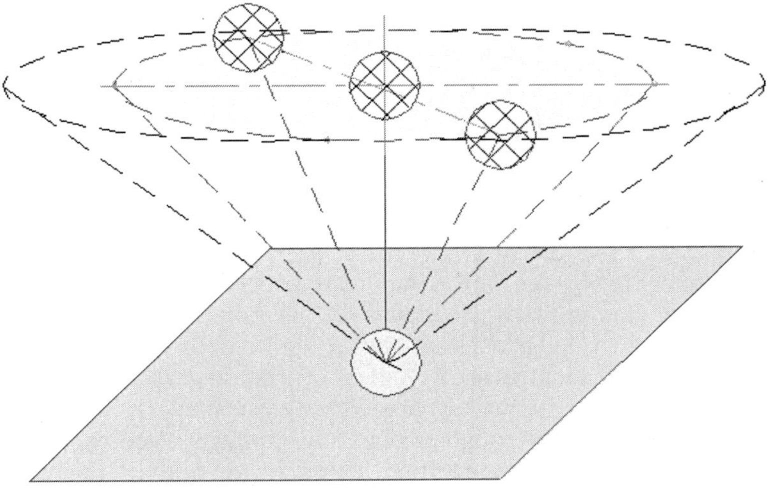

Figure 6. Sketch of precession and nutation of a polarized molecule on the plane

There is none of the physically reasoned statement that would disprove the assumption that any other polar molecule, radical or ion ought to behave in an electric field in a similar way. For instance, during frictional interaction of polytetrafluoroethylene, the molecular chain is destructed with formation of its fragments that can be considered as radicals. Active ends of the molecular fragments make up a chemical bond with the frictional surfaces with energy of about 690 eV. Meanwhile, such a molecule tends to take up a perpendicular position, which is impeded by the frictional heat being emitted in the area of dynamic contact of the friction assembly, and by the intermolecular action (Nikolsky et al., 1988). Exposure to an external electric field changes the motion of a polytetrafluoroethylene molecule. As a result of precession and nutation processes, such a molecule becomes stable in its vertical position. The precession and nutation lead to the formation of an additional electron cloud, parallel to the friction surface plane.

Presence of the Coulomb forces acting between the precessing molecules helps quite obviously to additionally organize their structural order, and resists to the disorganizing action from the temperature fields (Fig. 6). There being generated a structurally ordered molecular layer that is identical to a surfing film in terms of its action and structure.

3. SURFING FILMS

The ultra low values of the friction coefficient recorded in the selective transfer mode are connected with appearance of surfing films on the friction paths. The "surfing" structures appear due to complex physical and chemical processes, and complexing processes become of a special significance among them (Polyakov, 1988). Complexing during friction occurs when the processes associated with triboplasma and postplasmatic transformation phase of the aggregate state of the tribounit surface layer matter, are directly involved. In the moment of closest approach of the rough edges of the conjugate bodies, the surface layer is cleaned from impurities, oxide films, and in the places of dislocation outcrops, resulting from the structural thermal activation, the surface atoms have additional bonds. As almost all metals are electron acceptors, and most additives or tribodestruction products of friction materials have electron donors within their structure, then the latter can act as ligands. This creates favorable conditions for the formation of complex compounds. In the wearlessness condition, it is typical that chemical bond occurs between lubricant ligands and metal surface atoms that have available bonds. In this case a particle on the surface can interact with several lattice atoms at once. Consequently, the total interaction will be very strong. The chemically active components of the lubricant – ligands – form metal-containing complexes with three-dimensional or two-dimensional structure at the dislocation outcrop areas or contact places. Formation of a coordination bond in a complex, leads to weakening or rupture of the metal atomic bonds in the complex with the metal atoms in the friction body lattice. The bond between the metal and complex becomes adsorptive. The complex is now capable to move across the surface and combine with other complexes in film islands (Lubimov & Ryzhikov, 2001). In further acts of contacting between the rough surfaces, the complexes may mechanically come off the surface transferring to the volume of the lubricant. This creates dynamic stability of the surfing structures, with which complex formation and decay processes run at the same speed.

To minimize the frictional forces, the complex compound needs not only to develop on the friction surface but also to take a certain position. This positioning was named tribocoordination. Our studies conducted with a scanning tunnel microscope showed that the surfing films have quasi-crystalline structure, which is different from the surface structure where they appear. The important factor for the generation of a monomolecular and multimolecular highly oriented (coordinated) layer is the tribochemical interaction of the polar molecules with the solid body surface, which then provides easiest sliding of the lubricant on the boundary layer surface. This effect leads to the transition of the tribounit to the hydrodynamic friction condition with ultra low friction coefficient values inherent to it. A.S. Akhmatov has grounded the connection between the friction forces and structural organization of the lubricating layers' molecules. He has shown that ordered, pseudocrystalline structure of the lubricating layer adjacent to the solid surface is characterized by the friction modes transient to the hydrodynamic friction, and being accompanied with low values of the friction factor. For such a condition, it is typical to have oriented position of the filiform molecules. In this regard, specifically the surfing film's structure meets the friction moment minimization requirements most of all. Fig. 7 shows photographs of the friction surface conjugate with the epoxy-filled composite containing copper tetraethylthiuramdisulfide complex, which were made using a scanning tunnel microscope.

The film from Fig. 7 is fixed on steel at a voltage of 0.1V and 0.5 mA current; one can see that it represents the ordered atomic lattice with a distance between the atoms of 1.9Å. Increased vertical scanning up to 15Å (Fig. 7b), also gives us the ordered atomic grid with 1.9Å distance between the atoms, which resembles a crystal lattice. When larger areas are used (Fig. 7c), apparently the topography can be described as chaotic. Obtaining the friction surface images with the scanning tunnel microscope was possible thanks to the small thickness of the oxide layer. The thickness of metal oxide on the rubbing specimens' surface did not increase (in spite of their contact with atmospheric oxygen), because the oxide layer was covered with the surfing film layer. That is exactly why the topography of surfaces shown in Fig. 6 reflects mainly the morphology of the surfing film. Images of different surface areas in different scales given for each specimen, apparently for the first time allow establishing the relation between the large-scale structure of the surfing film surface, and the structure of the same film at the atomic level. Thus, comparing the images from Fig. 6, the conclusion can be made that the ordered position of the surfing film atoms, evident at the

22×22Å area, leads to the formation of long molecular chains oriented mainly in one direction. While obtaining an image in a larger scale, separate molecules and especially atoms become invisible. And the topography itself now starts reflecting the sliding pattern of tribounit surfaces.

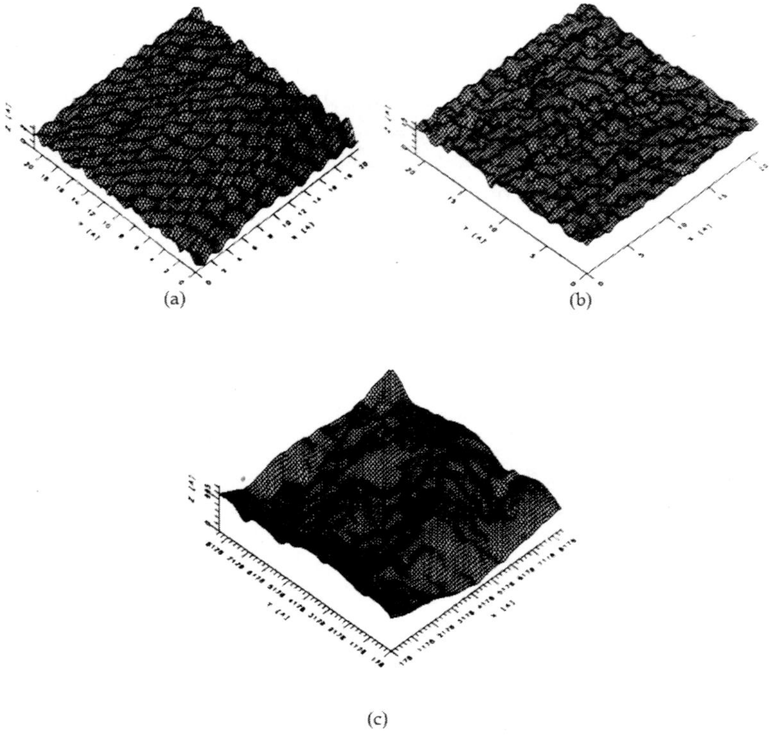

Figure 7. Surfing film area made of copper complexes

It can be taken for granted that the substances structured as surfing films, regardless whether they are liquid or plastically viscous in mass, with the same temperature and load, in the boundary state on the friction surface take on elastic forms, turning into a different quasi-crystalline aggregate state. Such lubricating layer has a higher stability, and is not squeezed out of the frictional contact area under external forces. Exposure to an electric field facilitates the emergence of a lubricating film on the friction surface, structural orientation of which resembles the tribo-coordinated surfing film, consisting of highly oriented molecules that have chemical bond with the metal surface. The orientation of such films is additionally maintained due to Coulomb forces, which provides them with additional stability and decreases the friction factor and material wear by many times.

4. CONTROL OF FRICTIONAL PROCESSES WITH AN ELECTRIC FIELD

The impact of an electric field action on a frictional contact for various materials of contacting bodies was studied using a side friction machine with the "plane - three pins" testing scheme. Schematic diagram for the friction assembly of the testing machine is shown in Fig. 8. The frictional parameters to be recorded are the linear wear and mass wear of the specimens, friction factor, bulk temperature of the specimens and lubricant.

The cartridge clip with the fixed indenters is held down to the counterbody at the set pressure. The specimens are loaded with the help of the lever mechanism. The machine design provides for the capability to alter the rotational speed of the indenters from 0.5 m/s to 1.2 m/s. The friction factor is determined based on the measured friction torque, recorded by the strain gages fixed on the beam 6 using the half-bridge circuit. The beam is connected to the friction assembly via the rigid constraint that prevents specimen 1 located in the metal cup 5 from turning. The cup 5 is installed on the thrust bearing. The friction assembly is electrically insulated from the machine frame with the help of the textolite pad 7. The friction torque is recorded by the pointer indicator of the instrument TMM-48. The uncertainty of such measurement is 1 - 1.5% which ensures high accuracy of the obtained results. Mass wear of the specimens is determined on the laboratory balance with 0.5 mg measurement error.

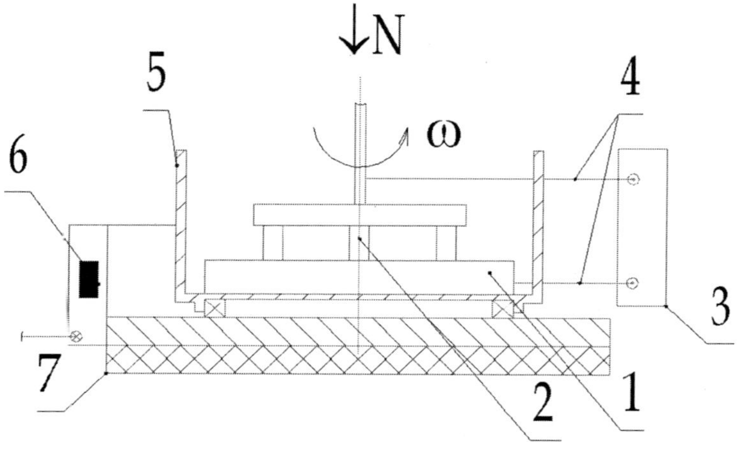

Figure 8. Schematic diagram of the friction assembly to study the impact of an electric field upon the tribounit frictional parameters: 1) specimen; 2) counterbodies; 3) external source of electric energy; 4) electrodes providing the potential to the friction assembly; 5) testing cup for lubricants; 6) beam with resistance strain gages; 7) electrical insulating (dielectric) pad.

4.1. Application Of An Electric Field To Friction Assemblies Working In Boundary Lubrication Conditions

Tests were carried out on the friction pair: "Bronze (БрО10Ц2, GOST 613-79) – Steel 45 (GOST 1050-88) in the lubricant medium - POLADYNE 10W-30 oil manufactured by Irvine, the US. Friction parameters: sliding velocity $\vartheta = 1$ m/s, contact pressure P = 0.5 MPa. A series of tests was conducted without an electric field, as well as tests where a positive potential was created on the steel counterbody, and then on the rubbing bronze indenters. The friction coefficient was selected as a measurand. The measurement results are shown in Fig. 9.

In our opinion, the significant decrease of the friction coefficient is driven by the intensifying effect of the electric field on the adhesive forces of the metal surface. The fact of the intensification of the adhesive forces under the action of the electric field was proved by experimental weighing (with an analytical balance) of the lubricant (adhesive), left on the steel counterbody surface after the friction. Relation of the adhesive mass and the magnitude and sign of the potential of the steel surface is shown in Figure 10. Along with insignificant amount of oil retained by the surface, it also has some amount of transferred bronze with the hardness that is smaller than that of steel.

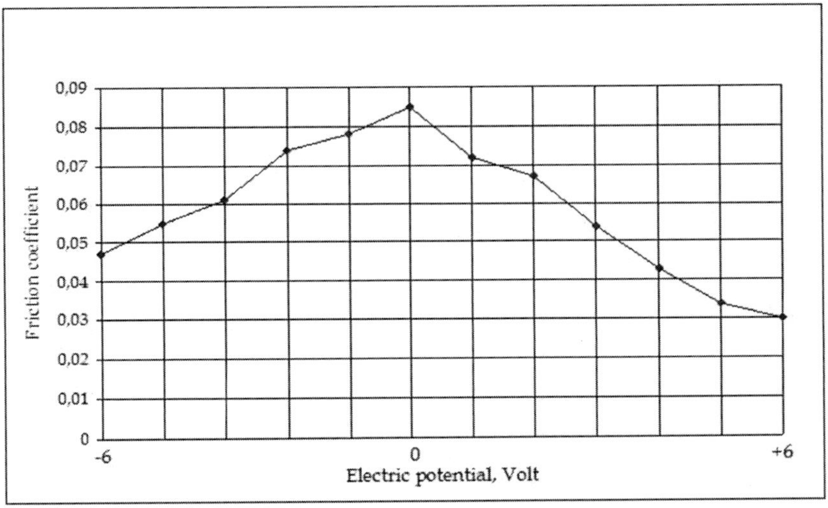

Figure 9. Relation between the friction coefficient of the "bronze-steel" conjugation and the steel surface potential's value and sign

Thus, the improved antifrictional characteristics observed in the tribounit are explained by the strengthened lubricant adhesion to the friction surfaces, which impedes its mechanical removal from the frictional contact area, and by the higher stability of the lubricating films. Also, higher chemical activity of the lubricant should be expected. Increased chemical activity of the present-day lubricants is one of the ways to achieve higher performance of friction assemblies. Oil's degree of activity can be visually monitored by the representative darkening after tests. Moreover, oil's reactivity can be evaluated by the decreased photo-emf and light flux, passing through the oil medium. The photo-emf measurement results for the tested oil samples, for each series, are shown in Fig. 11. It is evident that the oil taken after the tests where the positive potential had been applied to the steel surface is more chemically active. This can be explained by the fact that in this test, the recorded value of friction factor was minimal, as well as the frictional heating of the surfaces and the volume temperature of the lubricant.

The oil has been depleted by the wear products and thermal decomposition products of its components. Visually it is lighter and therefore, weakens the light flux passing though it at a smaller extent.

Notable is total correlation of the dependencies shown in Fig. 9, 10, 11 which confirms the positive impact of the used electric field on the lubricant properties.

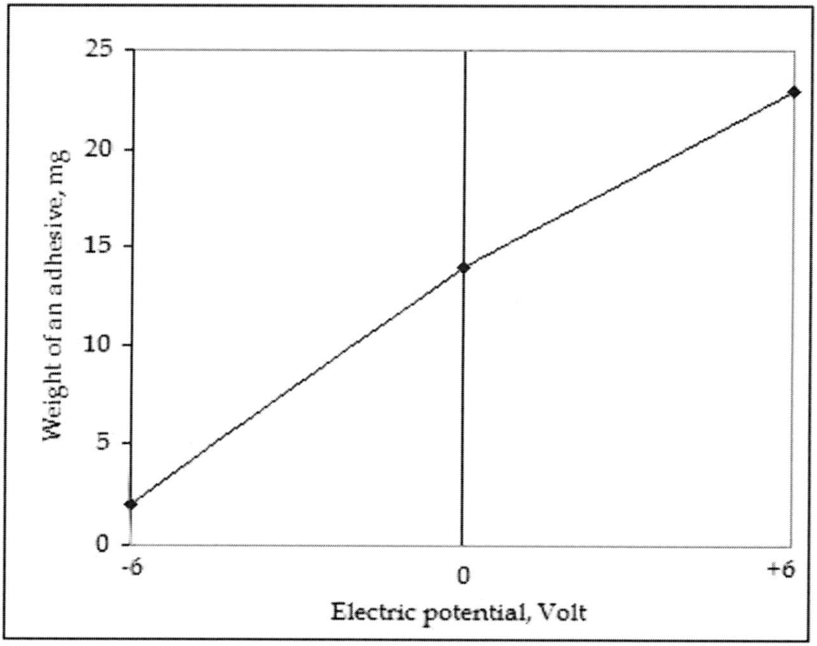

Figure 10. Relation between the adhesive mass M (mg) and the magnitude and sign of the steel surface potential

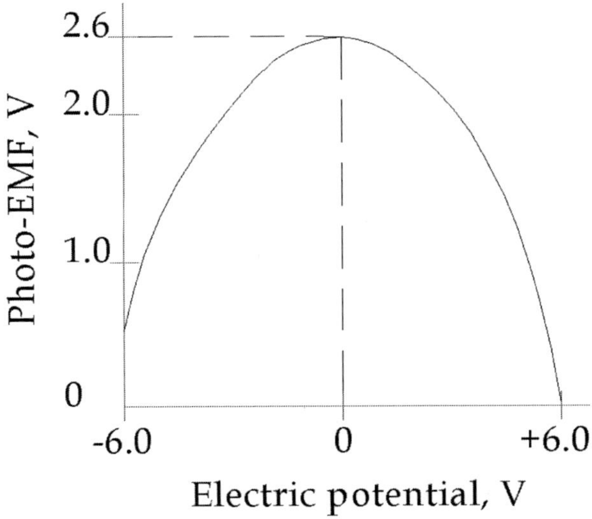

Figure 11. Relation between the photo-emf and the magnitude and sign of the steel surface potential

4.2. Application Of An Electric Field To Metal-Polymeric Friction Assemblies

Tests to evaluate the impact of an electric filed on metal-polymer conjugations with different combinations of polymer materials were carried out under the action of the load-and-speed factor ([pϑ-factor) equal to 0.5 MPaxm/s, the test time was t = 2 hours. Intensity of wear of polymer materials is affected not only by the electrical field magnitude but by its direction, too (Fig. 12). This relates to the charge build-up on the metal surface in the metal-polymer friction pairs, and also to the fact that the metal surface has either cathodal or anodal polarization, depending on the rubbing polymer nature. Therefore, electric field application may lead to the surface depolarization, which in its turn will either increase or decrease the friction transfer of the polymer material.

More detailed research has determined that depending on the magnitude and direction of the external electric field, not only the extent of wear of the polymer material changes, but so do the wear conditions (Fig. 13), and the appearance of wear particles.

At the same time, it is evident that with the overall improvement of antifrictional properties of the metal-polymer tribounit, observed when a positive potential is applied to the metal counterbody, also the running-in time is observed to reduce. The plot of relation between the linear wear of the PTFE specimen and the test time for the friction pair "PTFE-steel" under an electric field, can serve as illustration to the previous statement (Fig. 14).

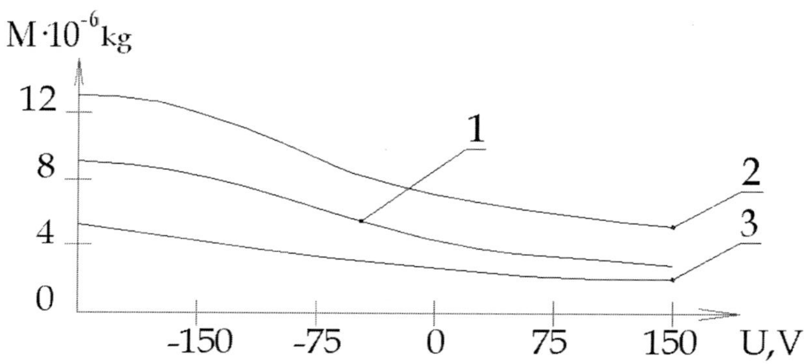

Figure 12. Relation between the mass wear M and the magnitude and sign of the potential applied to the metal counterbody 1) polyethylene; 2) polypropylene; 3) polyamide 6.

It is evident that the superimposed electrical field is capable to not only decrease or increase the specimen wear by 6 times, but also to decrease the running-in time for the material. This is explained by the fact that, with a certain pattern of filed imposition on the metal countersurface, third bodies are increasingly generated which concurs with optimization of the frictional properties of the friction assembly and transfer to the steady state of wear. Therefore, the action of an electrical field leads to the improved tribotechnical parameters of friction assemblies. Characteristic feature of metal-polymer couplings is that it is possible, without fear of electrical breakdown, impose tens of volts on the friction assembly. As all negative issues associated with electroerosion damage of the surface taking place when an electrical breakdown occurs, depend not upon the magnitude of the applied voltage, but upon the breakdown current generated in the frictional contact. Having considered that, we developed and patented an electronic device for the dynamic control of friction and material wear processes, which was named "Electronic regulator of friction" (ERF).

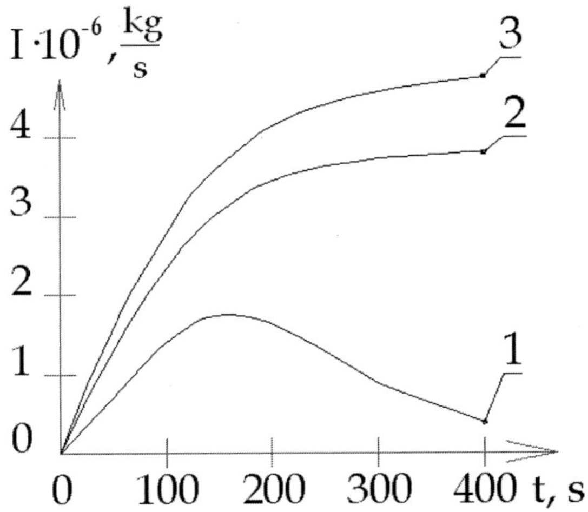

Figure 13. Relation between the mass wear rate I and the friction time t for the friction pair "PTFE – copper": 1) +100V on metal; 2) 0V; 3) –100V on metal.

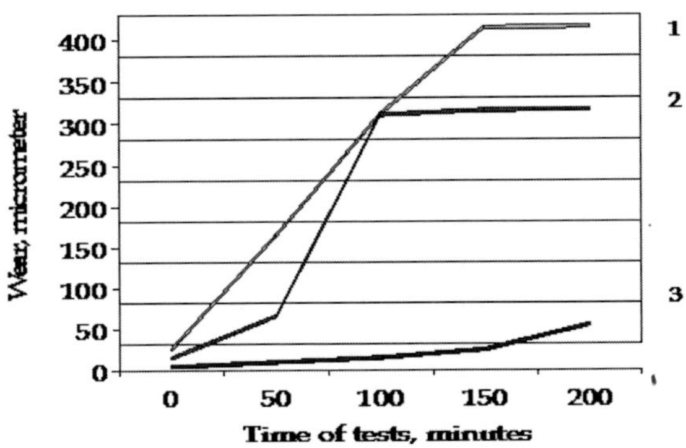

1) Heavy wear condition: -150V potential is applied to the steel counterbody; 2) Friction with no exposure to an electric field; 3) Minimal wear condition: +150V potential is applied to the steel counterbody.

Figure 14. Relation between the running-in time for polymer material and the friction conditions.

4.3. Electronic Regulator Of Friction

To control and regulate friction and wear processes of materials, it is necessary to provide a reference information channel which on the one hand, would transmit information about frictional processes, and on the other hand, would easily convert this information into a signal that controls the electrical field applied to the frictional contact area. The electronic friction regulator converts this reference signal into an electrical field with a certain voltage and direction. Frictional heating value of a friction pair can be selected as a reference signal. The heat Q emitted during friction is known to be a function of load N, friction coefficient f, sliding velocity ϑ, and friction time t:

$$Q = F(N; f; \vartheta; t) \tag{10}$$

In general terms, a change in tribological characteristics of a friction assembly reflects in the intensity of heating of the friction area. The less the contacting bodies are heated - the less apparent are the dilatation processes, the less is the bulk temperature of the lubricant, the better are the conditions to maintain the carrying capacity of the lubricating layer, stability of the lubricating protective films structure, and the less is the number of surface damages. Therefore, the increased heat release during friction may be identical to the deteriorated antifrictional parameters of a tribosystem. To prevent the wear of friction materials at high temperatures, surface active and chemically active additives are introduced into lubricants. Surface active additives ensure minimization of wear at relatively low temperatures – up to 300 °C, and chemically active additives decrease the wear and friction at higher temperatures – up to 800÷1200 °C. However, an additive added to a lubricant gradually deteriorates in the process of work, which entails its lower concentration in the friction area, and therefore the additive's modifying properties lower, too. Besides this, additional introduction of different types of additives into a lubrication material may disturb the thermodynamic, kinetic or mechanical compatibility of the base oil components. The maximum possible fuel savings (%), depending on an engine type and conditions of their operation, determined by the efficiency of the used motor oils and their additives, are shown in Table 2 (Matveevskyi et al., 1989):

Table 2. Fuel savings for vehicle engines

Engine Type	Short Runs		Middle Runs	Long Runs
	Cold Start	Town Conditions		
Gasoline	7,5	6,0	2,2	1,8
Diesel	5,8	4,8	2,5	3,0

Advantage of adaptive electronic lubricating devices over additives to lubricants is that you enable the electronic module only one time and then its positive action upon mechanisms' and machines' friction assemblies is continuous and unflagging, while the intensity of an additive's action decreases during operation process. The developed device in terms of its function may propose an alternative to quite a wide range of additives to lubricants, and has no restrictions related to the design and process features of friction assemblies. The device was developed for heavy loaded friction assemblies. The device was tested on the propulsion systems of passenger elevators, the electric drives of kettle mixers at lubricant factories, but the widest use this device has got in internal combustion engines. It was determined that this device connected to a power unit improves the power efficiency of engines for vehicles and drives of fixed machines. This happens due to the decreased friction forces in the mechanisms. Friction factors of the assemblies working in the boundary lubrication conditions decrease by 15%, and working with a lubricant – by up to 40%. Efficiency of the utilization of the Electronic Friction Regulator is determined by the load and speed performance of am operated friction assembly, as well as by the initial and maximal intensity of the electromagnetic field created by the device. The device itself has comparatively small dimensions (45 x 45 x 25 mm, its weight does not exceed 150 g). It is built as a stand-alone unit, and thanks to this it is adapted to the structures of machine's mechanisms without any design changes. At this stage of the accomplished work, the device is operable within the temperature range from minus 50 °C to plus 90 °C and has a service life no less than 15,000 hours. Application of an "ERF" in the friction assemblies of machines, at a certain intensity and direction of an electromagnetic field, will allow to:

- suppress the frictional electrification effect catalyzing development of corrosion and oxidation processes on the contacting surfaces, and metal anodic dissolution;
- improve lubricant adhesion to the surface being protected thus increasing the lubricant's protective functions;

- decrease a possibility for the hydrogen wear of the friction surface.

Figure 15. General view of the test bed

Table 3. Load characteristic at n = 2000 rpm

Base Case												
Condition No.	Me, nm	Ne, kW	Gt, kg/hr	ge, kg/ kW-hr	ηe	ηm	Po, bar	Tdg, ⁰C	CO,%	CH, ppm	NO, ppm	CO_2, %
1	2	3	4	5	6	7	8	9	10	11	12	13
1	20,16	4,22	1,99	0,471	0,174	0,543	2,8	383	0,102	93	1563	11,61
2	40,32	8,44	2,81	0,333	0,246	0,714	2,7	425	0,073	112	2687	11,43
3	60,48	12,67	3,71	0,293	0,279	0,797	2,6	480	0,067	110	3046	11,93
4	80,63	16,89	4,84	0,286	0,286	0,847	2,5	560	0,643	114	2420	12,28
5	83,15	17,42	6,15	0,353	0,232	0,852	2,5	540	6,441	177	596	9,58
Final condition, with the Electronic Regulator of Friction in 5 motor hours												
1	2	3	4	5	6	7	8	9	10	11	12	13
1	20,16	4,22	1,87	0,443	0,185	0,543	2,5	395	0,096	76	1710	11,96
2	40,32	8,44	2,67	0,317	0,258	0,713	2,5	421	0,070	97	2812	11,69
3	60,48	12,67	3,43	0,271	0,302	0,797	2,4	460	0,062	105	3675	11,73
4	80,63	16,89	4,62	0,273	0,299	0,845	2,4	555	0,585	95	2556	12,43
5	84,67	17,73	5,98	0,337	0,243	0,854	2,3	535	6,664	155	554	9,39

Table 4. Load characteristic at n = 3000 rpm

Base case												
Condition No.	Me, nm	Ne, kW	Gt, kg/hr	ge, kg/ kW-hr	ηe	ηm	Po, bar	Tdg, ⁰C	CO,%	CH, ppm	NO, ppm	CO_2, %
1	**2**	**3**	**4**	**5**	**6**	**7**	**8**	**9**	**10**	**11**	**12**	**13**
1	20,33	6,39	2,96	0,463	0,177	0,503	2,8	502	0,213	82	2100	12,47
2	40,66	12,77	4,26	0,334	0,245	0,677	2,7	548	0,093	88	3487	12,56
3	60,98	19,16	5,38	0,281	0,291	0,765	2,6	596	0,089	97	3937	12,37
4	81,31	25,55	7,22	0,282	0,290	0,818	2,6	656	0,629	101	2981	13,12
5	97,58	30,65	9,81	0,320	0,256	0,848	2,8	610	7,171	189	602	9,95
Final condition, with the Electronic Regulator of Friction in 5 motor hours												
1	**2**	**3**	**4**	**5**	**6**	**7**	**8**	**9**	**10**	**11**	**12**	**13**
1	20,33	6,39	2,89	0,453	0,181	0,503	2,6	506	0,128	52	2211	12,51
2	40,66	12,77	4,10	0,321	0,255	0,677	2,6	537	0,097	73	3480	12,67
3	60,98	19,16	5,23	0,273	0,299	0,765	2,6	585	0,099	83	4013	12,61
4	81,31	25,55	6,90	0,270	0,303	0,818	2,6	650	0,785	90	2889	13,15
5	97,58	30,65	9,92	0,324	0,253	0,848	2,6	610	7,553	177	536	9,68

This entails the increased service lifetime of the machine mechanisms, as well as their better environmental performance. The device design and method of its connection to a car power plant are presented in the papers by (Lubimov et al., 2008; Lubimov et al. 2009b). There were conducted two series of tests of different designs of the device. Below are shown the device photograph and the results of its trials on the test bed simulating internal combustion engine's operation.

As you can see from the picture, the dimensions of the Electronic Regulator of Friction (blue colored device indicated with the arrow) attached to the machine body are small in comparison to the machine overall sizes.

Table 5. Mechanical loss moment, Nm

Rotation frequency, rpm	Initial condition		Outcome	
	Closed throttle	Open throttle	Closed throttle	Open throttle
300	26,0	23,0	24,0	22,0
500	17,5	17,0	17,0	16,5
800	16,5	12,5	16,0	13,0
1000	16,5	12,5	16,0	12,5
1500	18,0	13,5	17,5	13,0
2000	19,0	14,0	18,0	14,0
2500	21,0	16,0	20,0	16,0
3000	24,0	18,5	21,0	17,5

Table 6. Averaged effects, %, against the engine base case condition according to the results of the first series of tests, considering the changes for all the conditions

Effective Power	Fuel Consumption	Effective Efficiency	CO	CH	NO_x
0,840	4,344	4,621	18,90	19,05	- 6,53

Symbols in the table: n – rotation frequency for the engine crankshaft; Me – net troque; Ne – effective power; Gt – fuel consumption per hour; ge – specific fuel consumption; ηe – effective efficiency; ηm – mechanical efficiency; Po – oil pressure; Tdg – temperature of discharge gases; CO – carbon oxide content in the engine discharge gases; CO2 – carbon dioxide content in the engine discharge gases; NO – nitrogen oxide content in the engine discharge gases; CH – residual hydrocarbons content the engine discharge gases.

From the given data, one can see that the enabled device decreases the fuel consumption and mechanical loss moment; increases the power and efficiency of the machine engine, decreases harmful contamination in the discharge gases, as well as their temperature. The observed effects were recorded after 5 motor hours of the experimental unit operation. After the device had been disconnected and the simulation machine had been restarted, the controlled variables again decreased down to the values corresponding to the base case condition of the engine. The next series of tests were aimed to study the impact of an electromagnetic field with 60 MHz frequency pulse action on the engine performance. For this purpose, the design of the Electronic Regulator of Friction was modified with some slight increase of its dimensions. The results of the development tests of this design version at n = 3000 rpm, as well as the averaged effects are shown in Table 7 and summary table 8.

At this moment, we can not provide the definite explanation for the increased content of nitrogen oxides in the discharge gases. One of the draft versions of reasoning is the ability of an electromagnetic field to inhibit generation of soot depositions on the friction surfaces of the piston-cylinder group of an internal combustion engine. Under the action of an electric field the working elements of the car engine are cleaned. Indirectly, this is proved by photographs of oil filters taken from the machines during operational tests of the device (Fig. 16). In the presented figure, the upper image is a new filter of the VAZ 21099 car that has run 1500 km with ERF, the lower image is the condition of the oil filter after the scheduled oil replacement after 10,000 km. The upper filter is filled with particulate depositions of the soot nature. It seems that under the action of an electromagnetic filed, the above mentioned depositions were removed from the cylinders' wall surfaces and retained by the oil filter.

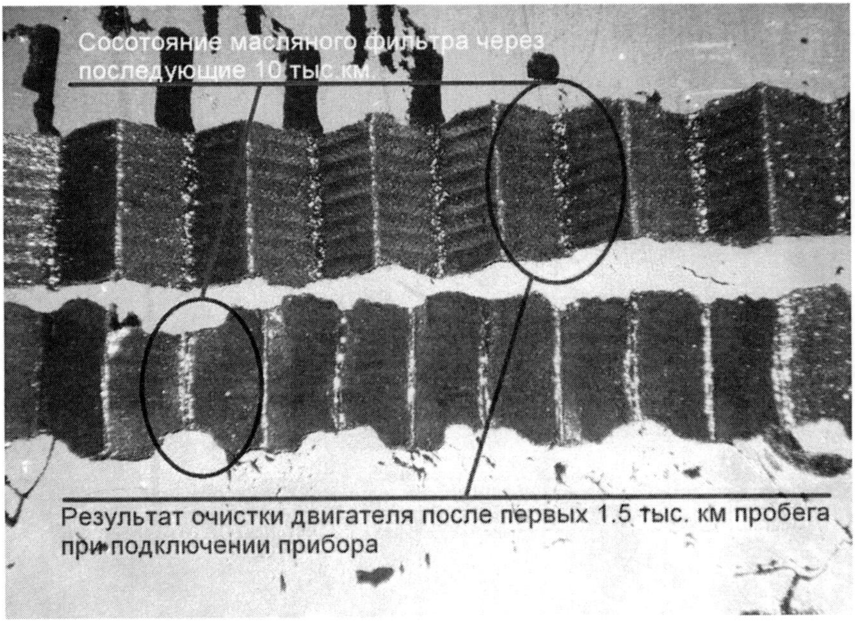

Figure 16. Condition of the oil filters with the utilization of the ERF (above) and without it (below)

The important feature of the developed device is that it can be used both for the improvement of the lubrication conditions of friction assemblies and for controlled tightening of their frictional parameters. This feature allows to use the device for faster running-in time of new machines and mechanisms in the conditions minimizing the initial wear, and it can be helpful for machine engineering companies to reduce time and energy consumption when testing their fabricated products.

Table 7. Load characteristic at n = 3000 rpm

Condition No.	Me, nm	Ne, kW	Gt, kg/hr	ge, kg/kW-hr	ηe	ηm	Po, bar	Tdg, °C	CO,%	CH, ppm	NO, ppm	CO_2, %
Base case												
1	**2**	**3**	**4**	**5**	**6**	**7**	**8**	**9**	**10**	**11**	**12**	**13**
1	20,33	6,39	2,90	0,455	0,180	0,503	2,7	494	0,135	55	2059	12,60
2	40,66	12,77	4,19	0,328	0,250	0,677	2,7	536	0,092	78	3421	12,67
3	60,98	19,16	5,28	0,276	0,297	0,765	2,6	585	0,099	95	4017	12,71
4	81,31	25,55	6,98	0,273	0,299	0,818	2,7	650	0,644	97	2987	13,32
5	98,59	30,97	9,98	0,322	0,254	0,849	2,8	612	7,356	174	574	9,96
Final condition, with the Electronic Regulator of Friction (60 MHz) in 5 motor hours												
1	**2**	**3**	**4**	**5**	**6**	**7**	**8**	**9**	**10**	**11**	**12**	**13**
1	20,33	6,39	2,79	0,436	0,187	0,503	2,7	506	0,132	55	2255	12,56
2	40,66	12,77	3,95	0,309	0,265	0,677	2,6	538	0,095	74	3562	12,64
3	60,98	19,16	4,99	0,261	0,314	0,765	2,6	582	0,100	86	4066	12,59
1	**2**	**3**	**4**	**5**	**6**	**7**	**8**	**9**	**10**	**11**	**12**	**13**
4	81,31	25,55	6,68	0,262	0,313	0,818	2,6	661	0,675	87	2988	13,21
5	98,59	31,23	9,38	0,303	0,270	0,849	2,7	608	8,024	163	488	9,48

Table 8. Averaged effects, %, against the engine base case condition according to the results of the second series of tests, considering the changes for all the conditions

Effective Power	Fuel Consumption	Effective Efficiency	CO	CH	NOx
1,392	5,036	5,233	21,14	24,78	-4,43

5. CONCLUSION

Today, research of the fine mechanisms of friction which are responsible for change of materials' antifriction parameters, and the impact of external force fields (electric, magnetic) on them, are underway. At this stage, a thesis has been defended which discusses the main principles of transformation of a matter aggregate state during friction, and impact of electromagnetic fields on lubricating action of oils and greases in metal couplings. The design of the Electronic Friction Regulator is being improved, and there being checked possible ways of its connection to vehicle engines with the purpose of finding the best option providing the maximal improvement of vehicle/machine power plant performance. Also, the research is underway to study possible negative consequences from utilization of this device. However, over the three years of monitoring of the technical condition of the machines operated with the electronic device, no negative aspects from its utilization were found.

REFERENCES

1. A. S. Akhmatov, 1963 Molekulyarnaya fizika granichnogo treniya (Molecular Physics of Boundary Friction), Fiziko-matematicheskaya literatura, Moscow, USSR.

2. D. H. Buckley, 1986 Poverkhnostnye javlenija pri adgezii i frikcionnom vzaimodejjstvii (Surface Effects in Adhesion, Friction, Wear and Lubrication), Mashinostroenie, Moscow, USSR.

3. Belyi, V.A.(1988). Problema sozdanija kompozicionnykh materialov i upravlenie ikh frikcionnymi svojjstvami (Creation of Composites and Control of their Frictional Properties). Trenie i iznos, Vol.3, No.3, (June 1988), pp. 384-396, ISSN 0202-4977

4. R. W. Cahn, 1968 Fizicheskoye metallovedenie (Physical Metallurgy), 3Mir, Moscow, USSR.

5. V. Cholakov, 1986 Nobelevskie premii. Uchenye i otkrytija (Nobel Prizes. Scientists and Discoveries), Mir, Moscow, Russia.

6. D. N. Garkunov, 2001 Tribotekhnika, iznos i bezyznosnost' (Tribotechnics, Wear and Wealessness), Moscow Agricultural Academy, 5-94327-004-3Russia.

7. Garnovskyi, A.D.; Rjabukhin, Yu.I. & Kuzharov, A.S. (1984). Koordinacionnaja khimija (Coordination Chemistry). Trenie i iznos, Vol.5, No.6, (December 1984), pp. 1011-1033, ISSN 0202-4977

8. V. L. Ginsburg, 1967 Rasprostranenie ehlektromagnitnykh voln v plazme (Electromagnetic Wave Propagation in Plasma), Fiziko-matematicheskaja literatura, Moscow, USSR.

9. M. Godet, 1984The body approach. A mechanical view of wear, Wear, 100 437 452 0043-1648

10. V. A. Goldade, V. A. Struk, S. S. Peseckyi, 1993 Ingibitory iznashivanija metallopolimernykh system (Wear Inhibitors of Metal-polymeric Systems), Khimija, 5-72450-876-1Russia.

11. G. Heinicke, 1987 Tribokhimija (Tribochemistry), Mir, Moscow, USSR.

12. S. Hawking, R. Penrose, 2007 Priroda prostranstva i vremeni (The Nature of Space and Time), Amfora, 978-5-36700-590-5St.-Petersburg, Russia.

13. Kadolich, Zh.V.; Pinchuk, L.S. & Anisov, A.P. (2001). Issledovanie poverkhnosti trenija polimernykh implantantov tazobedrennykh sustavov (Studying Friction Surface of Polymeric Implants of Hip Joints), Trenie i iznos, Vol.22, No.1, (February 2001), pp. 78-83, ISSN 0202-4977

14. Kostetsky, B.I. (1980). O roli vtorichnykh struktur v formirovanii mekhanizmov trenija, smazochnogo dejjstvija i iznashivanija (About the Role of Secondary Structures in Formation of Friction Mechanisms, Lubricating Action and Wear). Trenie i iznos, Vol.1, No.4, (August 1980), pp. 622-637, ISSN 0202-4977

15. L. D. Landau, E. M. Lifshitz, 2001 Teoreticheskaja fizika (Theoretical Physics), 6Gidrodinamika" (Hydrodynamics),Fiziko-matematicheskaja literatura, 5-92210-121-8Russia.

16. Lubimov, D.N.; et al. (1990). Ehlektromagnitnye spektry i fiziko-khimija metallopolimernogo frikcionnogo kontakta (The Electromagnetic Spectra and Physicochemistry of a Metal-polymeric Frictional Contact). Trenie i iznos, Vol.11, No.6, (December 1990), pp. 1084-1086, ISSN 0202-4977

17. Lubimov, D.N.; et al. (1992). Kinetika formirovanija i struktura smazochnykh sloev pri izbiratel'nom perenose v metallopolimernykh tribosoprjazhenijakh (The Kinetics of Formation and The Structure of Lubricating Layers during Selective Transfer in Metal-polymeric Tribounits). Trenie i iznos, Vol.13, No.3, (June 1992), pp. 496-500, ISSN 0202-4977

18. D. N. Lubimov, P. N. Kozachenko, A. E. Ivanov, 1994Ehlektropolevojj "serfing-ehffekt" (Electrofield "Surfing Effect"), Sovershenstvovanie tekhniki, tekhnologii i problemy ehkologii proizvodstv (Improvement of Technology and Machinery, Manufacturing

Environmental Issues), 3Shakhty Technology Institute for Consumer Services, Shakhty, Russia.

19. D. N. Lubimov, V. V. Levkin, A. E. Ivanov, 1998Nakhozhdenie kontaktnojj temperatury pri pomoshhi diffuzionno-ehnergeticheskikh predstavlenijj o trenii (Finding of Contact Temperature Using Diffusion and Energy Notions of Friction), Oborudovanie i tekhnologii byta i uslug (Equipment and Technologies of Housekeeping and Services), 29 104 106Shakhty, Russia.

20. D. N. Lubimov, V. A. Ryzhikov, 2001 Osnovy teorii trenija (Friction Theory Fundamentals), South Russian State Technical University, 5-88998-220-6Russia.

21. D. N. Lubimov, N. K. Vershinin, 2001Vlijanie frikcionnogo vzaimodejjstvija poverkhnostejj metallopolimernogo tribosoprjazhenija na ikh ehlektricheskie i khimicheskie svojjstva. (Impact of Frictional Interaction of the Metal-polymeric Tribounit Surfaces on Their Electrical and Chemical Properties). Sostojanija i perspektivy razvitija vostochnogo Donbassa (Conditions and Prospects of Eastern Donbass Development), pp. 274 278 5-88998-252-4

22. D. N. Lubimov, V. A. Ryzhikov, 2006 Fiziko-khimicheskie processy pri trenii (Physicochemical Processes in Friction), South Russian State Technical University, 5-88998-394-6Russia.

23. D. N. Lubimov, A. V. Kozhemyachenko, K. N. Dolgopolov, 2007Vlijanie postplazmennykh processov na formirovanie smazochnykh plenok na poverkhnostjakh trenija (Influence of Postplasmatic Processes on the Formation of Lubricating Films on Friction Surfaces), In: Bytovaja tekhnika, tekhnologija i tekhnologicheskoe oborudovanie predprijatijj servisa i mashinostroenija (Household Equipment, Technology and Process Equipment of Service and Engineering Companies), pp. 86 87South Russia State University of Economics and Service, 978-5-93834-302-3Shakhty, Russia.

24. Lubimov, D.N.; et al. (2007b). Issledovanie fizicheskikh svojjstv triboplazmy, kak osobogo agregatnogo sostojanija veshhestva (Study of the Physical Properties of Triboplasma As a Special Aggregate State of Matter), Mezhdunarodnaja nauchnaja konferencija Problemy razvitija estestvennykh, tekhnicheskikh i

social'nykh sistem (International Scientific Conference. Development Issues of Natural, Technical and Social Systems), Part 2, pp.38-43, ISBN 5-88040-047-6, Anton, Taganrog, Russia, April, 2007

25. D. N. Lubimov, et al. 2008Osnovnye principy i rezul'taty ispol'zovanija "Ehlektronnogo reguljatora trenija" dlja povyshenija ehkspluatacionnykh pokazatelejj dvigatelejj transportnykh sredstv (Main Principles and Results of Operation of the "Electronic Friction Regulator" to Imporve the Performance of Vehicle Engines), In: Intellektika, logistika, sistemologija (Intellectics, Logistics, Systemology), pp. 146 149Chelyabinsk Center of Scientific and Technical Information, 978-5-94218-081-2Chelyabinsk, Russia.

26. D. N. Lubimov, K. N. Dolgopolov, N. M. Bai, 2009Ehlektrodinamika poljarizacii gorjuche-smazochnykh materialov ot vneshnego istochnika (The Electrodynamics of Polarization of Fuels and Lubricants from an External Source), In: Ehkonomika i proizvodstvo (Economics and Production), pp. 124 130Chelyabinsk Center of Scientific and Technical Information, 978-5-94218-088-1Chelyabinsk, Russia.

27. D. N. Lubimov, et al. 2009bEhlektronnyjj reguljator trenija (ustrojjstvo i rezul'taty stendovykh ispytanijj) (Electronic Friction Regulator. Design and Development Testing Results), In: Ehkonomika i proizvodstvo (Economics and Production), pp.118 123Chelyabinsk Center of Scientific and Technical Information, 978-5-94218-088-1Chelyabinsk, Russia.

28. Matveevskyi et al. 1989 Smazochnye materialy: Antifrikcionnye i protivoiznosnye svojjstva. Metody ispytanijj (Lubricants. Antifriction and Antiwear Properties. Testing Methods), Mashinostroenie, Moscow, USSR.

29. Morozov, S.V.; Novoselov, K.S. & A. Heym, A.K. (2008). Ehlektronnyjj transport v grafene (Electron Transport in Graphene), Uspekhi fizicheskikh nauk, Vol.178, No.7, pp. 776-780, ISSN: 0042-1294 (Print), 1996-6652 (Online

30. Nikolsky, A.V.; Kozakov, A.T. & Kravchenko, V.I. (1988). Dinamika izmenenija khimicheskogo sostojanija poverkhnostejj trenija metallopolimernogo soprjazhenija v processe frikcionnogo vzaimodejjstvija (The Change Trend of the Friction Surface Chemical Condition of a Metal-polymeric Tribounit during Frictional

Interaction Process). Trenie i iznos, Vol.9, No.5, (October 1988), pp. 860-869, ISSN 0202-4977

31. L. S. Pinchuk, V. A. Goldade, A. V. Makarevich, 2004 Ingibirovannye plastiki (Inhibited Plastics), Metal-Polymer

32. Research Institute of National Academy of Science of Belarus, ISBN 985-647727-1, Gomel, Belarus.

33. Pinchuk, L.S. & Chernyakova, Yu.M. (2007). Sinovial'nyjj sustav kak "umnyjj" uzel trenija (Synovial Articulation as a "Smart Friction Assembly"), Trenie i iznos, Vol.28, No.4, (August 2007), pp. 410-417, ISSN 0202-4977

34. Pogosyan, A.K. & Oganesyan, K.V. (1986). Javlenie frikcionnogo perenosa: osnovnye zakonomernosti i metody issledovanija (Phenomenon of Frictional Transfer: Main Patterns and Study Methods). Trenie i iznos, Vol.7, No.6, (December 1986), pp. 998-1007, ISSN 0202-4977

35. A. A. Polyakov, 1988Samoorganizacija struktury izbiratel'nogo perenosa (Self-organization of Selective Transfer Structure), In: Dolgovechnost' trushhikhsja detalejj mashin (Durability of Rubbing Parts of Machines), 3Mashinostroenie, Moscow, USSR.

36. A. A. Polyakov, 1990Terminy izbiratel'nogo perenosa (Selective Transfer Terms), In: Dolgovechnost' trushhikhsja detalejj mashin (Durability of Rubbing Parts of Machines, 4Mashinostroenie, Moscow, USSR.

37. A. M. Prokhorov, 1992 Fizicheskaja ehnciklopedija (Physical Encyclopedia), 3Bol'shaja Rossijjskaja ehnciklopedija (Big Russian Encyclopedia), 5-85270-019-3Russia.

38. Ryzhikov V.A. & Dolgopolov K.N. (2005). Ehlektrokhimicheskie processy v uzlakh trenija (Electrochemical Processes in Friction Assemblies). IV Mezhdunarodnaja nauchno-prakticheskaja konferencija Problemy sinergetiki v tribologii, triboehlektrokhimii, materialovedenii i mekhatronike (IV International Scientific Workshop Conference. Synergetics Problems in Tribology, Triboelectrochemistry, material science and mechatronics), pp. 7-11, South Russian State Technical University, ISBN 5-88998-629-5, Novocherkassk, Russia, November 4, 2005

39. P. V. Sysoev, M. M. Bliznets, A. K. Pogosyan, 1990 Antifrikcionnye ehpoksidnye kompozity v stankostroenii (Antifrictional Epoxy-filled Composites in Machine Engineering), Nauka i tekhnika, 5-34300-536-5Belarus.

40. E. L. Shvedkov, D. Rovinsky, Ya, 1979 Slovar'-spravochnik po treniju, iznosu i smazke detalejj mashin (Glossary for Friction, Wear, and Machine Parts Lubrication), Naukova dumka, Kiev, USSR.

41. K. N. Voinov, 2010 Tribologija: mezhdunarodnaja ehnciklopedija. Istoricheskaja spravka, terminy, opredelenija (Tribology: International Encyclopedia (Background, Terms, Definitions), 1Anima, 978-5-99020-642-7St.-Petersburg, Russia.

6

RECENT ADVANCES IN THE SYNTHESIS AND BIOMEDICAL APPLICATIONS OF NANOCOMPOSITE HYDROGELS

Umile Gianfranco Spizzirri [1], Manuela Curcio [1], Giuseppe Cirillo [1,*], Tania Spataro [1], Orazio Vittorio [2,3], Nevio Picci [1], Silke Hampel [4], Francesca Iemma [1] and Fiore Pasquale Nicoletta [1]

[1] Department of Pharmacy, Health and Nutritional Sciences, University of Calabria, I-87036 Rende, Italy.

[2] Children's Cancer Institute Australia, Lowy Cancer Research Centre, University of New South Wales, Sydney, 2052, Australia

[3] Australian Centre for Nanomedicine, University of New South Wales, Sydney, 2052, Australia

[4] Leibniz Institute for Solid State and Materials Research, PF 270116, D-01171 Dresden, Germany

ABSTRACT

Hydrogels sensitive to electric current are usually made of polyelectrolytes and undergo erosion, swelling, de-swelling or bending in the presence of an applied electric field. The electrical conductivity of many polymeric materials used for the fabrication of biomedical devices is not high enough to achieve an effective modulation of the functional properties, and thus, the incorporation of conducting materials (e.g., carbon nanotubes and nanographene oxide) was proposed as a valuable approach to overcome this limitation. By coupling the biological and chemical features of both natural and synthetic polymers with the favourable properties of carbon nanostructures (e.g., cellular uptake, electromagnetic and magnetic behaviour), it is possible to produce highly versatile and effective nanocomposite materials. In the present review, the recent advances in the synthesis and biomedical applications of electro-responsive nanocomposite hydrogels are discussed.

KEYWORDS

carbon nanotubes; graphene; composite hydrogels; electro-responsive

1. ELECTRO-RESPONSIVE HYBRID HYDROGELS IN DRUG DELIVERY

Smart hydrogels are functional materials able to modulate equilibrium swelling as a function of a change in the external environment, such as pH, ionic strength, temperature, pressure, light, electro-magnetic and sound fields [1,2,3,4,5,6]. Over the last few decades, they have emerged as valuable tools for applications in biomedicine, mainly as drug delivery matrices and tissue engineering scaffolding materials [7,8].

An external electric field is a useful approach to activate/modulate the release of therapeutics, since the availability of equipment for controlling the magnitude of current, the duration of electric pulses, the intervals between pulses, *etc.*, allows the application of such a stimulus to be precisely handled [9,10].

The response of electro-responsive hydrogels to an external electric field can be erosion, swelling, de-swelling or actuation of the structure according to different mechanisms (Coulombic, electro-osmotic, electro-chemical and dynamic enrichment/depletion phenomena) [11] and to several parameters, including the experimental setup, the gel composition (charge density, swelling degree, nature of monomers, cross-linkers and pendant chains), pH, current magnitude, ionic strength, presence of drugs, *etc.* [12]. In particular, a low polyion concentration results in hydrogel de-swelling, while swelling is recorded at higher concentrations [13].

A key limitation for the preparation of effective electro-sensitive releasing devices is related to the low electrical conductivity of most biologically-significant polymers [12]. The incorporation of suitable functional additives containing a spatially-extended π bonding system represents a simple approach to overcome this limitation [14,15].

Among the different additives, carbon nanostructures, such as carbon nanotubes (CNT) and graphene (GP), are emerging as innovative and valuable materials to fabricate electro-responsive hydrogel systems [16,17] (Figure 1).

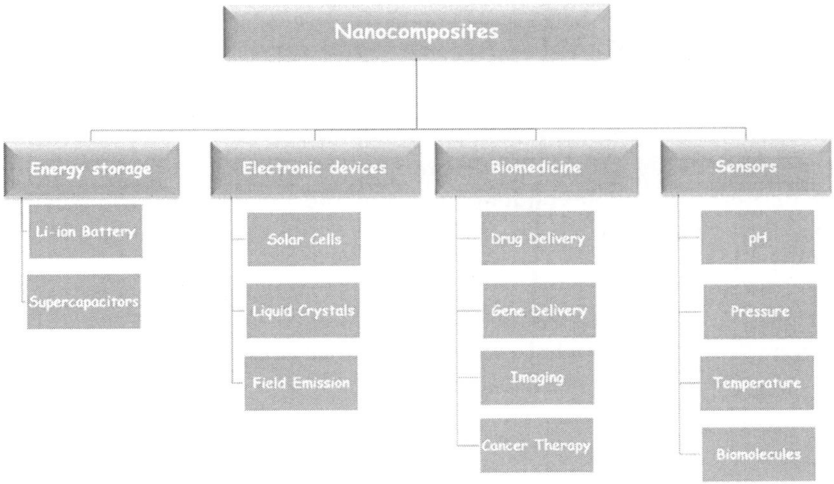

Figure 1. Technological applications of carbon nanohybrids.

CNT and GP are allotropes of carbon resulting from variations in covalent bonding between carbon atoms. Each allotrope has distinctive physical and chemical properties owing to the unique spatial arrangement that carbon atoms adopt [18].

The introduction of a small amount of carbon nanostructures into a polymer system allows the enhancement of some polymer properties, such as photo-conductivity [19], electro-conductivity [20], water solubility [21], mechanical and optical properties [22].

2. CARBON NANOTUBES

Carbon nanotubes (CNT) consist of a hexagonal arrangement of sp^2-hybridized carbon atoms (the C–C distance is about 1.4 Å) [23,24].

They can be envisioned as cylinders composed of rolled-up graphite planes with diameters in the nanometre scale [25,26]. The cylindrical nanotube usually has at least one end capped with a hemisphere of a fullerene structure [27].

Depending on the process for CNT fabrication, there are two types of CNT [28,29]: single-walled CNT (SWNT) and multiwalled CNT (MWCNT). SWCNT consist of a rolled single graphene layer, whereas MWCNT consist of two or more concentric cylindrical shells of graphene sheets coaxially arranged around a central hollow core with van der Waals forces acting between adjacent layers. The interlayer separation of the graphene layers of MWCNT is approximately 0.34 nm on average, each one forming an individual tube and the outer diameter ranging from 2.5 to 100 nm, while for SWCNT, this value ranges from 0.6 to 2.4 nm [30].

According to the rolling angle of the graphene sheet, CNT have three different chiralities: armchair, zigzag and chiral (Figure 2).

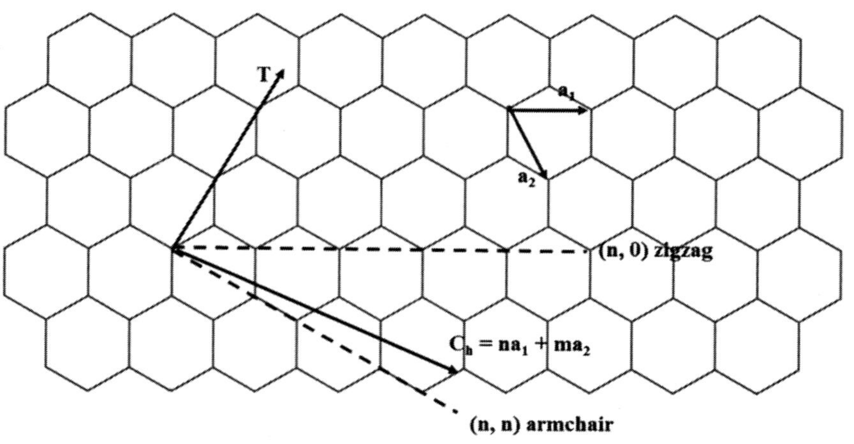

Figure 2. Chirality features of carbon nanotubes.

The tube chirality is defined by the chiral vector, $\mathbf{C}_h = n\mathbf{a}_1 + m\mathbf{a}_2$, where the integers (n, m) are the number of steps along the unit vectors ($\mathbf{a}_1$ and $\mathbf{a}_2$) of the hexagonal lattice [31]. Using this (n, m) naming scheme, the three types of orientation of the carbon atoms around the nanotube circumference are specified. If $n = m$, the nanotubes are called "armchair". If $m = 0$, the nanotubes are called "zigzag". Otherwise, they are called "chiral". The chirality of nanotubes has a significant impact on their transport properties, particularly on the electronic properties. For a given (n, m) nanotube, if $(2n + m)$ is a multiple of three, then the nanotube has a metallic behaviour; otherwise, the nanotube is similar to a semiconductor. For MWCNT, since they contain a multi-layer of graphene and each layer can have different chiralities, the prediction of their physical properties is more complicated than that of SWCNT [23].

Within the realm of biotechnology, CNT have been utilized for many different applications [32], including platforms for ultrasensitive recognition of antibodies [33], as nucleic acid sequencers [34] and as bioseparators [35], biocatalysts [36] and ion channel blockers [37] for facilitating biochemical reactions and biological processes [38]. Towards nanomedicine, CNT have been utilized as scaffolds for neuronal and ligamentous tissue growth for regenerative interventions of the central nervous system [39] and orthopaedic sites [40], substrates for detecting antibodies associated with human autoimmune diseases with high specificity [41,42] and carriers of contrast agent magnetic resonance imaging [43,44]. When coated with nucleic acids (DNA or RNA), vaccines and proteins, CNT have been shown as effective substrates for gene sequencing and as gene and drug delivery vectors to challenge conventional viral and particulate delivery systems [45,46,47,48,49,50,51,52].

It should be noted that the CNT outstanding reactivity, due to their enormous surface area and achieved by their infinitesimal size, serves as their best merit, on the one hand, but also as their worst attribute, on the other, especially when they enter humans and others living creatures' bodies [53,54,55,56], even if the exact mechanisms regulating their toxicity are still to be understood [57,58,59].

According to the WHO (World Health Organisation) definition, CNT possess a fibre-like structure, and the morphological similarity to asbestos fibres is the major concern in public health [60]. It was proven that exposing the mesothelial lining of the body cavity of mice (a surrogate for the mesothelial lining of the chest cavity) to long MWNT, an asbestos-like nanotube, length-dependent pathogenic processes are observed. These processes include inflammation and the formation of lesions known as granulomas [61]. In addition, CNT carcinogenic properties, causing mesothelioma at a high rate in intact male rats, have also been reported [62,63].

Since most of the toxicological concerns of CNT are related to their water insolubility or near insolubility [64], several studies have hence been performed aiming to prepare derivatives of highly functionalized CNT with reduced toxic effects [65]. CNT, in the pristine form, invariably exist as aggregates or bundles that are tightly bound by hydrophobic interactions between the sp^2 carbon tube shells [66].

When dealing with the functionalization of CNT, a distinction must be made between covalent and non-covalent functionalization. Covalent functionalization is based on the covalent linkage of functional entities onto the nanotube's carbon scaffold. It can be performed at the termini of the tubes or at their sidewalls. Direct covalent sidewall functionalization is associated with a change of hybridization from sp^2 to sp^3 and a simultaneous loss of conjugation. Defect functionalization takes advantage of chemical transformations of defect sites already present. Defect sites can be the open ends and holes in the sidewalls, terminated, for example, by carboxylic groups, and pentagon and heptagon irregularities in the hexagon graphene framework. Oxygenated sites, formed through oxidative purification, have also to be considered as defects [67,68].

A non-covalent functionalization is mainly based on supramolecular complexation using various adsorption forces, such as van der Waals' and π-stacking interactions. All of these functionalizations are exohedral derivatizations. A special case is the endohedral functionalization of CNT, *i.e.*, the filling of the tubes with atoms or small molecules [69].

Among others, surface modification of CNT may be realized either by covalent or non-covalent bonding (wrapping) of polymer molecules on the surface of CNT [70].

2.1. Non-Covalent Wrapping Techniques

Most of the CNT-hydrogel hybrids are synthesized by non-covalent wrapping techniques involving coating with polymeric materials [71]. The wrapping techniques benefit from being non-destructive and efficient [72], but some of the best dispersants are also toxic to living cells. Different biocompatible and bioactive polymeric materials have been proposed as valuable wrapping agents [73,74], including polymeric surfactant (e.g., Pluronic F127 [75]), oligopeptides and natural extracellular matrix proteins (e.g., gelatine, collagen, elastin and laminin) [76].

For preparing non-covalent CNT-hydrogel hybrids, the employed approaches involve the use of both pristine and preliminarily-functionalized CNT.

The first approach is based on the surface properties of CNT: the sp^2 bonded graphene structures at the sidewalls contain highly delocalized π electrons, which can form functionalized CNT with other π electron-rich compounds (e.g., conducting polymers) through π–π interactions. This organic functionalization method does not destroy the intrinsic structures of CNT and gives structurally intact CNT with functionalities, and thus, the unique properties of the CNT surface are transferred to the final nanohybrids. Recently, the potential interaction between the highly delocalized π-electrons of CNT and the π-electrons correlated with the lattice of the polymer skeleton has generated much interest and provided the motivation for studying the optical and electronic properties of composites of CNT and π-conjugated polymers [70].

On the other hand, a pre-functionalization could help with enhancing the interaction between CNT and polymer counterparts to optimize their dispersion in the hydrogel networks.

Many studies carried out on CNT/polymer composites [77] show that three main factors greatly influence the reinforcement of the CNT/polymer composite: (1) a good dispersion of the CNT in polymer matrix [78,79]; (2) a strong interfacial bonding between CNT and polymer [80]; and (3) a good alignment of the CNT in polymer matrix [81].

2.2. Covalent Functionalization of the CNT Surface

Alternative synthetic strategies consist of the preparation of a composite material where the CNT are covalently bound to the network.

Many techniques, including esterification, "click" chemistry, layer-by-layer self-assembly, radical polymerization, anionic coupling, supercritical CO_2-solubilized polymerization, γ-ray irradiation, reversible addition fragmentation chain-transfer polymerization, ring-opening polymerization and atom transfer radical polymerization, have been employed to functionalize CNT with polymers [82].

The main approaches for the fabrication of covalent CNT-polymer nanohybrids can be classified into "grafting from" and "grafting to" [83] (Figure 3).

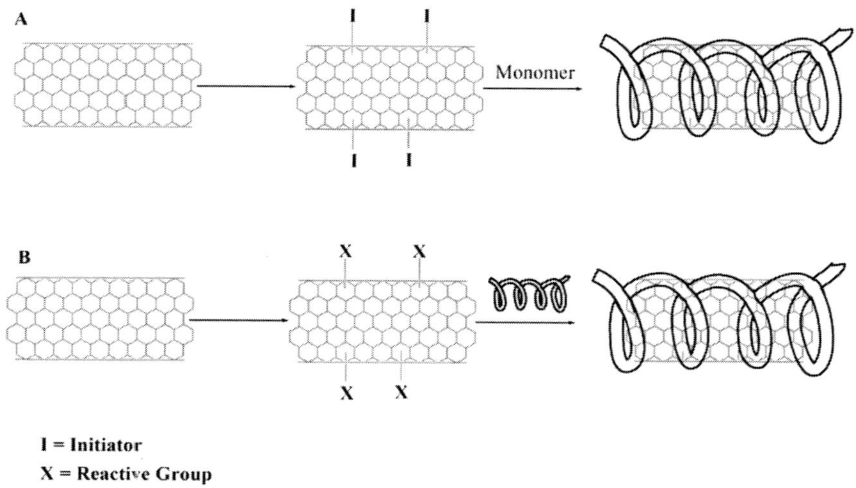

Figure 3. Schematic representation of the "grafting from" (**A**) and "grafting to" (**B**) approaches for CNT functionalization.

The "grafting to" approach involves pre-formed polymer chains reacting with the surface of either pristine or pre-functionalized CNT. The structure of CNT can be used to explain the rationale of this approach. The curvature of the carbon nanostructures imparts a significant strain upon the sp^2-hybridized carbon atoms, and as a consequence, the energy barrier required to convert these atoms to sp^3 hybridization is lower than that of the flat graphene sheets, making them susceptible to various addition reactions. The "grafting to" method onto CNT defect sites means that the readymade polymers with reactive end groups can react with the functional groups on the nanotube surfaces. An advantage of the "grafting to" method is that preformed commercial polymers of controlled molecular weight and polydispersity can be used. The main limitation of this technique is that initial binding of polymer chains sterically hinders the diffusion of additional macromolecules to the CNT surface, leading to a low grafting density. Furthermore, only polymers containing reactive functional groups can be used [84]. The "grafting from" approach involves the polymerization of monomers from surface-derived initiators on CNT. These initiators are covalently attached using the various functionalization reactions developed for small molecules. The advantage of the "grafting from" approach is that the polymer growth is not limited by steric hindrance, allowing high molecular weight polymers to be efficiently grafted. In addition, nanotube-polymer composites with quite high grafting density can be prepared. However, this method requires a strict control of the amounts of initiator and substrate, as well as accurate control of polymerization conditions. Moreover, the continuous π-electronic properties of CNT would be destroyed by acid oxidation (required for initiator linkage). As a result, compared to the "grafting from", the "grafting to" approach has much less alteration of the structure of CNT [85]. A further modification of the "grafting from" approach involves the direct polymerization of suitable monomers around the CNT surface, with the subsequent covalent addition to change the carbons from sp^2 to sp^3 hybridization [83].

2.3. Non-Covalent CNT-Hydrogels in Drug Delivery

As stated before, several wrapping polymers have been proposed for the preparation of non-covalent CNT composites (Table 1).

Solvent casting followed by melt mixing and coagulation methods were proposed to produce SWNT/poly(methyl methacrylate) nanocomposite fibres with enhanced elastic modulus, thermal stability and conductivity [86,87].

Pluronic F127 and gelatine were employed as wrapping materials in MWNT nanohybrids useful in the biomedical field [88]. MWNT/gelatine composites were prepared by dispersion of MWNT through ultrasonication in an aqueous medium with the anionic surfactant sodium dodecyl sulphate [89]. The key result of the study is the possibility to couple the advantages of both the nanostructure and protein. CNT act as reinforcing agents, considerably improving the structural and mechanical properties of the gelatine [90], while gelatine facilitates the dispersion of the carbon structures allowing, at the same time, the use of the final nanohybrid in biomedicine, where the protein is widely employed as a plasma expander, wound dressing, adhesive agent and controlled drug delivery system [91,92,93,94]. The water affinity of the nanohybrids was also found to be closely related to the CNT content: an increase in the CNT content resulted in a decrease in the swelling properties, while the stability and reversible behaviour upon exposure to direct current are considerably enhanced when compared to the protein alone, which underwent remarkable erosion processes.

Synthetic polymers can also be useful for the obtainment of composites for pharmaceutical and biomedical purposes. The coating of MWNT with poly(vinyl alcohol) allows the obtainment of versatile composites for different applications, including biosensors, artificial cartilage substitutes and drug delivery systems [95]. The formation of CNT/poly(vinyl alcohol) composite hydrogels without using chemical treatments of CNTs and chemical crosslinking agents in gel formation would enhance the mechanical and electrical properties of the hydrogels, meanwhile maintaining the good biocompatibility of both nanotubes and poly(vinyl alcohol) hydrogels.

Table 1. Electro-responsive carbon nanohybrids discussed in this review.

Application	Composition		Synthetic Approach	Ref.
	Carbon Portion	Polymeric Portion		
Conducting material	SWNT	PMMA	Melt processing	86
		PMMA	Coagulation	87
	CNT	PVP	Wrapping	96
	GO	PDMAEMA	ATRP	134
Biomedical device	MWNT	Pluronic F127	Wrapping	88
		Gelatine	Wrapping	88
		PVP/PVA	Wrapping	95
		P(AAm-*co*-EBA)	Thermal polymerization	100
	SWNT	PPy/PEGDA	Redox/Thermal polymerization	102
	CNT	P(NIPAAm-*co*-MEBA)	Dispersion	97
	GO	Cyclodextrin-dye complex	Mixing	140
		PS	PEP	154
	rGO	SAP and IBP	Mixing	144
	MWNT	PVA/PAA	Electrospinning	98
		IPN(PEO/PPETA)	Photo-polymerization	101
		P(Gelatine-*co*-NaMA-*co*-EBA)	Thermal polymerization	104
		P(MAA-*co*-EGDMA)	Thermal polymerization	107
		P(MAA-*co*-EBA)	Thermal polymerization	109–110
		P(Gelatine-*co*-AAm-*co*-EBA)	Photo-polymerization	111
		Cellulose	Mixing	113

Drug delivery system	CNT	PVA	Mixing	112
	GO/rGO	α-Cyclodextrin/Pluronic F-127	Mixing	141
	GO	KGM/SA	Mixing	142
		RGD-chitosan	Mixing	143
		Chitosan	Condensation	148
		PPy	Electro-polymerization	149
		PEI	Condensation	150
		PVCL	ATRP	152
		IPN(PEO/NIPAAm-*co*-MEBA)	Condensation	153
	rGO	PVA	Mixing	146
		JPT/PEGDE	Radical polymerization	155
Electrical sensor	MWNT	Gelatine	Wrapping	89–90
		C8-LPEI/EGDGE	Mixing	99
	SWNT	PEGDA	Photo-polymerization	103
Photosensitisers release	GO	HA	Condensation	151

CNT = carbon nanotubes; MWNT = multi-walled carbon nanotubes; SWNT = single-walled carbon nanotubes; GO = graphene oxide; GP = graphene; rGO = reduced graphene oxide; AAm = acrylamide; C8-LPEI = octyl modified linear polyethylenimine; EBA = *N,N'*-ethylenebisacrylamide; EGDMA = ethylene glycol dimethacrylate; EGDGE = ethylene glycol diglycidyl ether; HA= hyaluronic acid; IBP = hyperbranched polymer; IPN = interpenetrating polymer network; JPT= Jeffamine polyetheramine; KGM= konjac glucomannan; MAA = methacrylic acid; NaMA = sodium methacrylate; NIPAAm = *N*-isopropyl-acrylamide; MEBA = *N,N'*-methylenebisacrylamide; PDMAEMA = poly-2-(dimethylamino)ethyl-methacrylate; PEGDA = polyethylene glycol diacrylate; PEGDE = polyethylene glycol diglycidyl ether; PEI = poly-ethylenimine; PEO = polyethylene oxide; PMMA = polymethylmethacrylate; PPETA = polypentaerythritol triacrylate; PPy = polypyrrole; PS = polystyrene; PVA = polyvinyl alcohol; PVCL = poly-*N*-vinyl caprolactam; PVP = polyvinylpyrrolidone; SA= sodium alginate; SAP= superadsorbent polymer; ATRP = atom transfer radical polymerization; PEP = Pickering emulsion polymerization.

Poly(vinyl pyrrolidone) is another valuable polymer employed for the synthesis of hybrid hydrogels where CNT are uniformly dispersed. Furthermore, in this case, the strong interface between the nanotubes and matrix allows the preservation of the CNT pristine surface structure and the obtainment of exceptional mechanical and physical properties [96]. The surfactant poly(vinyl pyrrolidone) in cooperation with MWCNT demonstrated an important synergistic effect on the formation and performance of the composite hydrogels.

A different approach is based on the introduction of CNT into a thermally-sensitive poly(*N*-isopropylacrylamide) hydrogel matrix to fabricate hybrid hydrogels exhibiting an electric current sensitivity resulting from electro-thermal conversion. When an external electric field is applied, the electro-thermal conversion takes place to raise the local temperature of the composite beyond the lower critical solution temperature (around 32 °C) of poly(*N*-isopropylacrylamide) hydrogel. Here, an indirect and contact-type electric sensitivity is reached, while when polyelectrolyte hydrogels are employed, they always require being placed between two electrodes in a non-contact status [97].

As stated before, other strategies for the preparation of hybrid composites required a pre-functionalization of the CNT surface.

A preliminary surface modification of MWNT can be carried out by oxyfluorination to insert functional groups on their hydrophobic surface for improving the dispersion and the compatibility with a polymer matrix composed of poly(vinyl alcohol)/poly(acrylic acid) hydrogel. Thus, an electro-responsive transdermal drug delivery system was prepared by electrospinning [98]. A uniform distribution of the oxyfluorinated MWNT in the nanofibers was found to be crucial to the electro-responsive swelling and drug releasing behaviours. The amount of released drug increased by increasing the content of MWCNT, by using an oxyfluorination with higher oxygen content and by applying the highest electric voltage. The nanocomposite showed more than 80% cell viability, enough to be used as a non-toxic and biocompatible material.

In another study, Meredith *et al.* suggested the application of a CNT composite electrode hydrogel as a glucose biosensor and biofuel [99]. The mechanism at the basis of idea is the facile oxidation of nicotinamide cofactors, such as NADH, by azine dyes inserted inside the electrode. Azine dyes, carbon nanotubes, polymer hydrogels and enzymes were incorporated in composite electrodes in one single mixing step. The electro-polymerized composite electrodes were found to display the ability to electrocatalytically oxidize NADH more efficiently than a control electrode prepared without CNT.

2.4. Covalent CNT-Hydrogel in Drug Delivery

By inserting various amounts of MWNT (0.001, 0.002 and 0.01 wt %) in a radical polymerization of acrylamide and bisacrylamide, composite hydrogels with tuneable physico-chemical properties were synthetized and proposed for different applications, including artificial muscles, drug delivery, robotics and micropumps. The different amounts of CNT affect many properties, such as electro-responsive behaviour, water uptake and surface roughness, while not significantly influencing the electrical conductivity due to the lack of alignment of nanotubes in the polymeric network. In addition, water uptake depends on the drying method employed to recovery hydrogels [100].

A semi-interpenetrating polymer network, reinforced with MWNT and obtained by photo-polymerization of pentaerythritol triacrylate and polyethylene oxide, was proposed for the preparation of an electro-sensitive transdermal drug delivery system for ketoprofen [101]. The MWNT amount in the electrospun fibres confers remarkable electrical conductivity up to a MWCNT/PEO ratio of 10 wt %, while no further significant increase is recorded above this value.

The materials show a high cell viability (over 85%) on mouse fibroblasts, confirming their high biocompatibility. The *in vitro* drug diffusion experiments, tested by vertical Franz-type diffusion cells, were performed at various electric voltages (0, 5, 10 and 15 V) using both hybrid hydrogels and blank materials not containing CNT. For both materials, the amount of released drug significantly increases within the applied voltage. Furthermore, the release is greater for the hybrid than for the blank fibres, because the electrical stimulation results in a partial dissolution of polyethylene oxide in the semi-interpenetrating polymeric network, creating empty spaces, which represent the route for the drug diffusion.

An alternative methodology to improve the electrical conductivity and mechanical strength of the hydrogels without significant reduction in the swelling properties of SWNT composites was obtained through the interfacial polymerization of SWNT, polypyrrole and poly(ethylene glycol) diacrylate [102]. The ternary SWNT/polypyrrole (PPy)/ polyethylene glycol diacrylate (PEGDA) hydrogel possesses improved electrochemical performance, enhanced mechanical toughness, compactness and homogeneity of the surface structure than the binary hybrid hydrogel of SWNT/PEGDA previously reported by the same authors [103]. This conductive composite hydrogel could be applied in the field of implantable devices, including neural prosthetic devices, biosensors and bioelectronic arrays.

Spherical hybrid hydrogels composed of gelatine and MWNT were synthesized by emulsion polymerization in the presence of sodium methacrylate and *N,N'*-ethylenebisacrylamide and proposed as drug delivery microspheres for the electro-responsive release of diclofenac sodium salt [104]. Emulsion polymerization allows producing spherical particles with a narrow size distribution and dispersing designated functional molecules (e.g., MWNT) into the polymerization feed. With this technique, the gelatine can act as a macromolecular co-crosslinker in the synthesis of spherical gels (average diameters around 100 μm) suitable for delivering a therapeutic in response to pH and/or temperature modifications without any preliminary modification of the protein structure [105,106].

This straightforward procedure allows obtaining composite materials in a one-step process showing considerable advantages from both a synthetic and an application point of view; it is a single step and easily scalable process, and the intrinsic properties of either CNT or protein are not affected, but totally retained in the final composite.

The swelling properties confirm the polyelectrolyte behaviour of hydrogels, and the application of an external electric field (at pH 7.4) caused a significant reduction of their swelling degree as a consequence of a built-in osmotic pressure. Drug release experiments demonstrated the ability of the composites to control the release of drug over time. The electric stimulation resulted in a further increase of the release (+20%) in MWNT-containing materials.

The synthesis of molecularly-imprinted electro-responsive nanohybrids was reached by precipitation polymerization of methacrylic acid and ethylene glycol dimethacrylate in the presence of CNT [107]. Diclofenac sodium salt was employed as a template molecule, and its release profile was investigated to elucidate the ability of nanohybrids to release the template in response to 20-V DC voltage.

An interesting upgrade of the application of hybrid hydrogels as electro-responsive delivery systems is the possibility to reach pulsatile release profiles by the on/off switching of an applied electric field [108]. Electro-responsive MWNT-poly(methacrylic acid) hydrogel hybrids synthetized by *in situ* radical polymerization were proposed for "on demand" drug delivery [109]. Experiments, carried out employing radio-labelled sucrose as the model hydrophilic drug molecule showed a pulsatile release profile upon the on/off application of the electric field, and a drug release from the hybrid gels up to 70% was achieved after 80 min. The release of a model drug upon the application of a DC electric field was demonstrated *in vivo*following sub-cutaneous implantation of the hybrid hydrogels in a mouse model. This drug delivery device produces greater amounts of released drug from gel with a high MWNT content; however, the amount of drug released between each cycle of on/off electrical stimulation was not reproducible and decreased dramatically after the second electrical stimulation. Producing an adherent hybrid hydrogel is not trivial, as the reversible swelling and de-swelling of the polymer matrix upon electrical exposure can lead to appreciable levels of shear stress, which can result in structural damage. The same authors also studied the mechanisms by which MWNT improve the electrical, mechanical and thermal properties of the composite [110]. The presence of MWNT within the hydrogel matrix affected the mechanical properties of the hydrogel by decreasing the pore size and therefore the swelling/de-swelling of the gels. The damage to the hybrid gel surfaces after electrical stimulation and the loss of the pulsatile release profile at high cross-linker concentrations suggested that the mechanism of drug release involved a compression effect and magnified the stress on the polymeric network as a result of the electrical properties of MWNT.

Electro-responsive hybrid hydrogels, able to precisely modulate the release of therapeutics by the application of an external voltage, were proposed by the covalent insertion of MWNT into acrylamide-*co*-ethylene/bis-acrylamide hydrogels [111]. In this study, MWNT act as functional elements, responsible for both drug interaction and the electro-responsivity. The release/retention profile was closely related to the charge, with anionic drugs (diclofenac sodium salt) released at 12 V from the negatively-charged MWNT outer surface, and cationic drugs (ciprofloxacin) were retained at 12-V conditions and quickly released in off conditions (0 V). Hydrogel properties were found to depend strongly on the MWNT content, with the possibility to reach an on demand delivery by switching the external voltage on/off.

Different kinds of CNT hybrid materials were proposed as an electric interface between neural tissue and electrodes in the development of implantable devices for continuous monitoring and functional stimulation of the central nervous system in terms of electro-activity, biocompatibility and long-term stability for chronical application. To engineer an interface possessing these merits, a polymeric hydrogel based on poly(ethylene glycol) diacrylate and SWNT was employed to fabricate a hybrid hydrogel via a covalent anchoring strategy, *i.e.*, self-assembly of cysteamine followed by Michael addition between cysteamine and poly(ethylene glycol) diacrylate [103]. According to the authors, this hybrid hydrogel provides a favourable biomimetic microenvironment for cell attachment and growth due to its inherent biocompatibility and good adhesion to the electrode substrate.

More recently, Choi *et al.* synthesized CNT-incorporated polyvinyl alcohol (PVA)-based hydrogels by an electro-click reaction, which was controlled by an electrochemically-generated Cu(I) catalyst [112]. Results showed that CNT-incorporated hydrogel films were deposited faster than hydrogel films in the absence of CNT, as the embedded carbon nanotubes provided a larger electrochemically-active area. The release behaviour of tetracycline was enhanced under negatively-biased potential at pH 8 due to charge–charge repulsion.

Novel hybrid hydrogels composed of sodium alginate, bacterial cellulose and MWNT were synthesized using $CaCl_2$ as a crosslinking agent and proposed as a pH and electric field dual-stimulus responsive drug delivery system [113]. The releasing profile of the drug from the hybrid hydrogels was found to be dependent on the applied electric current strength at neutral pH and selective to the pH value of the surrounding acidic or alkaline environment, with little difference in the presence of the applied electric voltage. Drug release was mainly affected by the pH-induced protonation and deprotonation of sodium alginate and was enhanced under electric voltage application.

3. GRAPHENE

The atomic structure of graphite is characterized by the multiple stacking of one atom-thick sheets formed by carbon atoms arranged in a hexagonal lattice. The isolated two-dimensional crystal structures made of single atomic layers of graphite are called "graphene". The existence of single graphene sheets had been discussed in theory more than 50 years ago [114]. Yet, the existence of two-dimensional, atomically-thin crystal materials was considered physically impossible [115]. In 2004, a single sheet of graphene was isolated and characterized by Novoselov and Geim [116]. Since then, research on graphene has been increasing almost exponentially, attracting the interest of various scientific fields [117].

Interest in monoatomic graphene sheets also drew attention to other carbon-based materials that are now examined under a different light. The most notable of them is graphene oxide, *i.e.*, graphene sheets derivatized with oxygen-containing functional groups. By analogy to graphite and graphene, graphene oxide is the "building block" of graphite oxide. More particularly, it is the result of graphite's oxidation under acidic conditions, first described by Brodie *et al.* more than 150 years ago [118]. Today, the most popular method for the production of graphene oxide is based on the principle first introduced by Hummers and Offeman (commonly referred to as the Hummers method) that involves the oxidation of graphite by potassium permanganate in concentrated sulphuric acid [119].

Graphene is composed almost entirely of sp^2-hybridized carbon atoms and their electrons participate in aromatic conjugated domains. In graphite, van der Waals forces keep graphene sheets tightly together. Even in the case of single graphene sheets synthesized *de novo* (for example, through chemical vapour deposition of carbohydrates on a metal catalyst), the apolar nature of the carbonaceous material makes them highly hydrophobic [120]. On the other hand, the oxidation of graphite to graphite oxide loosens its firmly-packed graphene sheets. This occurs due to the random introduction of carbonyls, hydroxyls and epoxides on the planar surfaces and edges of the carbon sheets. Graphene oxide sheets can then be exfoliated from graphite oxide particles through ultra-sonication [121].

Graphene oxide has important potential applicability in drug delivery due to the presence of functional groups (epoxy, hydroxyl, carboxylic groups), conjugation systems, large surface area, low cytotoxic effects and low costs [122]. The presence of these functionalities decorating the basal planes and edges of GP layers significantly revises the van der Waals interactions between graphene sheets, thus conferring desirable dispersibility [123]. Furthermore, these functional groups, together with the high surface area and π-conjugated structure of GP, allow conjugation with a number of substances through both covalent and non-covalent modification techniques [124].

3.1. Graphene-Hydrogel Hybrids

Graphene can be functionalized via non-covalent binding, such as π–π stacking, cation-π and van der Waals interactions [125,126,127]. In contrast to covalent modifications, non-covalent ones are considered to have less impact on the conjugated structure and the mechanical and electrical properties of graphene.

Recently, a considerable number of works aimed to enhance the properties of graphene by chemical modification, obtaining new graphene derivatives [128,129,130]. The grafting of polymer chains onto graphene is proposed as an efficient way to improve the properties of graphene, such as its solubility [131], interfacial interactivity with target matrix [132] and electronic properties [133].

Atom transfer radical polymerization is one of the first used approaches to graft initiator molecules onto GO sheets and, subsequently, to synthesize *in situ* poly[2-(dimethylamino)ethyl methacrylate] chains on their surface [134]. As reported for CNT, this approach is unable to control the molecular weight and architecture of a resultant polymer, and the reaction conditions are also severely restricted by graphene solubility in a solvent [135]. Thus, also in the GP case, the "grafting to" approach seems to be a promising and simple way to functionalize GP with well-defined polymer chains, and several polymer-grafted GP nanocomposites have been synthesized using different coupling techniques [136,137,138,139].

3.2. Non-Covalent Graphene-Hydrogel Hybrids in Drug Delivery

A valuable approach for the preparation of hybrid hydrogels composed of GO and GO derivatives is the non-covalent inclusion into supramolecular structures, like cyclodextrins. This confers high elastic behaviour at elevated temperature to the resulting supramolecular hydrogels [140]. A key example of this concept was developed by a preliminary stabilization of GO in solution by the surfactant Pluronic F-127 and a subsequent inclusion into α-cyclodextrins. By this approach, hydrogels for potential systemic administration were prepared, with the key advantages of improving the ability of hydrogels to bind doxorubicin molecules and better controlling the release in comparison to a native hydrogel prepared in the absence of GO [141].

A similar approach involves the insertion of GO into composite hydrogels based on natural polysaccharides, such as konjac glucomannan and sodium alginate [142]. As expected, the GO nanosheets were found to significantly influence the micromorphology, swelling ability and drug loading of the composite hydrogel. Interestingly, *in vitro* release studies showed that the burst release phenomenon could be avoided and that an excellent pH-sensitive system could be achieved.

Efficient GO-hydrogel hybrids have been synthesized by the non-covalent coating of GO with cyclic RGD-modified chitosan. The nanohybrid was proposed as a drug delivery system for the targeted delivery of doxorubicin to hepatocellular carcinoma, with the GO counterpart conferring efficient loading properties, while the GO-chitosan hydrogen bonding interactions allow the selective drug release in the acidic tumour environment due to the pH-responsive behaviour. Furthermore, the nanocomposite is able to recognize hepatoma cells and to promote drug uptake by the cells, especially for those overexpressing integrins, as confirmed by cellular uptake and proliferation studies [143].

A similar non-covalent approach was employed for the preparation of a flexible, electrically-conducting hydrogel based on commercial superabsorbent polymer and hyperbranched polymer. Here, the incorporation of reduced graphene oxide (rGO) resulted in excellent electrical self-healing properties and water-absorption reusability [144].

Yi Wang and co-workers reported on the fabrication of flexible self-supporting hybrid films composed of GO intercalated with Mg–Al layered double hydroxide via a solvent evaporation process and proposed it as a tool for the delivery of benzylpenicillin [145]. The key finding of this study is the possibility to tailor the drug release by modification of the film composition. More interestingly, the results of antibacterial tests highlight that hybrid films can inhibit bacterial growth with key advantages from an application point of view.

Hybrid hydrogel membranes composed of reduced graphene oxide nanosheets and a poly(vinyl alcohol) matrix were proposed as an electrically-responsive drug release system for the anaesthetic drug lidocaine hydrochloride by Heng-Wen Liu *et al.* [146]. The presence of rGO in the nanohybrids was found to act as a physical barrier to inhibit the drug release, while the exposure to an electrical stimulus highly enhances the release.

3.3. Covalent Graphene-Hydrogel Hybrids in Drug Delivery

He *et al.* immobilized different kinds of functional moieties or polymers onto GP sheets via nitrene chemistry to obtain highly-engineered composite materials [147]. As a result, the functionalized graphene nanosheets, electrically conductive and dispersible in different solvents, remained in individually separated single layers, while the introduced functional groups can be further modified by differently chemical reactions for the fabrication of different kinds of graphene oxide-based hybrids.

Kumar Rana *et al.* prepared a covalent nanohybrid composed of chitosan and GO by condensation reaction [148]. Specifically, the carboxylic acid groups of GO were converted into acyl chlorides, and then, the conjugation to chitosan was achieved. The composite was proposed as a delivery device for ibuprofen and 5-fluorouracil with high efficiency due to its ability to penetrate into the cells. The releasing profile was found to be closely dependent on the ionization properties of the two drugs, with the release reaching a faster rate when the drugs are in the ionized form. Electrospinning methods were employed by Weaver *et al.* [149] for the preparation of GO nanohybrids able to act as drug delivery systems for the controlled release of the anti-inflammatory drug dexamethasone in response to electrical stimulation. The key finding of the study was the possibility to finely modulate the properties, such as the stability and dosing, of the GO nanocomposite release platform, which is proposed as a valuable tool for advanced drug delivery technologies. More in detail, polypyrrole was employed as the conducting element, while the GO nanosheets acted as nanocarriers, improving the amount of drug loaded into and released from the nanocomposite film. The authors stated that the system can be tuned according to various needs, making it a valuable tool for both therapeutic and research applications. The nanocomposite film, indeed, exhibited a linear release profile persisting over several hundred stimulations, and this is a clear indication that it could be used for long-term drug-release applications requiring repeated dosing over time.

The binding properties of the GO surface were explored for the preparation of a gene delivery vector [150]. In this approach, low-molecular weight branched polyethylenimine was employed as a cationic gene carrier, with GO acting as a fluorescence reagent probe. According to the authors' hypothesis, the excellent photoluminescence activities of the GO-based nanoconstruct will definitely merit further attention in the development of more sophisticated carrier systems that could serve both as a gene delivery vector and bioimaging tool.

GO can also act as a high loading effector for photosensitiser agents, as proven by Li *et al.* [151]. Here, a hyaluronic acid-GO conjugate was proposed as a cancer cell target and photoactivity switchable nanoplatform for photodynamic therapy. The conjugate was prepared through a multi-step synthetic strategy involving at first the chemical conjugation of modified hyaluronic acid and fractionated GO sheets (below 100 nm to enhance releasing efficiency). Subsequently, a photosensitiser was loaded into the GO sheet via π–π stacking and hydrophobic interactions. The photoactivity of the photosensitiser adsorbed on the nanocarriers was mostly quenched in aqueous solution to ensure biocompatibility, but was quickly recovered after the release from nanocarriers upon cellular uptake. The photodynamic therapy was remarkably improved, resulting in efficiencies 10-times higher than that of the free photosensitiser.

The loading properties of the GO surface via π–π stacking and hydrophobic interactions were also exploited for the preparation of pH-sensitive drug delivery devices with poly-*N*-vinyl caprolactam acting as the polymer counterpart conferring high solubility and stability in water and physiological solutions to GO [152]. GO was inserted via atom transfer radical polymerization, and the resulting composite was proposed for the intracellular delivery of the anticancer drug camptothecin with high releasing efficiency. The composite showed a high loading ratio, and the loaded drug is quickly released at a reduced pH (typical of the tumour micro-environment). In addition, the presence of poly-*N*-vinyl caprolactam allows a targeted and temperature-dependent delivery rate. The authors demonstrated that the pure nanocargo alone was almost non-toxic, whereas the nanohybrid showed strong potency against cancer cells, due to a cellular-uptake mechanism of energy-dependent endocytosis.

Thermo-sensitive releasing behaviour by GO nanohybrids was also reached by the covalent immobilization of thermo-responsive polymer nanoparticles on functionalized GO nanosheets. In the synthetic procedure, thermo-sensitive nanoparticles were first synthesized by free radical polymerization, and GO nanosheets were non-covalently modified with a bi-functional linker able to provide reactive sites for the subsequent binding to the polymer. Then, Adriamycin was loaded with high efficiency and the whole system tested in suitable mouse models with interesting results [153].

In a different approach, the Pickering emulsion polymerization method was proposed for the fabrication of core-shell-structured polystyrene microspheres containing GO units [154]. Here, GO shows the double activity of coating the polymer microspheres and imparting electro-responsive properties to the polymer composite suspension under an applied electric field.

An electro-conductive hydrogel system was prepared by incorporation of rGO into Jeffamine polyetheramine and polyethylene glycol diglycidyl ether by a one-step polymerization method [155]. The obtained hybrid hydrogels showed enhanced mechanical and electrical properties depending on rGO content and were investigated as a delivery device for methyl orange. A significant reduction in the passive release of the dye was observed by incorporating rGO, while upon electrical stimulation, the release rate and dosage could be tuned through variation of the percentage *w*/*w* of rGO, as well as of the polarity and amplitude of the applied electric potential.

ACKNOWLEDGMENTS

The work was financially supported by University of Calabria funding.

AUTHOR CONTRIBUTIONS

All of the authors generated the ideas, performed the literature search, and contributed to the organization, writing, and editing of this paper.

CONFLICTS OF INTEREST

The authors declare no conflict of interest.

REFERENCES

1. Qiu, Y.; Park, K. Environment-sensitive hydrogels for drug delivery. Adv. Drug Deliv. Rev. 2014, 64, 49–60.

2. Lee, S.C.; Kwon, I.K.; Park, K. Hydrogels for delivery of bioactive agents: A historical perspective. Adv. Drug Deliv. Rev. 2013, 65, 17–20.

3. Yucel, T.; Lovett, M.L.; Kaplan, D.L. Silk-based biomaterials for sustained drug delivery. J. Control. Release 2014, 190, 381–397.

4. Tanaka, T.; Nishio, I.; Sun, S.-T.; Ueno-Nishio, S. Collapse of gels in an electric field. Science 1982, 218, 467–469.

5. Tanaka, T.; Sun, S.-T.; Hirokawa, Y.; Katayama, S.; Kucera, J.; Hirose, Y.; Amiya, T. Mechanical instability of gels at the phase transition. Nature 1987, 325, 796–798.

6. Suzuki, A.; Tanaka, T. Phase transition in polymer gels induced by visible light. Nature 1990, 346, 345–347.

7. Sharpe, L.A.; Daily, A.M.; Horava, S.D.; Peppas, N.A. Therapeutic applications of hydrogels in oral drug delivery.Exp. Opin. Drug Deliv. 2014, 11, 901–915.

8. Liang, Y.; Kiick, K.L. Heparin-functionalized polymeric biomaterials in tissue engineering and drug delivery applications. Acta Biomater. 2014, 10, 1588–1600.

9. Pillay, V.; Tsai, T.-S.; Choonara, Y.E.; Du Toit, L.C.; Kumar, P.; Modi, G.; Naidoo, D.; Tomar, L.K.; Tyagi, C.; Ndesendo, V.M.K. A review of integrating electroactive polymers as responsive systems for specialized drug delivery applications. J. Biomed. Mater. Res. A 2014, 102, 2039–2054.

10. Bajpai, A.K.; Shukla, S.K.; Bhanu, S.; Kankane, S. Responsive polymers in controlled drug delivery. Prog. Polym. Sci.2008, 33, 1088–1118.

11. Glazer, P.J.; van Erp, M.; Embrechts, A.; Lemay, S.G.; Mendes, E. Role of pH gradients in the actuation of electro-responsive polyelectrolyte gels. Soft Matter 2012, 8, 4421–4426.

12. Murdan, S. Electro-responsive drug delivery from hydrogels. J. Control. Release 2003, 92, 1–17.

13. Jin, S.; Gu, J.; Shi, Y.; Shao, K.; Yu, X.; Yue, G. Preparation and electrical sensitive behavior of poly (N-vinylpyrrolidone-co-acrylic acid) hydrogel with flexible chain nature. Eur. Polym. J. 2013, 49, 1871–1880.

14. Tiitu, M.; Hiekkataipale, P.; Kainen, J.H.; Makela, T.; Ikkala, O. Viscoelastic and electrical transitions in gelation of electrically conducting polyaniline. Macromolecules 2002, 35, 5212–5217.

15. Chikar, J.A.; Hendricks, J.L.; Richardson-Burns, S.M.; Raphael, Y.; Pfingst, B.E.; Martin, D.C. The use of a dual PEDOT and RGD-functionalized alginate hydrogel coating to provide sustained drug delivery and improved cochlear implant function. Biomaterials 2010, 33, 1982–1990.

16. Indermun, S.; Choonara, Y.E.; Kumar, P.; Du Toit, L.C.; Modi, G.; Luttge, R.; Pillay, V. Patient-controlled analgesia: Therapeutic interventions using transdermal electro-activated and electro-modulated drug delivery. J. Pharm. Sci.2014, 103, 353–366.

17. Yan, F.; Zou, Y.; Wang, M.; Dai, L.; Zhou, X.; Chen, L. Synthesis and application of the fluorescent carbon dots. Prog. Chem. 2014, 26, 61–74.

18. Falcao, E.H.L.; Wudl, F. Carbon allotropes: Beyond graphite and diamond. J. Chem. Technol. Biotechnol. 2007, 82, 524–531.

19. Yoshino, K.; Yin, X.H.; Muro, K.; Kiyomatsu, S.; Morita, S.; Zakhidov, A.A.; Noguchi, T.; Ohnishi, T. Marked enhancement of photoconductivity and quenching of luminescence in poly(2,5-dialkoxy-p-phenylene vinylene) upon C60 doping. Jpn. J. Appl. Phys. 1993, 32, L357–L360.

20. Wang, C.; Guo, Z.-X.; Fu, S.; Wu, W.; Zhu, D. Polymers containing fullerene or carbon nanotube structures. Prog. Polym. Sci. 2004, 29, 1079–1141.

21. Ruoff, R.S.; Tse, D.S.; Malhotra, R.; Lorents, D.C. Solubility of C60 in a variety of solvents. J. Phys. Chem. 1993, 97, 3379–3383.

22. Dalton, A.B.; Collins, S.; Munoz, E.; Razal, J.M.; Ebron, V.H.; Ferraris, J.P.; Coleman, J.N.; Kim, B.G.; Baughman, R.H. Super-tough carbon-nanotube fibres. Nature 2003, 423, 703.

23. Ma, P.-C.; Siddiqui, N.A.; Marom, G.; Kim, J.-K. Dispersion and functionalization of carbon nanotubes for polymer-based nanocomposites: A review. Composites A 2010, 41, 1345–1367.

24. Iijima, S. Helical microtubules of graphitic carbon. Nature 1991, 354, 56–58.

25. Bethune, D.S.; Kiang, H.; De Vries, S.; Gorman, G.; Savoy, R.; Vazquez, J.; Beyers, R. Cobalt-catalysed growth of carbon nanotubes with single-atomic-layer walls. Nature 1993, 363, 605–607.

26. Dresselhaus, M.S.; Dresselhaus, G.; Saito, R. Physics of carbon nanotubes. Carbon 1995, 33, 883–891.

27. Zhang, W.; Zhang, Z.; Zhang, Y. The application of carbon nanotubes in target drug delivery systems for cancer therapies. Nanoscale Res. Lett. 2011, 6, 555.

28. Thostenson, E.T.; Ren, Z.F.; Chou, T.W. Advances in the science and technology of carbon nanotubes and their composites: A review. Compos. Sci. Technol. 2001, 61, 1899–1912.

29. Yakobson, B.I.; Avouris, P. Mechanical properties of carbon nanotubes. Top. Appl. Phys. 2001, 80, 287–327.

30. Huang, P.; Zhang, C.; Xu, C.; Bao, L.; Li, Z. Preparation and characterization of near-infrared region absorption enhancer carbon nanotubes hybrid materials. Nano Biomed. Eng. 2010, 2, 225–230.

31. Al-Haik, M.; Hussain, M.Y.; Garmestani, H. Adhesion energy in carbon nanotube-polyethylene composite: Effect of chirality. J. Appl. Phys. 2005, 97, 074306.

32. Ke, P.C.; Qiao, R. Carbon nanomaterials in biological systems. J. Phys. Condens. Matter 2007, 19, 373101.

33. Lv, Y.; Tan, T.; Svec, F. Molecular imprinting of proteins in polymers attached to the surface of nanomaterials for selective recognition of biomacromolecules. Biotechnol. Adv. 2013, 31, 1172–1186.

34. Qiu, W.; Xu, H.; Takalkar, S.; Gurung, A.S.; Liu, B.; Zheng, Y.; Guo, Z.; Baloda, M.; Baryeh, K.; Liu, G. Carbon nanotube-based lateral flow biosensor for sensitive and rapid detection of DNA sequence. Biosens. Bioelectron. 2015,64, 367–372.

35. Matlock-Colangelo, L.; Baeumner, A.J. Functionalized electrospun nanofibers as bioseparators in microfluidic systems. Lab Chip 2012, 12, 2612–2620.

36. Rather, J.A.; Pilehvar, S.; De Wael, K. Polycyclodextrin and carbon nanotubes as composite for tyrosinase immobilization and its superior electrocatalytic activity towards butylparaben an endocrine disruptor. J. Nanosci. Nanotechnol. 2015, 15, 3365–3372.

37. He, Z.; Corry, B.; Lu, X.; Zhou, J. A mechanical nanogate based on a carbon nanotube for reversible control of ion conduction. Nanoscale 2014, 6, 3686–3694.

38. Krieg, T.; Sydow, A.; Schröder, U.; Schrader, J.; Holtmann, D. Reactor concepts for bioelectrochemical syntheses and energy conversion. Trends Biotechnol. 2014, 32, 645–655.

39. Fabbro, A.; Prato, M.; Ballerini, L. Carbon nanotubes in neuroregeneration and repair. Adv. Drug Deliv. Rev. 2013, 65, 2034–2044.

40. Gupta, A.; Woods, M.D.; Illingworth, K.D.; Niemeier, R.; Schafer, I.; Cady, C.; Filip, P.; El-Amin, S.F. Single walled carbon nanotube composites for bone tissue engineering. J. Orthopaed. Res. 2013, 31, 1374–1381.

41. Abolmaali, S.S.; Tamaddon, A.M.; Dinarvand, R. A review of therapeutic challenges and achievements of methotrexate delivery systems for treatment of cancer and rheumatoid arthritis. Cancer Chemother. Pharmacol. 2013, 71, 1115–1130.

42. Delogu, L.G.; Magrini, A.; Bergamaschi, A.; Rosato, N.; Dawson, M.I.; Bottini, N.; Bottini, M. Conjugation of antisense oligonucleotides to PEGylated carbon Nanotubes enables efficient knockdown of PTPN22 in T lymphocytes.Bioconjug. Chem. 2009, 20, 427–431.

43. Wang, J.T.-W.; Cabana, L.; Bourgognon, M.; Kafa, H.; Protti, A.; Venner, K.; Shah, A.M.; Sosabowski, J.K.; Mather, S.J.; Roig, A.; et al. Magnetically decorated multiwalled carbon nanotubes as dual MRI and SPECT contrast agents. Adv. Funct. Mater. 2014, 24, 1880–1894.

44. Singh, R.K.; Patel, K.D.; Kim, J.-J.; Kim, T.-H.; Kim, J.-H.; Shin, U.S.; Lee, E.-J.; Knowles, J.C.; Kim, H.-W. Multifunctional hybrid nanocarrier: Magnetic CNTs ensheathed with mesoporous silica for drug delivery and imaging system. ACS Appl. Mater. Interfaces. 2014, 6, 2201–2208.

45. Vittorio, O.; Brandl, M.; Cirillo, G.; Spizzirri, U.G.; Picci, N.; Kavallaris, M.; Iemma, F.; Hampel, S. Novel functional cisplatin carrier based on carbon nanotubes-quercetin nanohybrid induces synergistic anticancer activity against neuroblastoma in vitro. RSC Adv. 2014, 4, 31378–31384.

46. Chakrabarti, M.; Kiseleva, R.; Vertegel, A.; Ray, S.K. Carbon nanomaterials for drug delivery and cancer therapy. J. Nanosci. Nanotechnol. 2015, 15, 5501–5511.

47. Zhang, J.; Li, X.; Huang, L. Non-viral nanocarriers for siRNA delivery in breast cancer. J. Control. Release 2014, 190, 440–450.

48. Castro Nava, A.; Cojoc, M.; Peitzsch, C.; Cirillo, G.; Kurth, I.; Fuessel, S.; Erdmann, K.; Kunhardt, D.; Vittorio, O.; Hampel, S.; et al. Development of novel radiochemotherapy approaches targeting prostate tumor progenitor cells using nanohybrids. Int. J. Cancer 2015.

49. Chen, Y.; Chen, H.; Shi, J. Inorganic nanoparticle-based drug codelivery nanosystems to overcome the multidrug resistance of cancer cells. Mol. Pharm. 2014, 11, 2495–2510.

50. Biju, V. Chemical modifications and bioconjugate reactions of nanomaterials for sensing, imaging, drug delivery and therapy. Chem. Soc. Rev. 2014, 43, 744–764.

51. Kaufmann, A.; Kunhardt, D.; Cirillo, G.; Hampel, S.; Schwenzer, B. Functionalized carbon nanotubes as transporters for antisense oligodeoxynucleotides. J. Mater. Chem. B 2014, 2, 7000–7008.

52. Cirillo, G.; Hampel, S.; Spizzirri, U.G.; Parisi, O.I.; Picci, N.; Iemma, F. Carbon nanotubes hybrid hydrogels in drug delivery: A perspective review. BioMed Res. Int. 2014, 2014, 825017.

53. Cirillo, G.; Hampel, S.; Klingeler, R.; Puoci, F.; Iemma, F.; Curcio, M.; Parisi, O.I.; Spizzirri, U.G.; Picci, N.; Leonhardt, A.; et al. Antioxidant multi-walled carbon nanotubes by free radical grafting of gallic acid: New materials for biomedical applications. J. Pharm. Pharmacol. 2011, 63, 179–188.

54. Sakamoto, Y.; Nakae, D.; Fukumori, N.; Tayama, K.; Maekawa, A.; Imai, K.; Hirose, A.; Nishimura, T.; Ohashi, N.; Ogata, A. Induction of mesothelioma by a single intrascrotal administration of multi-wall carbon nanotube in intact male Fischer 344 rats. J. Toxicol. Sci. 2009, 34, 65–76.

55. Tsukahara, T.; Matsuda, Y.; Haniu, H. The role of autophagy as a mechanism of toxicity induced by multi-walled carbon nanotubes in human lung cells. Int. J. Mol. Sci. 2014, 16, 40–48.

56. Sakamoto, Y.; Nakae, D.; Hagiwara, Y.; Satoh, K.; Ohashi, N.; Fukamachi, K.; Tsuda, H.; Hirose, A.; Nishimura, T.; Hino, O.; et al. Document Serum level of expressed in renal carcinoma (ERC)/mesothelin in rats with mesothelial proliferative lesions induced by multi-wall carbon nanotube (MWCNT). J. Toxicol. Sci. 2010, 35, 265–270.

57. Madl, A.K.; Plummer, L.E.; Carosino, C.; Pinkerton, K.E. Nanoparticles, lung injury, and the role of oxidant stress.Annu. Rev. Physiol. 2014, 76, 447–465.

58. Du, J.; Wang, S.; You, H.; Zhao, X. Understanding the toxicity of carbon nanotubes in the environment is crucial to the control of

nanomaterials in producing and processing and the assessment of health risk for human: A review. Environ. Toxicol. Pharmacol. 2013, 36, 451–462.

59. Hristozov, D.R.; Gottardo, S.; Critto, A.; Marcomini, A. Risk assessment of engineered nanomaterials: A review of available data and approaches from a regulatory perspective. Nanotoxicology 2012, 6, 880–898.

60. Boczkowski, K.; Lanone, S. Respiratory toxicities of nanomaterials—A focus on carbon nanotubes. Adv. Drug Deliv. Rev. 2012, 64, 1694–1699.

61. Poland, C.A.; Byrne, F.; Cho, W.-S.; Prina-Mello, A.; Murphy, F.A.; Davies, G.L.; Coey, J.M.D.; Gounko, Y.; Duffin, R.; Volkov, Y.; et al. Length-dependent pathogenic effects of nickel nanowires in the lungs and the peritoneal cavity.Nanotoxicology 2012, 6, 899–911.

62. Li, Y.; Zhang, Y.; Yan, B. Nanotoxicity Overview: Nano-Threat to Susceptible Populations. Int. J. Mol. Sci. 2014, 15, 3671–3697.

63. Luanpitpong, S.; Wang, L.; Rojanasakul, Y. The effects of carbon nanotubes on lung and dermal cellular behaviors.Nanomedicine 2014, 9, 895–912.

64. Roy, M.; Sonkar, S.K.; Tripathi, S.; Saxena, M.; Sarkar, S. Non-toxicity of water soluble multi-walled carbon nanotube on Escherichia-Coli colonies. J. Nanosci. Nanotechnol. 2012, 12, 1754–1759.

65. Yu, J.; Grossiord, N.; Koning, C.E.; Loos, J. Controlling the dispersion of multi-wall carbon nanotubes in aqueous surfactant solution. Carbon 2007, 45, 618–623.

66. Yurekli, K.; Mitchell, C.A.; Krishnamoorti, R. Small-angle neutron scattering from surfactant-assisted aqueous dispersions of carbon nanotubes. J. Am. Chem. Soc. 2004, 126, 9902–9903.

67. Hirsch, A.; Vostrowsky, O. Functionalization of carbon nanotubes. Top. Curr. Chem. 2005, 245, 193–237.

68. Tasis, D.; Tagmatarchis, N.; Bianco, A.; Prato, M. Chemistry of carbon nanotubes. Chem. Rev. 2006, 106, 1105–1136.

69. Grobert, N. Carbon nanotubes—Becoming clean. Mater. Today 2007, 10, 28–35.

70. Liu, J.Q.; Xiao, T.; Liao, K.; Wu, P. Interfacial design of carbon nanotube polymer composites: A hybrid system of noncovalent and covalent functionalizations. Nanotechnology 2007, 18, 165701.

71. Yang, B.X.; Pramoda, K.P.; Xu, G.Q.; Goh, S.H. Mechanical reinforcement of polyethylene using polyethylene-grafted multiwalled carbon nanotubes. Adv. Funct. Mater. 2007, 17, 2062–2069.

72. Gavrel, G.; Jousselme, B.; Filoramo, A.; Campidelli, S. Supramolecular chemistry of carbon nanotubes. Top. Curr. Chem. 2014, 348, 95–126.

73. Cirillo, G.; Vittorio, O.; Hampel, S.; Spizzirri, U.G.; Picci, N.; Iemma, F. Incorporation of carbon nanotubes into a gelatin-catechin conjugate: Innovative approach for the preparation of anticancer materials. Int. J. Pharm. 2013, 446, 176–182.

74. Spizzirri, U.G.; Hampel, S.; Cirillo, G.; Mauro, M.V.; Vittorio, O.; Cavalcanti, P.; Giraldi, C.; Curcio, M.; Picci, N.; Iemma, F. Functional Gelatin-Carbon Nanotubes Nanohybrids With Enhanced Antibacterial Activity. Int. J. Polym. Mater. 2015, 64, 439–447.

75. Kastrisianaki-Guyton, E.S.; Chen, L.; Rogers, S.E.; Cosgrove, T.; Van Duijneveldt, J.S. Adsorption of F127 onto Single-Walled Carbon Nanotubes Characterized Using Small-Angle Neutron Scattering. Langmuir 2015, 31, 3262–3268.

76. Tonelli, F.M.P.; Santos, A.K.; Gomes, K.N.; Lorençon, E.; Guatimosim, S.; Ladeira, L.O.; Resende, R.R. Carbon nanotube interaction with extracellular matrix proteins producing scaffolds for tissue engineering. Int. J. Nanomed.2012, 7, 4511–4529.

77. Yang, J.; Hu, J.; Wang, C.; Qin, Y.; Guo, Z. Fabrication and Characterization of Soluble Multi-Walled Carbon Nanotubes Reinforced P(MMA-co-EMA) Composites. Macromol. Mater. Eng. 2004, 289, 828–832.

78. Lillehei, P.T.; Kim, J.-W.; Gibbons, L.J.; Park, C. A quantitative assessment of carbon nanotube dispersion in polymer matrices. Nanotechnology 2009, 20, 325708.

79. Cadek, M.; Coleman, J.N.; Barron, V.; Hedicke, K.; Blau, W.J. Morphological and mechanical properties of carbon-nanotube-

reinforced semicrystalline and amorphous polymer composites. Appl. Phys. Lett. 2002, 81, 5123–5125.

80. Xu, X.J.; Thwe, M.M.; Shearwood, C.; Liao, K. Mechanical properties and interfacial characteristics of carbon-nanotube-reinforced epoxy thin films. Appl. Phys. Lett. 2002, 81, 2833–2835.

81. Xie, X.-L.; Mai, Y.-W.; Zhou, X.-P. Dispersion and alignment of carbon nanotubes in polymer matrix: A review. Mater. Sci. Eng. R 2005, 49, 89–112.

82. Spitalsky, Z.; Tasis, D.; Papagelis, K.; Galiotis, C. Carbon nanotube-polymer composites: Chemistry, processing, mechanical and electrical properties. Prog. Polym. Sci. 2010, 35, 357–401.

83. Cirillo, G.; Caruso, T.; Hampel, S.; Haase, D.; Puoci, F.; Ritschel, M.; Leonhardt, A.; Curcio, M.; Iemma, F.; Khavrus, V.;et al. Novel carbon nanotube composites by grafting reaction with water-compatible redox initiator system. Colloid Polym. Sci. 2013, 291, 699–708.

84. Cirillo, G.; Hampel, S.; Puoci, F.; Haase, D.; Ritschel, M.; Leonhardt, A.; Iemma, F.; Picci, N. Carbon nanotubes imprinted polymers hybrid materials for analytical applications. In Materials Science and Technology; Hutagalung, S.D., Ed.; InTech: Rijeka, Croatia, 2012; pp. 181–218.

85. Sakellariou, G.; Priftis, D.; Baskaran, D. Surface Initiated Polymerization from Carbon Nanotubes: Strategies and Perspectives. Chem. Soc. Rev. 2013, 42, 677–704.

86. Haggenmueller, R.; Commans, H.H.; Rinzler, A.G.; Fischer, J.E.; Winey, K.I. Aligned single-wall carbon nanotubes in composites by melt processing methods. Chem. Phys. Lett. 2000, 330, 219–225.

87. Du, F.; Fischer, J.E.; Winey, K.I. A coagulation method to prepare single-walled carbon nanotube/PMMA composites and their modulus, electrical conductivity, and thermal stability. J. Polym. Sci. B 2003, 41, 3333–3338.

88. Voge, C.M.; Johns, J.; Raghavan, M.; Morris, M.D.; Stegemann, J.P. Wrapping and dispersion of multiwalled carbon nanotubes improves electrical conductivity of protein–nanotube composite biomaterials. J. Biomed. Mater. Res. A 2013,101A, 231–238.

89. Haider, S.; Park, S.-Y.; Saeed, K.; Farmer, B.L. Swelling and electroresponsive characteristics of gelatin immobilized onto multi-walled carbon nanotubes. Sens. Actuators B 2007, 124, 517–528.

90. Li, H.; Qiang, W.D.; Lin, C.H.; Ling, L.B.; Zhen, L.G. A novel gelatin-carbon nanotubes hybrid hydrogel. Macromol. Biosci. 2003, 3, 720–724.

91. Ross-Murphy, S.B. Structure and rheology of gelatin gels: Recent progress. Polymer 1992, 33, 2622–2627.

92. Yang, X.J.; Zheng, P.J.; Cui, Z.D.; Zhao, N.Q.; Wang, Y.F.; Yao, K.D. Swelling behaviour and elastic properties of gelatin gels. Polym. Int. 1997, 44, 448–452.

93. Draye, J.P.; Delaey, B.; Van de Voorde, A.; Van Den Bulcke, A.; De Reu, B.; Schacht, E. In vitro and in vivobiocompatibility of dextran dialdehyde cross-linked gelatin hydrogel films. Biomaterials 1998, 19, 1677–1687.

94. Wan, Y.Z.; Wang, Y.L.; Yao, K.D.; Cheng, G.X. Carbon fiber-reinforced gelatin composites. II. Swelling behavior. J. Appl. Polym. Sci. 2000, 75, 994–998.

95. Huang, Y.; Zheng, Y.; Song, W.; Ma, Y.; Wu, J.; Fan, L. Poly(vinyl pyrrolidone) wrapped multi-walled carbon nanotube/poly(vinyl alcohol) composite hydrogels. Composites A 2011, 42, 1398–1405.

96. Misak, H.; Asmatulu, R.; Whitman, J.; Mall, S. High temperature cross-linking of carbon nanotube yarns using polyvinylpyrrolidone as a binding agent. J. Nanosci. Nanotechnol. 2015, 15, 2283–2288.

97. Wu, J.; Ren, Y.; Sun, J.; Feng, L. Carbon nanotube-coated macroporous poly(N-isopropyl-acrylamide) hydrogel and its electrosensitivity. ACS Appl. Mater. Interfaces 2013, 5, 3519–3523.

98. Yun, J.; Im, J.S.; Lee, Y.-S.; Kim, H.I. Electro-responsive transdermal drug delivery behavior of PVA/PAA/MWCNT nanofibers. Eur. Polym. J. 2011, 47, 1893–1902.

99. Meredith, M.T.; Giroud, F.; Minteer, S.D. Azine/hydrogel/nanotube composite-modified electrodes for NADH catalysis and enzyme immobilization. Electrochim. Acta 2012, 72, 207–214.

100. Kalisadan, V.; Sreekumar, S.; Parthasarathy, M. A green approach to electromechanical actuation based on lotus leaf-inspired nanocomposite hydrogels. Asian J. Sci. Res. 2014, 7, 262–272.

101. Im, J.S.; Bai, B.; Lee, Y.-S. The effect of carbon nanotubes on drug delivery in an electro-sensitive transdermal drug delivery system. Biomaterials 2010, 31, 1414–1419.

102. Xiao, Y.; He, L.; Che, J.J. An effective approach for the fabrication of reinforced composite hydrogel engineered with SWNTs, polypyrrole and PEGDA hydrogel. Mater. Chem. 2012, 22, 8076–8082.

103. He, L.; Lin, D.; Wang, Y.; Xiao, Y.; Che, J. Electroactive SWNT/PEGDA hybrid hydrogel coating for bio-electrode interface. Colloids Surf. B 2011, 87, 273–279.

104. Spizzirri, U.G.; Hampel, S.; Cirillo, G.; Nicoletta, F.P.; Hassan, A.; Vittorio, O.; Picci, N.; Iemma, F. Spherical gelatin/CNTs hybrid microgels as electro-responsive drug delivery systems. Int. J. Pharm. 2013, 448, 115–122.

105. Curcio, M.; Altimari, I.; Spizzirri, U.G.; Cirillo, G.; Vittorio, O.; Puoci, F.; Picci, N.; Iemma, F. Biodegradable gelatin-based nanospheres as pH-responsive drug delivery systems. J. Nanopart. Res. 2013, 15, 1581.

106. Curcio, M.; Puoci, F.; Spizzirri, U.G.; Iemma, F.; Cirillo, G.; Parisi, O.I.; Picci, N. Negative thermo-responsive microspheres based on hydrolyzed gelatin as drug delivery device. AAPS PharmSciTech 2010, 11, 652–662.

107. Puoci, F.; Hampel, S.; Parisi, O.I.; Hassan, A.; Cirillo, G.; Picci, N. Imprinted microspheres doped with carbon nanotubes as novel electroresponsive drug-delivery systems. J. Appl. Polym. Sci. 2013, 130, 829–834.

108. Kwon, I.C.; Bae, Y.H.; Kim, S.W. Heparin release from polymer complex. J. Control. Release 1994, 30, 155–159.

109. Servant, A.; Bussy, C.; Al-Jamal, K.; Kostarelos, K. Design, engineering and structural integrity of electro-responsive carbon nanotube-based hydrogels for pulsatile drug release. J. Mater. Chem. B 2013, 1, 4593–4600.

110. Servant, A.; Methven, L.; Williams, R.P.; Kostarelos, K. Electroresponsive polymer-carbon nanotube hydrogel hybrids for pulsatile drug delivery in vivo. Adv. Healthcare Mater. 2013, 2, 806–811.

111. Curcio, M.; Spizzirri, U.G.; Cirillo, G.; Vittorio, O.; Picci, N.; Nicoletta, F.P.; Iemma, F.; Hampel, S. On demand delivery of ionic drugs from electro-responsive CNT hybrid films. RSC Adv. 2015, 56, 44902–44911.

112. Choi, E.J.; Shin, J.; Khaleel, Z.H.; Cha, I.; Yun, S.-H.; Cho, S.-W.; Song, C. Synthesis of electroconductive hydrogel films by an electro-controlled click reaction and their application to drug delivery systems. Polym. Chem. 2015, 6, 4473–4478.

113. Shi, X.; Zheng, Y.; Wang, C.; Yue, L.; Qiao, K.; Wang, G.; Wang, L.; Quan, H. Dual stimulus responsive drug release under the interaction of pH value and pulsatile electric field for a bacterial cellulose/sodium alginate/multiwalled carbon nanotube hybrid hydrogel. RSC Adv. 2015, 5, 41820–41829.

114. Slonczewski, J.C.; Weiss, P.R. Band structure of graphite. Phys. Rev. 1958, 109, 272–279.

115. Venables, J.A.; Spiller, G.D.T.; Hanbucken, M. Nucleation and growth of thin films. Rep. Prog. Phys. 1984, 47, 399–459.

116. Novoselov, K.S.; Geim, A.K.; Morozov, S.V.; Jiang, D.; Zhang, Y.; Dubonos, S.V.; Grigorieva, I.V.; Firsov, A.A. Electric field in atomically thin carbon films. Science 2004, 306, 666–669.

117. Bitounis, D.; Ali-Boucetta, H.; Hong, B.-H.; Min, D.-H.; Kostarelos, K. Prospects and challenges of graphene in biomedical applications. Adv. Mater. 2013, 25, 2258–2268.

118. Brodie, B.C. On the atomic weight of graphite. Philos. Trans. R. 1859, 149, 249–259.

119. Shao, G.; Lu, Y.; Wu, F.; Yang, C.; Zeng, F.; Wu, Q. Graphene oxide: The mechanisms of oxidation and exfoliation. J. Mater. Sci. 2012, 47, 4400–4409.

120. Li, D.; Muller, M.B.; Gilje, S.; Kaner, R.B.; Wallace, G.G. Processable aqueous dispersions of graphene nanosheets. Nat. Nanotechnol. 2008, 3, 101–105.

121. Gao, H.; Duan, H. 2D and 3D graphene materials: Preparation and bioelectrochemical applications. Biosens. Bioelectron2015, 65, 404–419.

122. Acik, M.; Lee, G.; Mattevi, C.; Pirkle, A.; Wallace, R.M.; Chhowalla, M.; Cho, K.; Chabal, Y.J. The role of oxygen during thermal reduction of graphene oxide studied by infrared absorption spectroscopy. Phys. Chem. C 2011, 115, 19761–19781.

123. Altavilla, C.; Ciliberto, E. Inorganic nanoparticles: Synthesis, applications, and perspectives; CRC Press: Boca Raton, FL, USA, 2010.

124. Akhavan, O. The effect of heat treatment on formation of graphene thin films from graphene oxide nanosheets. Carbon2010, 48, 509–519.

125. Liu, J.Q.; Yang, W.R.; Tao, L.; Li, D.; Boyer, C.; Davis, T.P. Thermosensitive graphene nanocomposites formed using pyrene-terminal polymers made by RAFT polymerization. J. Polym. Sci. A 2010, 48, 425–433.

126. Li, X.; Wang, X.; Zhang, L.; Lee, S.; Dai, H. Chemically derived, ultrasmooth graphene nanoribbon semiconductors.Science 2008, 319, 1229–1232.

127. Liu, S.; Tian, J.Q.; Wang, L.; Li, H.L.; Zhang, Y.W.; Sun, X.P. Stable aqueous dispersion of graphene nanosheets: Noncovalent functionalization by a polymeric reducing agent and their subsequent decoration with Ag nanoparticles for enzymeless hydrogen peroxide detection. Macromolecules 2010, 43, 10078–10083.

128. Loh, K.P.; Bao, Q.; Ang, P.K.; Yang, J. The chemistry of graphene. J. Mater. Chem. 2010, 20, 2277–2289.

129. Dreyer, D.R.; Park, S.; Bielawski, C.W.; Ruoff, R.S. The chemistry of graphene oxide. Chem. Soc. Rev. 2010, 39, 228–240.

130. Lee, S.H.; Dreyer, D.R.; An, J.H.; Velamakanni, A.; Piner, R.D.; Park, S.; Zhu, Y.W.; Kim, S.O.; Bielawski, C.W.; Ruoff, R.S. Polymer brushes via controlled, surface-initiated atom transfer radical polymerization (ATRP) from graphene oxide. Macromol. Rapid Commun. 2010, 31, 281–288.

131. Veca, L.M.; Lu, F.S.; Meziani, M.J.; Cao, L.; Zhang, P.Y.; Qi, G.; Qu, L.W.; Shrestha, M.; Sun, Y.P. Polymer functionalization and solubilization of carbon nanosheets. Chem. Commun. 2009, 18, 2565–2567.

132. Fang, M.; Wang, M.K.G.; Lu, H.B.; Yang, Y.L.; Nutt, S. Covalent polymer functionalization of graphene nanosheets and mechanical properties of composites. J. Mater. Chem. 2009, 19, 7098–7105.

133. Elias, D.C.; Nair, R.R.; Mohiuddin, T.M.G.; Morozov, S.V.; Blake, P.; Halsall, M.P.; Ferrari, A.C.; Boukhvalov, D.W.; Katsnelson, M.I.; Geim, A.K.; et al. Control of graphene's properties by reversible hydrogenation: Evidence for graphene. Science 2009, 323, 610–613.

134. Yang, Y.; Wang, J.; Zhang, J.; Liu, J.; Yang, X.; Zhao, H. Exfoliated graphite oxide decorated by PDMAEMA chains and polymer particles. Langmuir 2009, 25, 11808–11814.

135. Paredes, J.I.; Villar-Rodil, S.; Martinez-Alonso, A.; Tascon, J.M.D. Graphene oxide dispersions in organic solvents.Langmuir 2008, 24, 10560–10564.

136. Makharza, S.; Cirillo, G.; Bachmatiuk, A.; Ibrahim, I.; Ioannides, N.; Trzebicka, B.; Hampel, S.; Rümmel, M.H. Graphene oxide-based drug delivery vehicles: Functionalization, characterization, and cytotoxicity evaluation. J. Nanopart. Res. 2013, 15, 2099.

137. Makharza, S.; Cirillo, G.; Bachmatiuk, A.; Vittorio, O.; Mendes, R.G.; Oswald, S.; Hampel, S.; Rummel, M.H. Size-dependent nanographene oxide as a platform for efficient carboplatin release. J. Mater. Chem. B. 2013, 1, 6107–6114.

138. Makharza, S.; Vittorio, O.; Cirillo, G.; Oswald, S.; Hinde, E.; Kavallaris, M.; Büchner, B.; Mertig, M.; Hampel, S. Graphene oxide—Gelatin nanohybrids as functional tools for enhanced carboplatin activity in neuroblastoma cells.Pharm. Res. 2015, 32, 2132–2143.

139. Servant, A.; Leon, V.; Jasim, D.; Methven, L.; Limousin, P.; Fernandez-Pacheco, E.V.; Prato, M.; Kostarelos, K. Graphene-based electroresponsive scaffolds as polymeric implants for on-demand drug delivery. Adv. Healthcare Mater. 2014, 3, 1334–1343.

140. Zo, H.J.; Lee, J.-S.; Song, K.-W.; Kim, M.; Lee, G.; Park, J.S. Incorporation of graphene oxide into cyclodextrin-dye

supramolecular hydrogel. J. Incl. Phenom. Macrocycl. Chem. 2014, 79, 357–363.

141. Hu, X.; Li, D.; Tan, H.; Pan, C.; Chen, X. Injectable graphene oxide/graphene composite supramolecular hydrogel for delivery of anti-cancer drugs. J. Macromol. Sci. Pure Appl. Chem. 2014, 51, 378–384.

142. Wang, J.; Liu, C.; Shuai, Y.; Cui, X.; Nie, L. Controlled release of anticancer drug using graphene oxide as a drug-binding effector in konjac glucomannan/sodium alginate hydrogels. Colloids Surf. B 2014, 113, 223–229.

143. Wang, C.; Chen, B.; Zou, M.; Cheng, G. Cyclic RGD-modified chitosan/graphene oxide polymers for drug delivery and cellular imaging. Colloids Surf. B. 2014, 122, 332–340.

144. Peng, R.; Yu, Y.; Chen, S.; Yang, Y.; Tang, Y. Conductive nanocomposite hydrogels with self-healing property. RSC Adv. 2014, 4, 35149–35155.

145. Wang, Y.; Zhang, D.; Bao, Q.; Wua, J.; Wan, Y. Controlled drug release characteristics and enhanced antibacterial effect of graphene oxide-drug intercalated layered double hydroxide hybrid films. J. Mater. Chem. 2012, 22, 23106–23113.

146. Liu, H.-W.; Hu, S.-H.; Chen, Y.-W.; Chen, S.-Y. Characterization and drug release behavior of highly responsive chip-like electrically modulated reduced graphene oxide-poly(vinyl alcohol) membranes. J. Mater. Chem. 2012, 22, 17311–17320.

147. He, H.K.; Gao, C. General approach to individually dispersed, highly soluble, and conductive graphene nanosheets functionalized by nitrene chemistry. Chem. Mater. 2010, 22, 5054–5064.

148. Rana, V.K.; Choi, M.-C.; Kong, J.-Y.; Kim, G.Y.; Kim, M.J.; Kim, S.-H.; Mishra, S.; Singh, R.P.; Ha, C.-S. Synthesis and drug-delivery behavior of chitosan-functionalized graphene oxide hybrid nanosheets. Macromol. Mater. Eng. 2011, 296, 131–140.

149. Weaver, C.L.; Larosa, J.M.; Luo, X.; Cui, X.T. Electrically controlled drug delivery from graphene oxide nanocomposite films. ACS Nano 2014, 8, 1834–1843.

150. Kim, H.; Namgung, R.; Singha, K.; Oh, I.-K.; Kim, W.J. Graphene oxide-polyethylenimine nanoconstruct as a gene delivery vector and bioimaging tool. Bioconjug. Chem. 2011, 22, 2558–2567.

151. Li, F.; Park, S.J.; Ling, D.; Park, W.; Han, J.Y.; Na, K.; Char, K. Hyaluronic acid-conjugated graphene oxide/photosensitizer nanohybrids for cancer targeted photodynamic therapy. J. Mater. Chem. B 2013, 1, 1678–1686.

152. Kavitha, T.; Kang, I.-K.; Park, S.-Y. Poly(N-vinyl caprolactam) grown on nanographene oxide as an effective nanocargo for drug delivery. Colloids Surf. B 2013, 115, 37–45.

153. Wang, H.; Sun, D.; Zhao, N.; Yang, X.; Shi, Y.; Li, H.; Su, Z.; Wei, G. Thermo-sensitive graphene oxide-polymer nanoparticle hybrids: Synthesis, characterization, biocompatibility and drug delivery. J. Mater. Chem. B 2014, 2, 1362–1370.

154. Kim, S.D.; Zhang, W.L.; Choi, H.J. Pickering emulsion-fabricated polystyrene-graphene oxide microspheres and their electrorheology. J. Mater. Chem. C 2014, 2, 7541–7546.

155. Mc Kenna, N.; Calvert, P.; Morrin, A.; Wallace, G.G.; Moulton, S.E. Electro-stimulated release from a reduced graphene oxide composite hydrogel. J. Mater. Chem. B 2015, 3, 2530–2537.

7

NANOPARTICLE-INTEGRATED HYDROGELS AS MULTIFUNCTIONAL COMPOSITE MATERIALS FOR BIOMEDICAL APPLICATIONS

Marco Biondi [1,†], Assunta Borzacchiello [2,†,*], Laura Mayol [1] and Luigi Ambrosio [2,3]

[1] Dipartimento di Farmacia, Università di Napoli Federico II, Via D. Montesano 49, 80131 Napoli, Italy

[2] Istituto per i Polimeri Compositi e Biomateriali (IPCB-CNR), P.le Tecchio 80, 80125 Napoli, Italy

[3] Dipartimento Scienze Chimiche e Tecnologie dei Materiali (DSCTM-CNR), P.le Aldo Moro 7, 00185 Roma, Italy

ABSTRACT

This review focuses on the most recent developments in the field of nanocomposite hydrogels intended for biomedical applications. Nanocomposite hydrogels are hydrated polymeric networks with a physically or covalently crosslinked three-dimensional (3D) structure swollen with water, in the presence of nanoparticles or nanostructures. A wide array of nanomaterials (polymeric, carbon-based, metallic, ceramic) can be incorporated within the hydrogel network to obtain reinforced nanocomposite hydrogels. Nanocomposites represent a new class of materials with properties absent in the individual components. In particular, the incorporation of nanomaterials within a polymeric hydrogel network is an attractive approach to tailor the mechanical properties of the hydrogels and/or to provide the nanocomposite with responsiveness to external stimuli.

KEYWORDS

nanocomposites; nanoparticles; hydrogels; biomedical applications; nanocomposite hydrogels

1. INTRODUCTION

Hydrogels represent a class of soft materials, of synthetic and/or natural origin, of particular interest for biomedical applications such as tissue engineering, regenerative medicine, and controlled drug delivery [1,2,3,4,5,6,7,8,9,10,11] thanks to their physical, chemical, and biological properties compatible with those of biological tissues [12,13,14,15,16]. Hydrogels are physically or chemically cross-linked natural or synthetic three-dimensional (3D) networks, which can be cast into various shapes and retain high amounts of water (up to 4000% of their dry weight), although they are hardly hydrosoluble. Their water retention properties mainly arise from the presence of hydrophilic groups, such as amido, amino, carboxyl, and hydroxyl, in the polymer chains; the swelling degree depends on the polymer composition and, in addition, the cross-link density and nature. The water content of hydrogels creates a highly porous structure, a soft and elastic consistency, and a low interfacial tension in contact with water or biological fluids. These features make hydrogel's properties closer to those of biological tissues than any other synthetic biomaterial. However, their application fields may be widened if a nanometric phase is embedded within the hydrogel's matrix [2]. Thus, recent advances in hydrogel technology have led to the development of nanocomposite hydrogels (NCHs), also named nanocomposites, for biomedical applications; currently there is significant and increasing research interest in their development (Table 1). NCHs are hydrated polymeric networks with a physically or covalently crosslinked 3D structure, swollen with water in the presence of nanoparticles or nanostructures, that can be covalently or non-covalently immobilized in the matrix (Figure 1).

Table 1. Number of publications per year dealing with nanocomposite hydrogels (NCHs). As can be seen, since 2005 the scientific community devoted increasing interest to NCHs in the biomedical field. PubMed was used as the biomedical bibliographic database.

Publication Date	2002	2003	2004	2005	2006	2007	2008	2009	2010	2011	2012	2013	2014	June 2015
Number of publications	1	1	2	1	7	11	18	21	27	29	47	75	58	36

Research trends are currently focused on the incorporation of many nanoparticulate systems such as carbon-based nanomaterials, polymeric nanoparticles, ceramic nanoparticles, and metal/metal-oxide nanoparticles into the hydrogel network so as to obtain NCHs [17]. The interaction of these nanosystems with the polymeric chains of the hydrogel structure results in the peculiar properties of the nanocomposite absent in the individual components [17,18,19]. Nanoparticle addition may reinforce the starting hydrogels and provide the NCHs with responsiveness to external stimuli such as mechanical, thermal, magnetic, and electric. In particular, the nature of the nanosystems incorporated in the starting hydrogel determines the kind of stimuli to which the NCH is responsive. Here we will discuss the NCH properties and applications, classifying them on the basis of the chemical nature of the embedded nanomaterials summarized in Figure 1.

2. CARBON-BASED NANOCOMPOSITE HYDROGELS

Carbon-based nanomaterials such as carbon nanotubes (CNTs) and graphene are being used to provide conventional hydrogels with improved mechanical and electrical properties [20]. In particular, both CNTs- or graphene-based NCHs (Figure 2) are being studied for applications such as actuators, biosensors, tissue engineering scaffolds, drug delivery, and biomedical devices [18,21].

CNTs exist in the multi-wall or single-wall forms. Generally, CNTs are hydrophobic because of their strong $\pi-\pi$ self-interactions, which in turn limits the interactions with the hydrogel network and favors the spontaneous formation of aggregates. These issues have been circumvented by surface modifications of CNTs by polar groups such as amines, hydroxyls, and carboxyls [22]. Due to their high electrical conductivity, NCHs reinforced with CNTs can be used to engineer electrically conductive tissues such as nerves, muscles, and cardiac tissues [23,24]. Graphene is a two-dimensional (2D) nanomaterial and is an excellent conductor of heat and electricity. Graphene solubility in physiological conditions and interaction with hydrophilic polymers can be ameliorated by its transformation into graphene oxide (GO) in an acidic milieu [25].

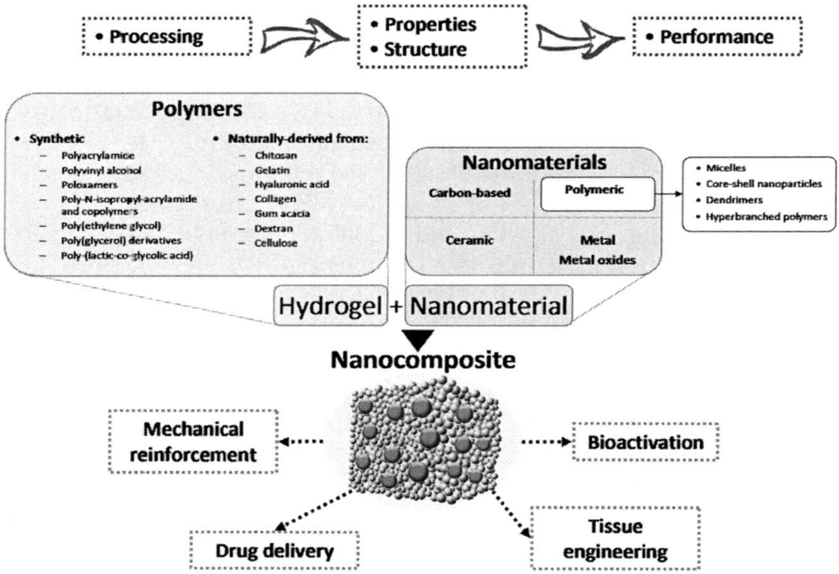

Figure 1. Processing parameters influence the structure and the properties of the NCH and, as a consequence, their performance in the medical field. Different polymers, both synthetic and naturally derived, can be used to produce the hydrogel, in which a variety of nanomaterials can be embedded. The resulting NCH can be used for a number of biomedical applications.

CNTs and GO, in addition to being physically dispersed in hydrogels, can be covalently conjugated to hydrogel polymer chains, thus promoting the transfer of the mechanical strength through the cross-linked network. For example, it has been estimated that the mechanical strength of a polymer matrix increases by an order of magnitude by introducing <1% of chemical bonds between CNTs and matrix [26].

Actually, the mechanical properties of a biomaterial are crucial and need to be properly tailored on the basis of the specific biological tissue to be replaced/repaired. Moreover, a scaffold for tissue engineering should be able to mimic the biological milieu of the extracellular matrix (ECM) so as to guide cell differentiation in a manner dependent on their stiffness. In this context, hybrid NCHs based on reinforced methacrylated gelatin (GelMA) have been recently produced as analogues of the ECM [27]. Multiwall COOH-functionalized CNTs were coated with GelMA by using the hydrophobic interactions between the polypeptide chains of GelMA and the sidewalls of the nanotubes. Moreover, it was also possible to photopattern the CNT-GelMA hybrid hydrogel and controllable dimensions and shapes could be attained. It was also shown that a CNT amount as low as 0.5% led to a threefold increase in the hydrogel tensile modulus [28]. Besides, CNT-GelMA hydrogels allowed a strong alignment of cardiomyocytes and the formation of tight intercellular junctions. It is interesting to note that, in the presence of CNTs, the external voltage needed to induce the cell beating was significantly reduced, thanks to the increase of electrical conductivity caused by CNTs. In more detail, a more stable spontaneous beating behavior was observed when cardiac tissues were cultured on CNT-GelMA and, furthermore, the beating rate was on average three folds higher than those measured from tissues cultured on CNT-free GelMA (~70 *vs.*23 beats per minute on Day 6).

In general, by functionalizating CNTs with polymers, it is possible to promote the dispersion of carbon-based nanosystems and their physical interactions with the surrounding hydrogel [29].

Carbon-based NCHs have also been exploited for controlled drug delivery applications. In this context, CNTs and GO were used to reinforce hydrogels made up of natural/synthetic materials such as carboxymethyl guar gum, poly(acrylamide) (PAAm), and poly(vinyl alcohol) (PVA). Those hydrogels, despite their recognized biocompatibility, biodegradability, and biological recognition, lack the mechanical strength to control the release rate of loaded drug(s). When required, carbon-based nanomaterials are also added to confer an electrically triggered drug release. As an example, multiwalled CNTs have been proposed to reinforce hybrid hydrogels based on carboxymethyl guar gum for the transdermal delivery of diclofenac sodium [30,31]. Drug release was found to be slower with an increasing CNT content within the NCH. Nonetheless, the low permeability of the skin restricts the utility of this approach and major research is devoted to develop methodologies to increase the delivery of drugs across this barrier.

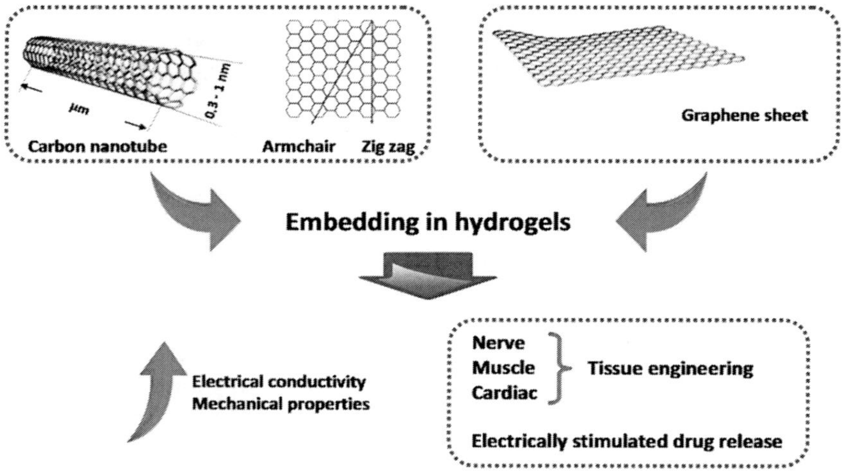

Figure 2. Nanocomposite hydrogels (NCHs) from carbon-based nanomaterials such as carbon nanotubes (CNTs) and graphene. CNTs exist in different atomic configurations (namely armchair and zig-zag) and architectures (single- and multi-walled) and can be chemically modified to enhance their hydrophilicity and, therefore, their interaction with the surrounding hydrogel. The addition/conjugation of CNTs and graphene derivatives provides NCHs with improved mechanical properties and electrical conductivity. For these reasons, NCHs embedding carbon-based nanomaterials can be potentially used for numerous applications, such as tissue engineering of electrically conductive tissues along with electrically-stimulated drug delivery.

In the context of electrically triggered drug delivery, PVA-based NCHs containing reduced graphene oxide (rGO) for the release of lidocaine, a hydrophilic drug, have been explored. The delivery of the drug is based on the migration of the electrically charged drug toward the oppositely charged electrode. Higher rGO content was correlated to a more negative charge of the rGO-PVA polymeric network and to a faster release rate of lidocaine [32].

In another recent study, GO sheets were functionalized by peroxidation induced by radiation, to obtain graphene peroxide (GPO). The resulting GPO was conjugated to PAAm gel to produce NCHs. The chemical conjugation of GPO caused an increase in tensile strength and elongation increased by 900% and 500% after the addition of 3 mg/mL of GPO, compared to conventional hydrogels [33]. By this approach it is in principle possible to obtain tissues able to withstand prolonged mechanical stresses.

In summary, the addition/conjugation of CNTs and graphene derivatives to hydrogel matrices allows us to obtain NCHs with improved mechanical and electrical properties. However, the actual usefulness of carbon-based nanomaterials as tissue substitutes, scaffolds for tissue engineering, or matrices for controlled drug delivery is yet to be ascertained because of concerns about their *in vivo* biocompatibility, which has not yet been fully assessed. Indeed, various studies have been performed on carbon-based materials with many cell lines [34,35,36,37], and contradictory results have been obtained. For example, GO could support the proliferation and adhesion of kidney cells, osteoblasts,, and human embryonic stem cells [38,39]; on the contrary, other studies revealed a concentration-dependent cytotoxic activity of GO against fibroblasts [40], and a genotoxic effect against mesenchymal stem cells (hMSCs) [41]. These puzzling findings have been partially explained considering that the oxygen groups, preparation methods of GO, size, charge, and the structural defects of graphene affect the *in vivo* and *in vitro* interactions along with the overall biocompatibility [25].

3. POLYMERIC NANOPARTICLE-BASED NANOCOMPOSITE HYDROGELS

Among the manifold nanomaterials that can be incorporated into hydrogels, a variety of polymeric nanoparticles have also been used, mainly aiming to endow the final NCH with controlled drug release ability, along with the mechanical reinforcement [42]. Actually, the concept of polymeric nanoparticles [43] includes a number of other different systems such as micelles [44], core-shell particles [45], dendrimers [46], and hyper-branched polymers [47] (Figure 3).

The very nature of conventional hydrogels hampers the loading of hydrophobic drugs and, in this regard, NCHs, incorporating polymeric nanoparticles, can help to overcome this issue. For example, a micelle-based, NCH-encapsulating erythromycin, a hydrophobic antibiotic drug, has been recently developed by Liu *et al.* [48]. The drug has been encapsulated in Pluronic F-127 diacrylate macromer micelles and the hydrogel has been obtained by photopolymerization under a low-intensity UV light. Pluronics are triblock copolymers made up of poly(ethylene oxide)–poly(propylene oxide)–poly(ethylene oxide) (PEO–PPO–PEO), displaying amphiphilic properties due to the presence of hydrophilic EO and hydrophobic PO segments on polymer backbone and therefore widely used as solubility enhancers in the drug delivery field [49]. The produced NCH allowed for efficient loading and a sustained release of the drug. In a similar vein, many chemotherapeutic drugs are sparingly water soluble, and this is related to their unwanted noxious side effects. For instance, paclitaxel has been loaded within a thermosensitive micelle-hydrogel hybrid system based on Pluronic F-127 and carboxymethyl chitosan to reduce the dangerousness of the drug and to improve its water solubility. The system, intended for local chemotherapy, was cross-linked with glutaraldehyde and *in vivo* studies revealed a decrease in the tumor progression rate and a reduction in side effects compared with free paclitaxel [50].

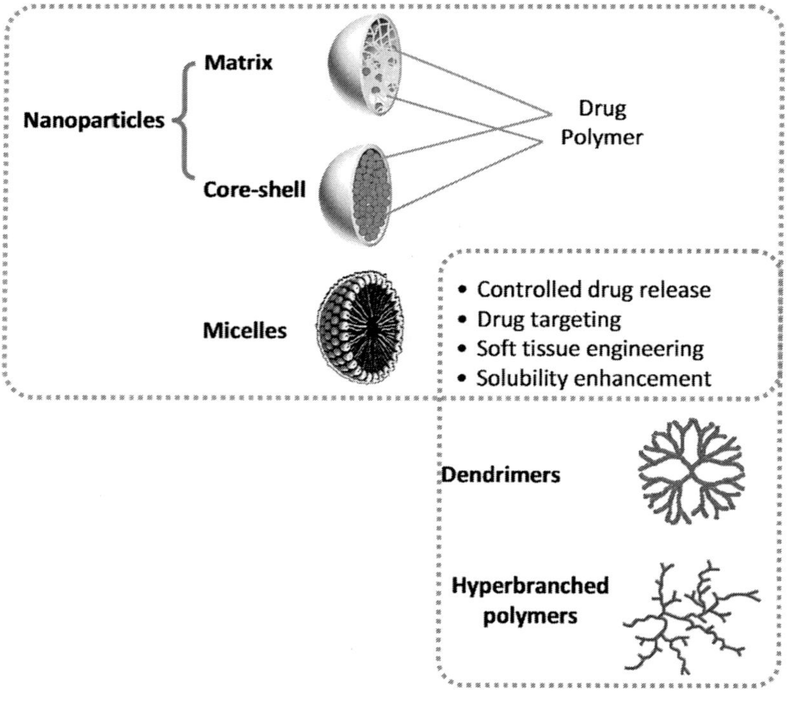

Figure 3. Nanocomposite hydrogels based on polymeric nanoparticles. Particle inclusion in hydrogels allows an increase of hydrogel mechanical properties, along with the possibility to control drug release rate. Amphiphilic (macro)molecules such as micelles, dendrimers, and hyperbranched polymers can also act as solubility enhancers of sparingly soluble drugs.

Additionally, dendritic polymers can encapsulate drug molecules into their interior cavities or form polymer–drug complexes/conjugates through a host–guest chemistry and, furthermore, they can be used to reinforce the hydrogel network by covalent or non-covalent interactions [51]. Among dendrimers, poly(amidoamine) (PAMAM), made of repetitively branched subunits of amide and amine, exhibit high biocompatibility probably because they resemble the chemical structure of globular proteins. In a recent study PAMAM nanoparticles have been physically integrated into collagen scaffolds leading to a significant increase of scaffold mechanical properties and to a subsequently enhanced proliferation of human conjunctival fibroblasts in the hydrogels [52]. In another example, hyper-branched polyesters with a globe-like nanostructure have been employed to produce NCHs by UV photopolymerization. They have been shown to possess a controlled porosity and could encapsulate dexamethasone acetate, a lipophilic drug, with high efficiency due to their globe-like amphiphilic nanostructure. The resulting NCHs allowed a one-week release of the drug, which cannot be attained with a conventional control hydrogel [53]. These NCHs can be applied as systems that require porous structures with controlled drug delivery properties, such as scaffolds for tissue engineering [54]. In this context, a hydrogel scaffold made from a tri-block copolymer consisting of a poly(ethylene glycol) (PEG) core and methacrylated poly(glycerol succinic acid) dendrimer terminal block has been evaluated for soft tissue engineering applications. Depending on dendrimer concentration, hydrogel stiffness and hydration/degradation features could be tailored. In particular, the obtained NCH displayed a transition from primarily elastic behavior (loss angle $\delta \sim 6°$) at the lowest macromer concentration to a more viscoelastic behavior (loss angle $\delta \sim 71°$) at higher macromer concentrations and, correspondingly, the compressive modulus increased from ~4 KPa to ~34 KPa. Furthermore, the obtained NCH proved suitable for cartilage repair since encapsulated chondrocytes were able to synthesize neo-cartilaginous material containing proteoglycans and type II collagen [55]. In a recent work, hyperbranched polyglycerols (HBP) have been crosslinked by a one-step gelation via biomimicking mineralization, without organic solvents or catalysts [56]. The resulting NCH formed a 3D mesoporous network, and their mechanical properties could be easily tailored by regulating the amount of the precursor. HBP solutions possess a very low viscosity and, therefore, HBP-based NCHs can also be formulated at high concentrations (tens of percent). These nanosystems are optically transparent in both the wet and the dried state and hold great promise for applications as optical devices.

4. CERAMIC NANOPARTICLE-BASED NANOCOMPOSITE HYDROGELS

Several advanced NCHs can be obtained by combining inorganic ceramic nanoparticles with natural or synthetic polymeric hydrogels. A wide range of bioactive nanoparticles, such as hydroxyapatite (HAP), synthetic silicate nanoparticles, bioactive glasses, silica, calcium phosphate, glass ceramic, and b-wollastonite, can be used to this aim (Figure 4) [57]. Ceramic nanoparticles can reinforce the hydrogel, taking advantage of their high mechanical strength; furthermore, because they are made of minerals with a crucial role in the normal homeostasis and turnover of human tissues, they can also provide the final NCH with favorable biological cues [58]. Both these features are compelling reasons for their use in the field of tissue engineering and regenerative medicine. As a matter of fact, these properties of ceramic nanoparticle-based NCHs are cardinal to fulfilling the conflicting requirements of hard tissue engineering and regenerative medicine applications. Indeed, widely used scaffolding materials of natural origin such as polysaccharides (e.g., chitosan, CHI, or hyaluronic acid, HA), despite their attractive features such as biodegradability/ biocompatibility, low toxicity, and low manufacture/disposal costs, generally lack mechanical and chemical stability and, therefore, their use as scaffolding materials as such is often compromised [59]. On the other hand, synthetic hydrogel materials such as PEG are bio-inert and, as such, are unable to impart an optimal milieu to endorse cell adhesion and tissue development; therefore, some bioactivation strategy must be taken in this sense [60]. In both cases, the addition of ceramic nanoparticles to obtain NCHs can be helpful to fulfill these needs.

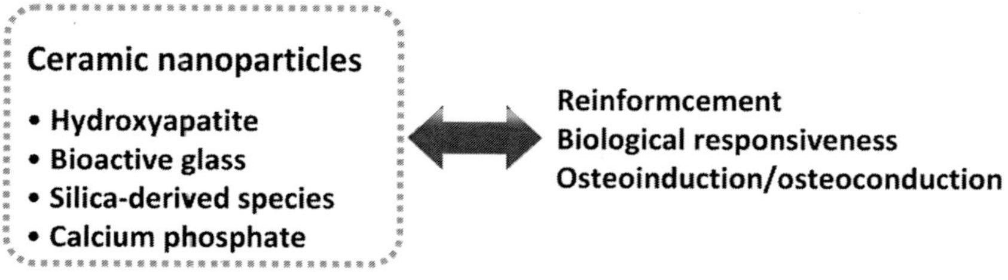

Figure 4. Nanocomposite hydrogels from ceramic nanoparticles. The inclusion of these nanomaterials allows a surprising reinforcement of the hydrogel. Furthermore, since most of these inorganic nanoparticles are made of minerals with a crucial role in the normal homeostasis of living tissues, they can provide the NCH with a biological responsiveness.

HA is a naturally occurring linear polysaccharide and a primary constituent of ECM of human connective tissues [61,62]. HA is involved in diverse *in vivo* functions such as arthritis joint lubrication and control of soft tissues' viscoelastic properties, and also in important cell functions such as cell motility and adhesion to the cell matrix [63,64]. To improve the mechanical performance of HA scaffolds, HA-based hydrogels have recently been reinforced with calcium and silica nanoparticles [65,66]. Bisphosphonate-functionalized HA hydrogels have been reinforced by reversible bonds with calcium phosphate nanoparticles [65]. The obtained NCHs displayed the capacity for self-healing as well as adhesiveness properties to mineral surfaces such as enamel and hydroxyapatite. Most importantly, these non-covalently cross-linked NCHs are surprisingly robust yet biodegradable upon extensive *in vitro* and *in vivo* testing and show bone interactive capacity evidenced by bone ingrowth into material remains. The second example of nanohybrid hydrogels consists of a cross-linked HA matrix including different amounts of silica-derived species. This inorganic filler phase controls the mechanical and swelling properties of cross-linked HA hydrogel, therefore making this NCH suitable for tissue engineering application, in which scaffold properties need to be modulated according to the specific tissue to be replaced.

Another natural polymer widely used in the biomedical field is CHI, because of its excellent biocompatibility, biodegradability, non-toxicity, and wound healing properties [67]. Aranaz and co-workers have recently reported a novel process method for the formation of NCH, namely ice segregation induced self-assembly (ISISA), based on urease-assisted hydrolysis of urea to produce CHI gels with a homogeneous and tailored 3D network structure. This process, applied to CHI solutions containing calcium phosphate salts, promoted the precipitation of nanoparticulate amorphous calcium phosphate and the gelation of CHI under mild conditions. The resulting NCH, with controlled properties, is suggested for its use as tissue engineering substrate [68]. In another report, calcium phosphate salts (CPS) and bone morphogenetic protein 2 (BMP2) have been immobilized, combined or alone, into CHI scaffolds by ISISA process. The obtained CHI-based NCH possessed the controlled release properties of BMP2, with a preserved osteoinductivity of the protein; furthermore, the NCH displayed interesting osteoconductivity features [69].

Among synthetic materials, PEG is also widely employed in tissue engineering thanks to its recognized biocompatibility. In a recent application, HAP was incorporated within a PEG matrix to confer elastomeric properties to the NCHs [70]. The addition of HAP to the polymeric network imparted elastomeric properties, enhanced mechanical strength, and improved the physiological stability of the NCH networks. For a starting 15% PEG hydrogel, the addition of HAP from 0% to 5% tripled the fracture stress and toughness of the obtained NCH, and its ultimate strain increased by 20%. Moreover, the addition of HAP resulted in enhanced osteoblast cell adhesion characteristics when compared with hydrogels made of PEG alone. Similar results were obtained when HAP was replaced with silica nanoparticles [71]. Due to their enhanced bioactivity and higher mechanical strength, these NCH networks (PEGHAP and PEG-Silica) can be used as injectable fillers for orthopedic applications [72].

5. METAL- AND METAL OXIDE-BASED NANOCOMPOSITE HYDROGELS

NCHs based on metal and metal oxide nanoparticles represent a promising new class of biomaterials since they possess several intriguing properties such as antimicrobial activity and responsiveness to electrical/magnetic stimuli and/or to light. Metallic nanoparticles mainly include noble metals such as platinum (Pt), gold (Au), and silver (Ag), or other metals such as cobalt (Co) and nickel (Ni), whereas metal oxide nanoparticles include iron oxide (Fe_3O_4, Fe_2O_3), titania (TiO_2), alumina (Al_2O_3), and zirconia (ZrO_2) [73].

Metallic nanoparticles can provide NCHs with antimicrobial activity since they are able to bind non-specifically to bacterial membranes, therefore inducing structural changes to bacteria, which allows for increased membrane permeability [74]. Moreover, metal/metal oxide nanoparticles stand out for their ferromagnetic and conducting/semi-conducting properties, thus providing the NCHs with electrical and magnetic properties that can be suitable for biomedical applications such as tissue regeneration.

In the context of antimicrobial materials, the intensified interest in NCHs with silver nanoparticles is due to their high antimicrobial effect, as recently reviewed by Dallas *et al.* [75]. Ag-based NCHs have been used as the functional coating in dental filling applications and also in wound and burn dressing to prevent infections. Different naturally occurring materials such as CHI, gum acacia, dextran, and gelatin, or synthetic materials such as PAAm, poly(acrylic acid) (PAA), *N*-isopropylacrylamide (NIPAAm), methyl methacrylate (MMA), and PVA have been used to incorporate metallic nanoparticles within the hydrogel matrix, thus obtaining NCHs with antimicrobial properties that are basically not harmful to healthy cells [76,77,78,79,80,81,82] Hydrogel-based substrates able to respond to magnetic and electrical fields are fundamental for the formation of tissues and in particular for those that require the propagation of electrical signals, such as nerves and muscles. For example, the incorporation of Au nanowires within alginate hydrogels has allowed us to improve the electrical conductivity between adjacent cardiac cells. Tissues grown on these composite matrices were thicker and better aligned than those grown on pristine alginate and, when electrically stimulated, the cells in these tissues synchronously contracted. The NCHs have shown an ability to be engineered as cardiac patches for treating damaged heart tissue after a heart attack [83]. Moreover, NCHs incorporating metal oxide nanoparticles can also enhance the bioactivity of hydrogels. In particular, nanoparticles of HAP and titania, entrapped within a polymeric matrix based on poly(L-lactic-*co*-glycolic acid) (PLGA), can enhance osteoblast adhesion and proliferation [84].

Besides using the metallic or metal-oxide based nanocomposite hydrogels for the abovementioned applications, they are also explored for biosensing, diagnostic and bioactuation applications, and stimuli-responsive controlled drug release [73]. In the last field, Gaharwar *et al.* demonstrated that an NCH based on magnetic nanoparticles, covalently conjugated with thermoresponsive hydroxypropyl cellulose (HPC), can remotely interact with external magnetic fields in order to obtain stimuli-responsive hydrogels. In particular, when an external magnetic field was applied, the nanoparticles incorporated within the hydrogel network generate heat, thus resulting in coil-to-globule transition of the polymer chains and in the release of therapeutic agents from the nanocomposite hydrogel [85]. The response of nanoparticles to electromagnetic stimuli has also been applied to the preparation of light responsive hydrogels, which can be useful for drug delivery purposes and biotissue repair. A photoresponsive hybrid hydrogel loaded with core–shell, lanthanide-doped upconverting nanoparticles (UCNPs) has been used to convert near-infrared (NIR) light into UV light. When the hydrogel is irradiated with 980 nm light, photodegradation and subsequent release of the embedded therapeutics occur (Figure 5) [85]. The stimulus of light is also very interesting because it is instantaneous and can be delivered rapidly, easily, and with high efficiency. A light-responsive material can be used for cell instruction and drug delivery purposes. Porous CHI substrates containing gold nanorods, which can absorb incoming laser light, have been tested as laser-activatable adhesives. Under a near-infrared light, nanoparticles' dispersion in CHI enables the activation of the polar groups of CHI strands, therefore enhancing tissue adhesion [87]. In a more recent example, similar NCHs based on CHI-containing gold nanorods were integrated with PCL–PEO–PCL micelles containing the drug to be released for on-demand release of a widely used antitumor drug, such as doxorubicin [88].

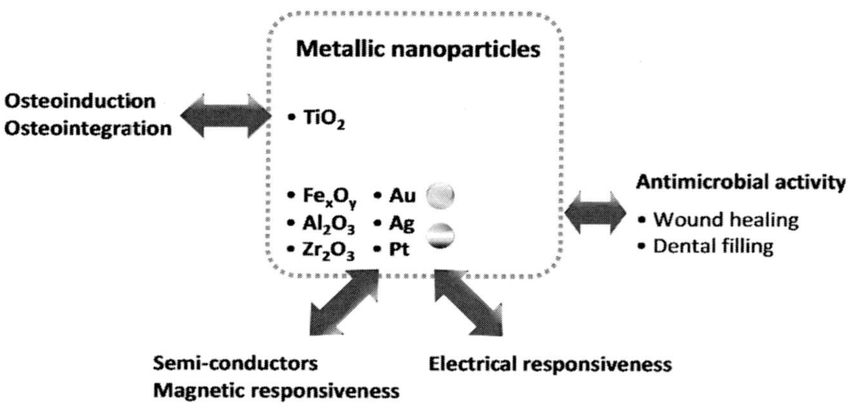

Figure 5. Nanocomposite hydrogels from metallic nanoparticles. The inclusion of metallic nanoparticles within hydrogels also allows us to obtain NCHs with electrical/magnetic responsiveness. In addition, noble metals such as silver also possess antimicrobial activity, and are therefore interesting for wound dressing.

6. FUTURE PERSPECTIVES

The incorporation of nanomaterials within polymeric hydrogels represents an attractive approach to tailor the mechanical properties of the hydrogels and/or to provide the NCH with responsiveness to mechanical, thermal, magnetic, and electric stimuli. Further studies have to be carried out to better understand the interactions, at different length scales, between the polymeric chains of the hydrogels and the nanophase and, in the case of drug delivery systems, between the interior part of the nanoparticles and the drug(s) loaded into them. In this context, the understanding of the relationship between structure and properties, at all scales from nano- to macro-, is crucial and will allow for custom design of NCHs' physico-chemical and electrical properties, so as to tailor them for specific applications.

Furthermore, the integration of suitable biological cues within the NCHs may provide them with biological features, thus leading to an increasingly detailed design of the biomaterial to be used in the field of cell/drug delivery and tissue engineering. Additionally, the proper combination of multiple phases within a NCH network could allow us to better mimic the structure and properties of native tissues and design increasingly sophisticated, stimuli-responsive NCHs. All these strategies may direct the development of the next generation of NCHs.

ACKNOWLEDGMENTS

The authors acknowledge the Italian Ministry of University and Research for providing financial support (Project: MIUR_PON Diateme).

AUTHOR CONTRIBUTIONS

All the authors contributed to literature survey, review organization and proof reading.

CONFLICTS OF INTEREST

The authors declare no conflict of interest.

REFERENCES

1. Biondi, M.; Ungaro, F.; Quaglia, F.; Netti, P.A. Controlled drug delivery in tissue engineering. Adv. Drug Deliv. Rev.2008, 60, 229–242.

2. Vashist, A.; Gupta, Y.K.; Ahmad, S. Recent advances in hydrogel based drug delivery systems for the human body. J. Mater. Chem. B 2014, 2, 147–166.

3. Biondi, M.; Indolfi, L.; Ungaro, F.; Quaglia, F.; La Rotonda, M.I.; Netti, P.A. Bioactivated collagen-based scaffolds embedding protein-releasing biodegradable microspheres: tuning of protein release kinetics. J. Mater. Sci. Mater. Med.2009, 20, 2117–2128.

4. Mollica, F.; Biondi, M.; Muzzi, S.; Ungaro, F.; Quaglia, F.; La Rotonda, M.I.; Netti, P.A. Mathematical modelling of the evolution of protein distribution within single PLGA microspheres: prediction of local concentration profiles and release kinetics. J. Mater. Sci. Mater. Med. 2008, 19, 1587–1593.

5. Ungaro, F.; Biondi, M.; d'Angelo, I.; Indolfi, L.; Quaglia, F.; Netti, P.A.; La Rotonda, M.I. Microsphere-integrated collagen scaffolds for tissue engineering: effect of microsphere formulation and scaffold properties on protein release kinetics. J. Control. Release 2006, 113, 128–136.

6. Mayol, L.; Biondi, M.; Russo, L.; Malle, B.M.; Schwach-Abdellaoui, K.; Borzacchiello, A. Amphiphilic hyaluronic acid derivatives toward the design of micelles for the sustained delivery of hydrophobic drugs. Carbohydr. Polym. 2014,102, 110–116.

7. Mayol, L.; Biondi, M.; Quaglia, F.; Fusco, S.; Borzacchiello, A.; Ambrosio, L.; La Rotonda, M.I. Injectable thermally responsive mucoadhesive gel for sustained protein delivery. Biomacromolecules 2011, 12, 28–33.

8. Mayol, L.; Quaglia, F.; Borzacchiello, A.; Ambrosio, L.; La Rotonda, M.I. A novel poloxamers/hyaluronic acid in situforming hydrogel for drug delivery: Rheological, mucoadhesive and in vitro release properties. Eur. J. Pharm. Biopharm. 2008, 70, 199–206.

9. Borzacchiello, A.; Mayol, L.; Ramires, P.A.; Pastorello, A.; Di, B.C.; Ambrosio, L.; Milella, E. Structural and rheological characterization of hyaluronic acid-based scaffolds for adipose tissue engineering. Biomaterials 2007, 28, 4399–4408.

10. Guarnieri, D.; Battista, S.; Borzacchiello, A.; Mayol, L.; de, R.E.; Keene, D.R.; Muscariello, L.; Barbarisi, A.; Netti, P.A. Effects of fibronectin and laminin on structural, mechanical and transport properties of

3D collageneous network. J. Mater. Sci. Mater. Med. 2007, 18, 245–253.

11. Maltese, A.; Borzacchiello, A.; Mayol, L.; Bucolo, C.; Maugeri, F.; Nicolais, L.; Ambrosio, L. Novel polysaccharides-based viscoelastic formulations for ophthalmic surgery: Rheological characterization. Biomaterials 2006, 27, 5134–5142.

12. Fisher, O.Z.; Khademhosseini, A.; Langer, R.; Peppas, N.A. Bioinspired materials for controlling stem cell fate. Acc. Chem. Res. 2010, 43, 419–428.

13. Slaughter, B.V.; Khurshid, S.S.; Fisher, O.Z.; Khademhosseini, A.; Peppas, N.A. Hydrogels in regenerative medicine.Adv. Mater. 2009, 21, 3307–3329.

14. Kloxin, A.M.; Kloxin, C.J.; Bowman, C.N.; Anseth, K.S. Mechanical Properties of Cellularly Responsive Hydrogels and Their Experimental Determination. Adv. Mater. 2010, 22, 3484–3494.

15. Borzacchiello, A.; Mayol, L.; Garskog, O.; Dahlqvist, A.; Ambrosio, L. Evaluation of injection augmentation treatment of hyaluronic acid based materials on rabbit vocal folds viscoelasticity. J. Mater. Sci. Mater. Med. 2005, 16, 553–557.

16. Borzacchiello, A.; Mayol, L.; Schiavinato, A.; Ambrosio, L. Effect of hyaluronic acid amide derivative on equine synovial fluid viscoelasticity. J. Biomed. Mater. Res. A 2010, 92, 1162–1170.

17. Annabi, N.; Tamayol, A.; Uquillas, J.A.; Akbari, M.; Bertassoni, L.E.; Cha, C.; Camci-Unal, G.; Dokmeci, M.R.; Peppas, N.A.; Khademhosseini, A. 25th anniversary article: Rational design and applications of hydrogels in regenerative medicine. Adv. Mater. 2014, 26, 85–123.

18. Goenka, S.; Sant, V.; Sant, S. Graphene-based nanomaterials for drug delivery and tissue engineering. J. Control. Release 2014, 173, 75–88.

19. Schexnailder, P.J.; Gaharwar, A.K.; Bartlett, R.L.; Seal, B.L.; Schmidt, G. Tuning cell adhesion by incorporation of charged silicate nanoparticles as cross-linkers to polyethylene oxide. Macromol. Biosci. 2010, 10, 1416–1423.

20. Cha, C.; Shin, S.R.; Annabi, N.; Dokmeci, M.R.; Khademhosseini, A. Carbon-based nanomaterials: Multifunctional materials for biomedical engineering. ACS Nano 2013, 7, 2891–2897.

21. Kuilla, T.; Bhadra, S.; Yao, D.H.; Kim, N.H.; Bose, S.; Lee, J.H. Recent advances in graphene based polymer composites. Prog. Polym. Sci. 2010, 35, 1350–1375.

22. Ma, P.C.; Siddiqui, N.A.; Marom, G.; Kim, J.K. Dispersion and functionalization of carbon nanotubes for polymer-based nanocomposites: A review. Compos. Part A-Appl. Sci. Manuf. 2010, 41, 1345–1367.

23. Mottaghitalab, F.; Farokhi, M.; Zaminy, A.; Kokabi, M.; Soleimani, M.; Mirahmadi, F.; Shokrgozar, M.A.; Sadeghizadeh, M. A Biosynthetic Nerve Guide Conduit Based on Silk/SWNT/Fibronectin Nanocomposite for Peripheral Nerve Regeneration. PLoS ONE. 2013, 8.

24. Huang, Y.C.; Hsu, S.H.; Kuo, W.C.; Chang-Chien, C.L.; Cheng, H.; Huang, Y.Y. Effects of laminin-coated carbon nanotube/chitosan fibers on guided neurite growth. J. Biomed. Mater. Res. Part A 2011, 99A, 86–93.

25. Makharza, S.; Cirillo, G.; Bachmatiuk, A.; Ibrahim, I.; Ioannides, N.; Trzebicka, B.; Hampel, S.; Rummeli, M.H. Graphene oxide-based drug delivery vehicles: Functionalization, characterization, and cytotoxicity evaluation. J. Nanopart. Res. 2013, 15.

26. Frankland, S.J.V.; Caglar, A.; Brenner, D.W.; Griebel, M. Molecular simulation of the influence of chemical cross-links on the shear strength of carbon nanotube-polymer interfaces. J. Phys. Chem. B 2002, 106, 3046–3048.

27. Shin, S.R.; Bae, H.; Cha, J.M.; Mun, J.Y.; Chen, Y.C.; Tekin, H.; Shin, H.; Farshchi, S.; Dokmeci, M.R.; Tang, S.; et al. Carbon nanotube reinforced hybrid microgels as scaffold materials for cell encapsulation. ACS Nano 2012, 6, 362–372.

28. Shin, S.R.; Jung, S.M.; Zalabany, M.; Kim, K.; Zorlutuna, P.; Kim, S.B.; Nikkhah, M.; Khabiry, M.; Azize, M.; Kong, J.; et al. Carbon-nanotube-embedded hydrogel sheets for engineering cardiac constructs and bioactuators. ACS Nano 2013,7, 2369–2380.

29. Li, C.X.; Mezzenga, R. Functionalization of Multiwalled Carbon Nanotubes and Their pH-Responsive Hydrogels with Amyloid Fibrils. Langmuir 2012, 28, 10142–10146.

30. Liu, X.; Kruger, P.; Maibach, H.; Colditz, P.B.; Roberts, M.S. Using skin for drug delivery and diagnosis in the critically ill. Adv. Drug Deliv. Rev. 2014, 77, 40–49.

31. Giri, A.; Bhowmick, M.; Pal, S.; Bandyopadhyaya, A. Polymer hydrogel from carboxymethyl guar gum and carbon nanotube for sustained trans-dermal release of diclofenac sodium. Int. J. Biol. Macromol. 2011, 49, 885–893.

32. Liu, H.W.; Hu, S.H.; Chen, Y.W.; Chen, S.Y. Characterization and drug release behavior of highly responsive chip-like electrically modulated reduced graphene oxide-poly(vinyl alcohol) membranes. J. Mater. Chem. 2012, 22, 17311–17320.

33. Liu, J.Q.; Chen, C.F.; He, C.C.; Zhao, L.; Yang, X.J.; Wang, H.L. Synthesis of graphene peroxide and its application in fabricating super extensible and highly resilient nanocomposite hydrogels. ACS Nano 2012, 6, 8194–8202.

34. Liao, K.H.; Lin, Y.S.; Macosko, C.W.; Haynes, C.L. Cytotoxicity of graphene oxide and graphene in human erythrocytes and skin fibroblasts. ACS Appl. Mater. Interfaces 2011, 3, 2607–2615.

35. Chang, Y.L.; Yang, S.T.; Liu, J.H.; Dong, E.; Wang, Y.W.; Cao, A.N.; Liu, Y.F.; Wang, H.F. In vitro toxicity evaluation of graphene oxide on A549 cells. Toxicol. Lett. 2011, 200, 201–210.

36. Zhang, X.Y.; Hu, W.B.; Li, J.; Tao, L.; Wei, Y. A comparative study of cellular uptake and cytotoxicity of multi-walled carbon nanotubes, graphene oxide, and nanodiamond. Toxicol. Res. 2012, 1, 62–68.

37. Zhang, Y.B.; Ali, S.F.; Dervishi, E.; Xu, Y.; Li, Z.R.; Casciano, D.; Biris, A.S. Cytotoxicity effects of graphene and single-wall carbon nanotubes in neural phaeochromocytoma-derived PC12 cells. ACS Nano 2010, 4, 3181–3186.

38. Park, S.; Mohanty, N.; Suk, J.W.; Nagaraja, A.; An, J.H.; Piner, R.D.; Cai, W.W.; Dreyer, D.R.; Berry, V.; Ruoff, R.S. Biocompatible, Robust free-standing paper composed of a TWEEN/graphene composite. Adv. Mater. 2010, 22, 1736–1740.

39. Agarwal, S.; Zhou, X.Z.; Ye, F.; He, Q.Y.; Chen, G.C.K.; Soo, J.; Boey, F.; Zhang, H.; Chen, P. Interfacing live cells with nanocarbon substrates. Langmuir 2010, 26, 2244–2247.

40. Wang, K.; Ruan, J.; Song, H.; Zhang, J.L.; Wo, Y.; Guo, S.W.; Cui, D.X. Biocompatibility of Graphene Oxide. Nanoscale Rese. Lett. 2011, 6.

41. Akhavan, O.; Ghaderi, E.; Akhavan, A. Size-dependent genotoxicity of graphene nanoplatelets in human stem cells.Biomaterials 2012, 33, 8017–8025.

42. Merino, S.; Martin, C.; Kostarelos, K.; Prato, M.; Vazquez, E. Nanocomposite hydrogels: 3D polymer-nanoparticle synergies for on-demand drug delivery. ACS Nano 2015, 9, 4686–4697.

43. Brayden, D.J.; Cryan, S.A.; Dawson, K.A.; O'Brien, P.J.; Simpson, J.C. High-content analysis for drug delivery and nanoparticle applications. Drug Discov. Today 2015, 20, 942–957.

44. Salcher, A.; Nikolic, M.S.; Casado, S.; Velez, M.; Weller, H.; Juarez, B.H. CdSe/CdS nanoparticles immobilized on pNIPAm-based microspheres. J. Mater. Chem. 2010, 20, 1367–1374.

45. Li, G.L.; Mohwald, H.; Shchukin, D.G. Precipitation polymerization for fabrication of complex core-shell hybrid particles and hollow structures. Chem. Soc. Rev. 2013, 42, 3628–3646.

46. Schluter, A.D.; Halperin, A.; Kroger, M.; Vlassopoulos, D.; Wegner, G.; Zhang, B.Z. Dendronized polymers: Molecular objects between conventional linear polymers and colloidal particles. ACS Macro Lett. 2014, 3, 991–998.

47. Irfan, M.; Seiler, M. Encapsulation using hyperbranched polymers: From research and technologies to emerging applications. Ind. Eng. Chem. Res. 2010, 49, 1169–1196.

48. Liu, T.; Wu, T.; Liu, H.X.; Ke, B.; Huang, H.X.; Jiang, Z.Y.; Xie, M.Q. Ultraviolet-crosslinked hydrogel sustained- release hydrophobic antibiotics with long-term antibacterial activity and limited cytotoxicity. J. Appl. Polym. Sci. 2014,131.

49. Mayol, L.; Serri, C.; Menale, C.; Crispi, S.; Piccolo, M.T.; Mita, L.; Giarra, S.; Forte, M.; Saija, A.; Biondi, M.; et al. Curcumin loaded PLGA-poloxamer blend nanoparticles induce cell cycle arrest in mesothelioma cells. Eur. J. Pharm. Biopharm. 2015, 93, 37–45.

50. Ju, C.Y.; Sun, J.; Zi, P.; Jin, X.; Zhang, C. Thermosensitive micelles-hydrogel hybrid system based on poloxamer 407 for localized delivery of paclitaxel. J. Pharm. Sci. 2013, 102, 2707–2717.

51. Joshi, N.; Grinstaff, M. Applications of dendrimers in tissue engineering. Curr. Top. Med. Chem. 2008, 8, 1225–1236.

52. Zhong, S.; Yung, L.Y. Enhanced biological stability of collagen with incorporation of PAMAM dendrimer. J. Biomed. Mater. Res. A 2009, 91, 114–122.

53. Zhang, H.B.; Patel, A.; Gaharwar, A.K.; Mihaila, S.M.; Iviglia, G.; Mukundan, S.; Bae, H.; Yang, H.; Khademhosseini, A. Hyperbranched polyester hydrogels with controlled drug release and cell adhesion properties. Biomacromolecules2013, 14, 1299–1310.

54. Oral, E.; Peppas, N.A. Responsive and recognitive hydrogels using star polymers. J. Biomed. Mater. Res. Part A 2004,68A, 439–447.

55. Sontjens, S.H.M.; Nettles, D.L.; Carnahan, M.A.; Setton, L.A.; Grinstaff, M.W. Biodendrimer-based hydrogel scaffolds for cartilage tissue repair. Biomacromolecules 2006, 7, 310–316.

56. Postnova, I.; Silant'ev, V.; Kim, M.H.; Song, G.Y.; Kim, I.; Ha, C.S.; Shchipunov, Y. Hyperbranched polyglycerol hydrogels prepared through biomimetic mineralization. Coll. Surf. B Biointerfaces 2013, 103, 31–37.

57. Hench, L.L.; Polak, J.M. Third-generation biomedical materials. Science 2002, 295, 1014–1017.

58. Hoppe, A.; Guldal, N.S.; Boccaccini, A.R. A review of the biological response to ionic dissolution products from bioactive glasses and glass-ceramics. Biomaterials 2011, 32, 2757–2774.

59. Khan, F.; Ahmad, S.R. Polysaccharides and their derivatives for versatile tissue engineering application. Macromol. Biosci. 2013, 13, 395–421.

60. Zhu, J.M. Bioactive modification of poly(ethylene glycol) hydrogels for tissue engineering. Biomaterials 2010, 31, 4639–4656.

61. Barbucci, R.; Rappuoli, R.; Borzacchiello, A.; Ambrosio, L. Synthesis, chemical and rheological characterization of new hyaluronic acid-based hydrogels. J. Biomater. Sci.-Polym. Ed. 2000, 11, 383–399.

62. Lapcik, L.; Lapcik, L.; de Smedt, S.; Demeester, J.; Chabrecek, P. Hyaluronan: Preparation, structure, properties, and applications. Chem. Rev. 1998, 98, 2663–2684.

63. Borzacchiello, A.; Russo, L.; Malle, B.M.; Schwach-Abdellaoui, K.; Ambrosio, L. Hyaluronic acid based hydrogels for regenerative medicine applications. BioMed Res. Int. 2015, 2015.

64. Fusco, S.; Borzacchiello, A.; Miccio, L.; Pesce, G.; Rusciano, G.; Sasso, A.; Netti, P.A. High frequency viscoelastic behaviour of low molecular weight hyaluronic acid water solutions. Biorheology 2007, 44, 403–418.

65. Nejadnik, M.R.; Yang, X.; Bongio, M.; Alghamdi, H.S.; van den Beucken, J.J.; Huysmans, M.C.; Jansen, J.A.; Hilborn, J.; Ossipov, D.; Leeuwenburgh, S.C. Self-healing hybrid nanocomposites consisting of bisphosphonated hyaluronan and calcium phosphate nanoparticles. Biomaterials 2014, 35, 6918–6929.

66. Valles-Lluch, A.; Poveda-Reyes, S.; Amoros, P.; Beltran, D.; Monleon, P.M. Hyaluronic acid-silica nanohybrid gels.Biomacromolecules 2013, 14, 4217–4225.

67. Mayol, L.; De Stefano, D.; Campani, V.; De Falco, F.; Ferrari, E.; Cencetti, C.; Matricardi, P.; Maiuri, L.; Carnuccio, R.; Gallo, A.; et al. Design and characterization of a chitosan physical gel promoting wound healing in mice. J. Mater. Sci. Mater. Med. 2014, 25, 1483–1493.

68. Aranaz, I.; Gutierrez, M.C.; Ferrer, M.L.; del, M.F. Preparation of chitosan nanocompositeswith a macroporous structure by unidirectional freezing and subsequent freeze-drying. Mar. Drugs 2014, 12, 5619–5642.

69. Gutierrez, M.C.; Jobbagy, M.; Ferrer, M.L.; del Monte, F. Enzymatic synthesis of amorphous calcium phosphate—Chitosan nanocomposites and their processing into hierarchical structures. Chem. Mater. 2008, 20, 11–13.

70. Gaharwar, A.K.; Dammu, S.A.; Canter, J.M.; Wu, C.J.; Schmidt, G. Highly extensible, tough, and elastomeric nanocomposite hydrogels from poly(ethylene glycol) and hydroxyapatite nanoparticles. Biomacromolecules 2011, 12, 1641–1650.

71. Gaharwar, A.K.; Rivera, C.; Wu, C.J.; Chan, B.K.; Schmidt, G. Photocrosslinked nanocomposite hydrogels from PEG and silica

nanospheres: Structural, mechanical and cell adhesion characteristics. Mater. Sci. Eng. C Mater. Biol. Appl.2013, 33, 1800–1807.

72. Gaharwar, A.K.; Rivera, C.P.; Wu, C.J.; Schmidt, G. Transparent, elastomeric and tough hydrogels from poly(ethylene glycol) and silicate nanoparticles. Acta Biomater. 2011, 7, 4139–4148.

73. Schexnailder, P.; Schmidt, G. Nanocomposite polymer hydrogels. Coll. Polym. Sci. 2009, 287, 1–11.

74. Barani, H.; Montazer, M.; Samadi, N.; Toliyat, T. In situ synthesis of nano silver/lecithin on wool: Enhancing nanoparticles diffusion. Coll. Surf. B Biointerfaces 2012, 92, 9–15.

75. Dallas, P.; Sharma, V.K.; Zboril, R. Silver polymeric nanocomposites as advanced antimicrobial agents: Classification, synthetic paths, applications, and perspectives. Adv. Coll. Interface Sci. 2011, 166, 119–135.

76. Palza, H. Antimicrobial polymers with metal nanoparticles. Int J Mol Sci. 2015, 16, 2099–2116.

77. Ma, Y.Q.; Yi, J.Z.; Zhang, L.M. A facile approach to incorporate silver nanoparticles into dextran-based hydrogels for antibacterial and catalytical application. J. Macromol. Sci. Part A-Pure Appl. Chem. 2009, 46, 643–648.

78. Vimala, K.; Sivudu, K.S.; Mohan, Y.M.; Sreedhar, B.; Raju, K.M. Controlled silver nanoparticles synthesis in semi-hydrogel networks of poly(acrylamide) and carbohydrates: A rational methodology for antibacterial application.Carbohydr. Polym. 2009, 75, 463–471.

79. Bardajee, G.R.; Hooshyar, Z.; Rezanezhad, H. A novel and green biomaterial based silver nanocomposite hydrogel: Synthesis, characterization and antibacterial effect. J. Inorg. Biochem. 2012, 117, 367–373.

80. Mohan, Y.M.; Lee, K.; Premkumar, T.; Geckeler, K.E. Hydrogel networks as nanoreactors: A novel approach to silver nanoparticles for antibacterial applications. Polymer 2007, 48, 158–164.

81. Wei, Q.B.; Fu, F.; Zhang, Y.Q.; Tang, L. Preparation, characterization, and antibacterial properties of pH-responsive P(MMA-co-MAA)/silver nanocomposite hydrogels. J. Polym. Res. 2014, 21.

82. Juby, K.A.; Dwivedi, C.; Kumar, M.; Kota, S.; Misra, H.S.; Bajaj, P.N. Silver nanoparticle-loaded PVA/gum acacia hydrogel: Synthesis, characterization and antibacterial study. Carbohydr. Polym. 2012, 89, 906–913.

83. Dvir, T.; Timko, B.P.; Brigham, M.D.; Naik, S.R.; Karajanagi, S.S.; Levy, O.; Jin, H.W.; Parker, K.K.; Langer, R.; Kohane, D.S. Nanowired three-dimensional cardiac patches. Nat. Nanotechnol. 2011, 6, 720–725.

84. Liu, H.; Webster, T.J. Mechanical properties of dispersed ceramic nanoparticles in polymer composites for orthopedic applications. Int J Nanomedicine. 2010, 5, 299–313.

85. Gaharwar, A.K.; Wong, J.E.; Müller-Schulte, D.; Bahadur, D.; Richtering, W. Magnetic nanoparticles encapsulated within a thermoresponsive polymer. J Nanosci Nanotechnol. 2009, 9, 5355–5361.

86. Wang, C.; Cheng, L.; Liu, Z. Upconversion Nanoparticles for Photodynamic Therapy and Other Cancer Therapeutics. Theranostics 2013, 3, 317–330.

87. Matteini, P.; Ratto, F.; Rossi, F.; Centi, S.; Dei, L.; Pini, R. Chitosan films doped with gold nanorods as laser-activatable hybrid bioadhesives. Adv. Mater. 2010, 22, 4313–4316.

88. Matteini, P.; Martina, M.R.; Giambastiani, G.; Tatini, F.; Cascella, R.; Ratto, F.; Cecchi, C.; Caminati, G.; Dei, L.; Pini, R. Light-responsive nanocomposite sponges for on demand chemical release with high spatial and dosage control. J. Mater. Chem. B 2013, 1, 1096–1100.

8

SYNTHESIS AND PROPERTIES OF DISCONTINOUOSLY REINFORCED ALUMINUM MATRIX COMPOSITES

Dusan Bozic [1] and Biljana Dimcic[1]

[1] Institute of Nuclear sciences "Vinca" University of Belgrade, Serbia

1. INTRODUCTION

Due to the very high performance requests, a new type of, so called, "high tech" materials which are used in aircraft, rocket or automobile industry, need to have a good combination of properties under static and dynamic loading. One group of materials that can meet such requirements are metallic composites which consist of the metallic matrix (metal or alloy) reinforced whit whiskers or directionally oriented fibers of secondary phase.

Currently, there are many different composite materials available, enabling one to easily select the most suitable combination of metallic matrix and reinforcing phase, depending on the uniquely defined application requests. By selecting proper type of matrix and suitable chemical composition, shape and the amount of reinforcing phase, many composites with different mechanical properties can be produced. Potential application of such materials is very wide, but the limiting factor is still their quite high price due to the complexity of their production.

In the last two decades there was an intense development of one class of composites with aluminum alloy matrix best known as the discontinuously reinforced aluminum materials (DRA). These materials consist of aluminum alloy matrix reinforced by ceramic particles.

Although, DRA composites can be produced in many ways, powder metallurgy technique offers the best results.

This article presents results of both fundamental and development studies of sintered composite materials with the Al-Zn-Mg-Cu matrix and the SiC reinforcing phase, conducted in last couple of years. As this is the composite which is already commercially applied or on the verge of being applied, all gathered results are thoroughly analyzed from the aspects of its synthesis, microstructure and several mechanical and fracture properties.

2. MATRIX PROPERTIES OF SINTERED DRA MATERIALS

Mostly used alloys in the aircraft industry belong to the Al-Zn-Mg-Cu system. Usually in these alloys, traces of transient elements (below 1 mass %) such as: Fe, Si, Co and Ni can also be found. Commercially these alloys are known as 7000 or 7xxx class. One that has been used for this study is best known as CW67 alloy. Compared to the Aluminium alloys of the same composition produced by different techniques (such as melting and casting), sintered alloys have much better properties such as strength, fracture toughness and stress corrosion resistance (Tietz & Palmour, 1986). Mentioned advantages of these alloys can be related to the high amount of fine secondary precipitates and undiluted dispersed particles.

It is known that the phase particles which consist of elements such as Co or Fe+Ni are less prone to coarsening than the particles consisting of Cr, Ti or Mo (Tietz & Palmour, 1986). Precipitating phases Co_2Al_9 and $FeNiAl_9$, which are formed in the Al-Zn-Mg-Cu alloy, are very fine, ranging from 0.05-2 μm in size.Even finer particles can be produced by forming the dispersed oxide phase due to presence of oxygen (0.05-0.08 mass %) in this type of alloy.

Increased stress corrosion resistance of sintered DRA is favored by adequate grain morphology and presence of the higher amount of Co_2Al_9 precipitates in the structure. Presence of higher amount of Co has a very big influence on the higher fatigue resistance values of sintered alloys compared to casted ones (Tietz & Palmour, 1986). In order to obtain finer and more stable structure, a small amount of elements, such as Zr (0.2-0.8 mass %) can be added. Within casted alloys, amount of Zr cannot be higher than 0.12 mass %.

For further improvement of mechanical properties of sintered aluminum alloys, mechanical alloying can be applied (Tietz & Palmour, 1986). By performing mechanical alloying, multiple strengthening effects can be achieved: strengthening of the solid solution (commonly with Mg as an alloying element), precipitation strengthening and dispersed phase strengthening. It has been found that higher stress corrosion resistance of mechanically alloyed alloys of 550 MPa and high value of the Young's modulus of 76.5 GPa compared to other sintered alloys with values of 310 MPa for the stress corrosion resistance and 73GPa for the Young's modulus, make these alloys very interesting candidates for the further development and characterization.

Improvement of many properties (strength, Young's modulus, toughness etc.) of sintered Al-Zn-Mg-Cu alloys compared to the casted ones is achieved by using the contemporary powder metallurgy techniques. Production of alloy powders by rapid cooling process (rate of cooling in the range from 10^4 to 10^9 K/s), vacuum degassing of powders or compacts and powder compacting at elevated temperatures (hot pressing) are just a few of mentioned powder metallurgy techniques which can improve alloy properties (Figure 1).

An initial powder which is used for the production of sintered aluminium alloys can be made by several techniques, but mostly used one is pulverization. This technique is equally used in industry and smaller laboratories and there are several different types of it (Figure 2): gas atomization, ultrasound gas atomization, splat quenching, melt spinning *etc.* (Tietz & Palmour, 1986).

As it can be seen from the Fig.1, there are two mostly used compaction processes. Both processes consist of several steps. The first one starts with encapsulation, followed by powder degassing ending with compaction at higher temperatures (Tietz & Palmour, 1986). In the second process, powder is iso-staticaly pressed up to 70% of the theoretical density, degassed and only afterwards compacted at high temperatures up to theoretical density (Bozic et al., 1997). Both of these production processes are widely used, but the first one is more complicated (due to encapsulation/decapsulation), while the second one is less effective in removal of hydrogen and oxygen originating from the absorbed moisture in hydrated aluminum oxides. Degassing temperatures are in both cases in the range from 400 to 500 °C. Powder compaction at elevated temperatures (up to 500 °C) is done by hot pressing, hot iso-static pressing or hot extrusion which is mainly used for the production of the final products. By hot pressing at 450 °C during 2h and applied pressure of 35 MPa, a pore free compact can be produced (Fig. 3).

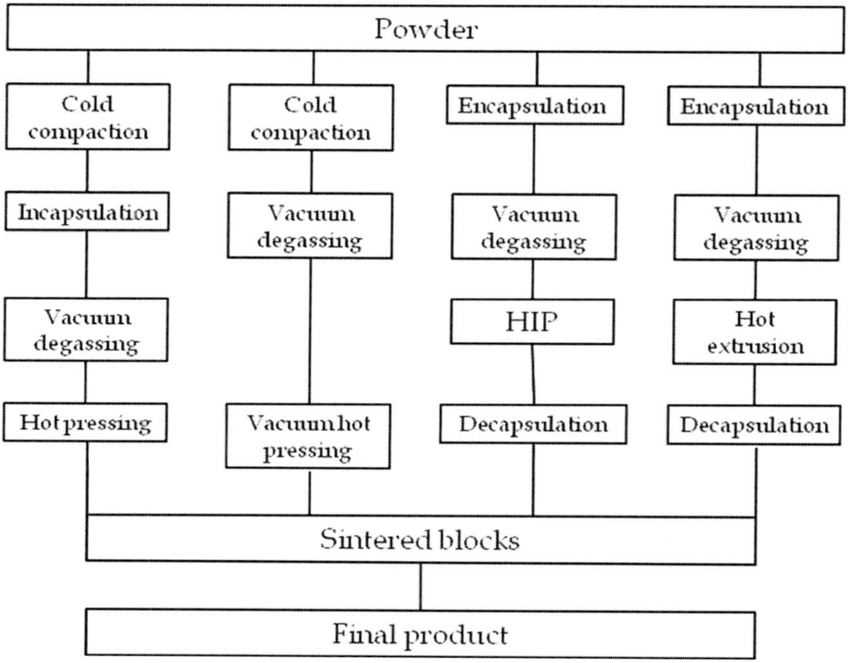

Figure 1. Schematic presentation of high strength aluminum alloys production techniques (**Tietz & Palmour, 1986**).

Primary particle boundaries can be seen in the microstructure of the hot pressed compacts (arrows inFig. 3), and most of the unwanted oxides which are deteriorating mechanical properties of the materials are located at these places (Gnjidic et al., 2001). Therefore, it is necessary to deform these materials by hot extrusion or rolling before use for the high risk applications (Bozic et al., 1997).

Since the Al-Zn-Mg-Cu alloy matrix is the precipitation strengthened alloy, it is important to perform adequate thermal treatment of the alloy after compaction. This treatment consists of the solution annealing at 475°c for 1h followed by water quenching and further ageing at 120°C for 24h or 160°C for 16h.

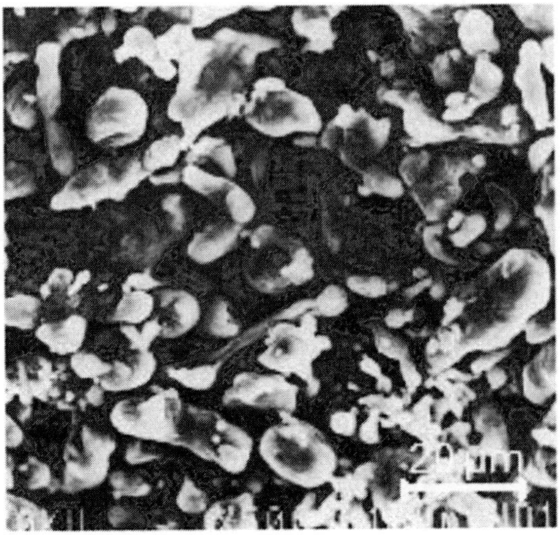

Figure 2. SEM. Aluminum alloy powder (CW67) produced by gas atomization.

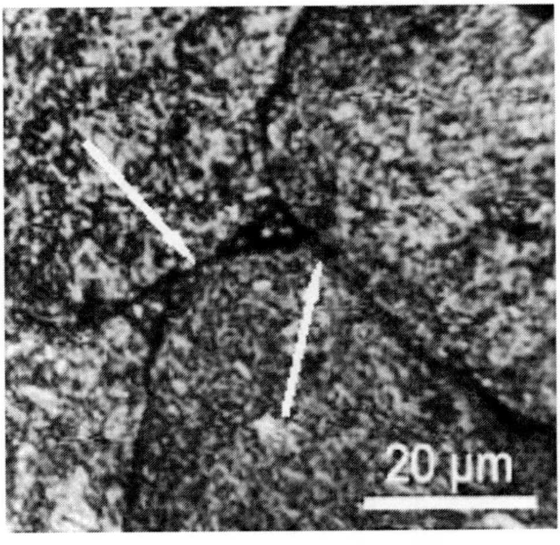

Figure 3. SEM. Microstructure of hot-pressed Al-Zn-Mg-Cu alloy. Arrows denote a few prior particle boundaries.

Mechanism of the thermally induced precipitation in Al-Zn-Mg-Cu alloys with the high Zn : Mg ratio can be presented in this way:

Saturated solid solution - Gunier-Preston (GP) zones - μ' ($MgZn_2$) - μ($MgZn_2$)

Gunier-Preston (GP) zones are precipitating from the saturated solid solution, and they have coherent interfaces with the parent phase. Semi coherent μ' ($MgZn_2$) phase has the monoclinic type of the crystal lattice, while the stable μ($MgZn_2$) phase has the hexagonal crystal structure. These phases are precipitating at different temperatures in the different time interval. When GP zones and the small amount of μ' ($MgZn_2$) are present in the structure, alloy exhibits the highest strength values. Addition of Cu above 1 mass % can also increase the strength (strengthening due to alloying) but the precipitating mechanism still remains the same. When there is more Cu in the chemical composition of the alloy, it can be involved in the precipitation by replacing the Zn atoms.

3. REINFORCING PHASE PROPERTIES OF SINTERED DRA MATERIALS

Selection of an appropriate composition, morphology and volume fraction of reinforcing phase is a matter of trade off between the requested mechanical properties and its economic value. Typical properties which have the biggest effect on the selection of the particular reinforcing phase are: Young's modulus value, wear resistance, fracture toughness, fatigue resistance and thermal expansion coefficient. A proper choice of the type and geometry of the reinforcing phase is very important for achieving the best combination of required properties and costs.Mostly used reinforcing phases for the production of discontinuously reinforced composites are: SiC in the shape of whiskers, particles of SiC or Al_2O_3, short Al_2O_3 or graphite fibers. Compared to fibers, particles and whiskers are easier for production; they are less expensive and have stable properties.

Whiskers are monocrystalline materials with a high length/width ratio (50 to 100). This shape of reinforcing phase is mostly used for composites that will be submitted to extrusion, rolling or forging. Unlike the particle reinforced composites, materials reinforced with whiskers are slightly less isotopic and their price is much higher.

Reinforcing phase in particle shape is not very expensive. Maximal/minimal particle dimension ratio is quite low (1 to 5). They have more isotropic properties compared to whiskers, resulting in better characteristics of composite materials. This property enables obtaining better characteristics of composite materials. It is important to know that this type of reinforcing phase is most commonly made by conventional methods used for the production of the metallic materials as well.

Technical SiC, which is mostly used as a particle reinforcing phase is often produced by reaction between SiO_2 and coke (in excess amount). SiC powder obtained in this way may still contain a small amount of un-reacted SiO_2 or free graphite (Fig. 4). Free graphite in a SiC powder can reduce the mechanical properties of the composites and therefore it has to be removed. Removal of this graphite can be done by thermal treatment of the powder at 900 °C for 2h (Gnjidic & Bozic, 1999).

Unlike composites produced by melting and casting that consist of particles with dimensions of hundreds of microns, composites produced by powder metallurgy techniques can use only much finer particles. Usual size of the SiC particles lies in the range from submicron size up to around 20 μm (Figs. 5-7).

As it can be seen from the images, SiC powder characteristics can differ depending on its particle shape, which can be: spherical (Figs. 5 and 6), polygonal (Fig. 6) or irregular with sharp edges (Fig. 7). In submicron powders, appearance of the agglomerates is common (Figs. 5).

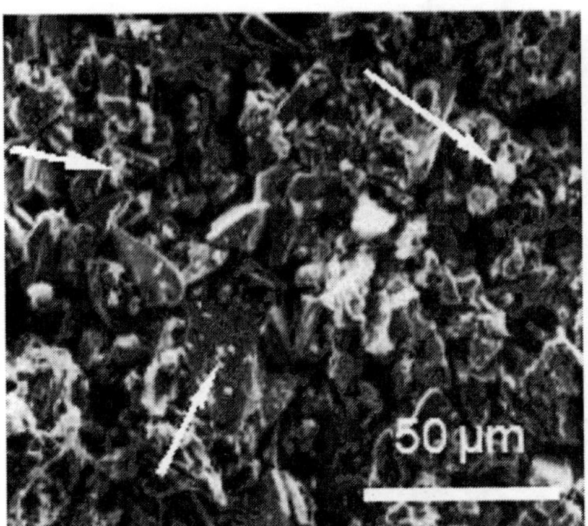

Figure 4. SEM. SiC powder particles before heat treatment. Arrows appointing free graphite.

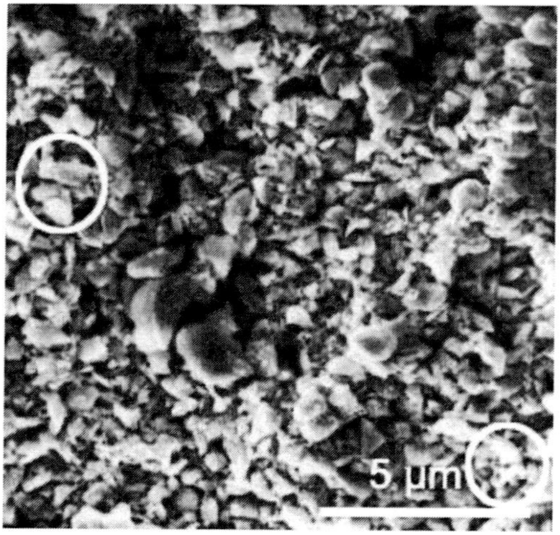

Figure 5. SEM. Size and shape of SiC powder particles (average size 0.7 μm). Circles denote agglomerates.

Figure 6. SEM. Size and shape of SiC particles (average size 10 μm).

4. MIXING PROCESS

Mixing or homogenization of powders is one of the most critical processes in production of DRA materials by powder metallurgy techniques. If it is done properly, particles of the reinforcing phase can be uniformly distributed in the composite metallic matrix which improves the structural and mechanical properties of these materials. The most important parameters that influence effectiveness of the mixing process are: mixing dish filling rate, rotation velocity and mixing time (German, 1994). Optimal values of these parameters are mostly confidential or patent protected. This means that for every combination of powders (shape, size, amount of reinforcing phase) and mixing dish type it is necessary to experimentally find optimal mixing conditions, *i.e.* optimal values of the previously mentioned parameters.

Figure 7. SEM. Different shapes of SiC particles (average size 15 μm).

One method for quantitative analysis of the SiC particle distribution homogeneity is based on the concept of homogeneity index, Q, which is a quantitative measure of the one component powder particle distribution homogeneity in the mixture of a few powders (Gray, 1973). It should be noted that this concept can only be valid under assumption that the volume fraction of the observed particles is equal to their surface fraction.

Standard deviation (α) of particle surface measurements is defined by Eq. (1):

$$\alpha = \frac{1}{N}\sum\left|A_i - A_f\right| \tag{1}$$

Where N is the number of repeated measurements, A_i is the individual area fraction and A_f is the mean value of all measured particle surfaces.

Equation (2) describes the case in which completely separated system appears (complete segregation) or, in other words, system in which the whole measuring surface belongs to the particles of just one powder component.

$$\alpha = 2A_f(1 - A_f) \tag{2}$$

By combining Eqs. (1) and (2), homogeneity index, Q, can be calculated as:

$$Q = \alpha / \alpha_{seg} \tag{3}$$

Values of homogeneity index vary from 0 (perfect uniform distribution of particles in the mixture) to 1 (complete non-uniformity of the powder particles).

Using this method, a homogeneity index for all mixtures can be determined by analyzing the eight randomly chosen measuring fields split into 16 measuring units. The size of the measuring unit can be calculated from the Eq. (4):

$$A = 1/N_A \tag{4}$$

Where N_A is the number of SiC particles in the measuring unit.

By using any type of software for the quantitative analysis of the microstructure, homogeneity of SiC particle distribution in metal matrix can be calculated (Mc Kimpson et al., 1999).

In our study, which is based on the data found in the literature (German, 1994), mixing dish rotation velocity was taken to be constant and equal 70 rpm for all regimes. Beside the rotation velocity, mixing time of 30 min was taken as a constant for the first five regimes as well. The amount of powders was different and it was increased by 5 vol. % for each following regime. In the second five regimes, mixing dish filling rate was kept constant while the time of mixing was changed. The selected mixing dish filling rate was the optimal value from the first five experimental regimes (Table 1).

Mixing of powders occurred due to 'diffusion" in the cylindrical dish without any mixing accelerators. Results of quantitative microstructural analysis showing the effect of the mixing dish filling rate on the homogeneity index values are presented in Fig. 8.

All mixtures were well mixed and differences between their homogeneity indexes were in the range of 15%. Dependence of homogeneity index on the mixing dish filling rate for three different volume fractions of SiC particles in the mixture exhibits the same trend. The optimal value was achieved for the mixture with the lowest amount of the reinforcing phase.

The resulting force that affects particles in powder mixture during mixing is a result of unified influences of centrifugal, gravitational and frictional forces.The latter one exhibits the predominant effect on the mixing (homogenization) results. It was found in this study that optimal mixing dish filling rate was the lowest one (20 vol. %). The reason for this might be the fact that inter-particle contact is minimal during mixing since the available length for the free fall of particles is maximal, and therefore, agglomerates are hard to form and the ones already formed can easily be destructed (Fig. 8). With increase of the mixing dish filling rate, amount of SiC powder also increases forming more agglomerates that need to be destroyed. Available length for the free fall of particles also decreases resulting in overall decrease in powder mixture homogeneity. Negative effect of agglomerate presence in the microstructure can also be increased with the increase of SiC particle amount. It is known that a very large number of parameters influence the efficiency of particle stacking in the volume of the powder (Gray, 1973). The most important parameters are shape, size, physical and chemical characteristics of the powder particles as well as the shape, size and material of the mixing dish and many more. The real influence of only one parameter is very hard to predict since it often overlaps with the influence of another one. In the case of heterogeneous system there is one additional parameter and that is the volume fraction of the second powder. Higher amount of additional powder decreases the efficiency of particle stacking but not very dramatically since homogeneity indexes do not differ much between the different analyzed mixtures (Fig. 8).

Table 1. Different mixing regimes for metallic-ceramic composites

Regime	Mixing dish filling rate (vol. %)	Mixing time (min)
I	20	30
II	25	30
III	30	30
IV	35	30
V	40	30
VI	30	60
VII	30	90
VIII	30	120
IX	30	150

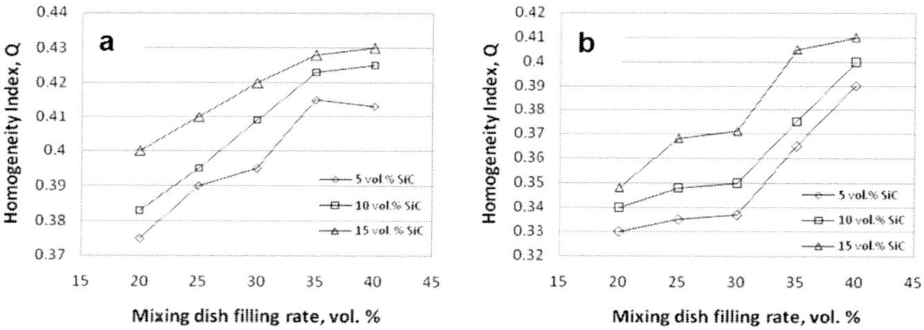

Figure 8. Plot of homogeneity index values *vs.* amount of the mixing dish filling rate for alloys with a) 0.7 μm and b) 15 μm sized reinforcing powder particles.

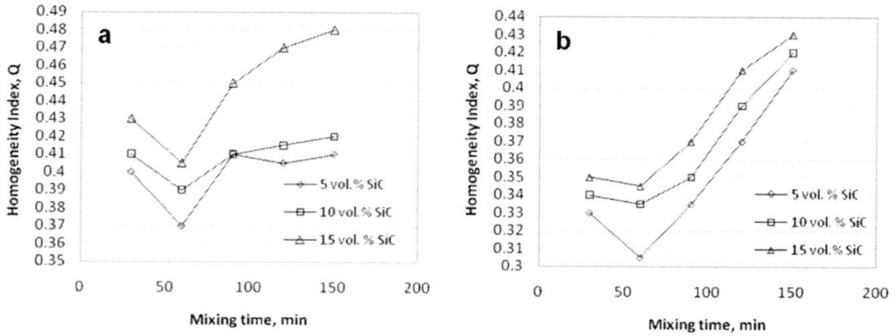

Figure 9. Plot of homogeneity index values *vs.* mixing time for alloys with a) 0.7 μm and b) 15 μm sized reinforcing powder particles.

It can be noticed from Fig. 9 that influence of the mixing time on the homogeneity index values is much higher than the influence of the mixing dish filling rate (Fig. 8). Values of homogeneity index increase up to 40% with the increase of mixing time compared to the optimal one.

Plots of homogeneity index values *vs.* mixing time, for the optimal value of mixing dish filling rate, exhibit the same trend for all mixtures. The best homogeneity is, like in the previous case (Fig. 8), obtained for the mixtures with 5 vol. % SiC. Mixing time of 60 min was optimal from the aspect of secondary (reinforcing) phase distribution. It is known that during longer mixing time the effect of particle "diffusion" is more pronounced (German, 1994). All plots display the increase of homogeneity index with increase of the amount of reinforcing phase in the mixture during constant mixing time. That kind of behavior is somewhat expected because of increase of the number of reinforcing phase particles that need to be mixed. Figs. 10 a-c and 11 a-c display the microstructures of samples with the optimal homogeneity which are produced under the optimal mixing conditions.Figs. 10 d and 11 d illustrate so called diffusion mixing as a result of prolonged mixing time. Scanning electron images clearly show clusters of SiC as well as areas without SiC presence.

SiC powder with medium size particles of 15 μm, used in this study is characterized by wider particle size distribution. Around 20 mass.% of this granulation contained particles larger than 20 μm, and a certain percent of 1-2 μm sized particles. Several authors (Flom and Arsenault, 1989) found that when a SiC particle size is 20 μm or larger, fracture process changes from mainly matrix controlled failure to particle cracking controlled. On the other hand, particles of around 1 μm are often prone to agglomeration during mixing. Therefore, even under optimal mixing conditions, segregation of particles, presence of surface defects and cracking of large particles can be expected with this kind of reinforcing material.

By sieving this fraction, a powder with the medium size SiC particles of 10 μm, characterized by the considerably narrower particle size distribution than the previous one can be extracted, Fig. 6.

Mixing of aluminum alloy and SiC powders was conducted under the same conditions as in the previous cases in this study. Dependence of the homogeneity index on the mixing dish filling rate and mixing time is shown in Figs. 12 and 13. Although the character of the plots is similar to the previous ones in Figs. 8 and 9, we can see some differences. In this case, higher degree of homogenization is achieved and better distribution of reinforcing particles is observed even for longer mixing times. Microstructures of such mixtures are presented in Fig. 14 a and b.

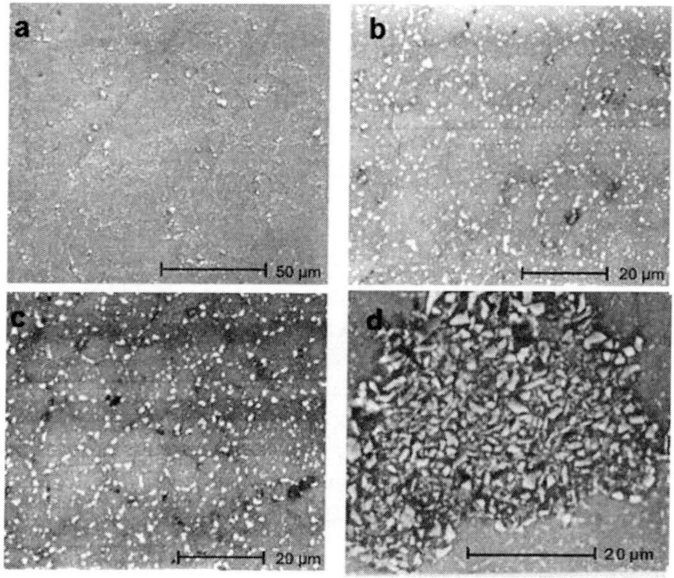

Figure 10. SEM. Composite microstructure (d_{SiC}=0.7 µm)-optimal homogeneity. a) 5 vol. % SiC, b) 10 vol. % SiC, c) 15 vol. % SiC, and example of prolonged mixing time effect, d) 15 vol. % SiC, 150 min mixing time.

Different values of mixing dish filling rates and mixing time have great influence on the SiC particle distribution in aluminum alloy and hence on the mechanical properties an mechanical behavior of the produced DRA material.

5. METHODS OF POWDER MIXTURE CONSOLIDATION

In order to make a fully dense product, powder mixture needs to be consolidated. Powder metallurgy consolidation techniques consist of three basic steps:

1) Cold isostatic pressing (CIP) - enables formation of compacts with density of 65-75% of the theoretical density;

2) Compacts degassing at high temperatures-removal of absorbed gases and moisture;

3) Hot or hot vacuum pressing, hot iso-static pressing or hot extrusion -production of compacts with theoretical density.

Existence of the open porosity in compacts after cold iso-static pressing enables further vacuum degassing. Degassing temperature is usually in the range from 400 to 500°C. Incomplete degassing results in detaining a large amount of oxides on the particle surface which cause weaker particle bonding in the following stages of consolidation process. In this kind of material, formation of cracks and fracture along the primary particle boundaries is likely to occur. On the other hand, too high degassing temperatures or very long degassing times are not very favorable since they can lead to decrease of the Zn amount in the composite matrix (in 7000 alloys).

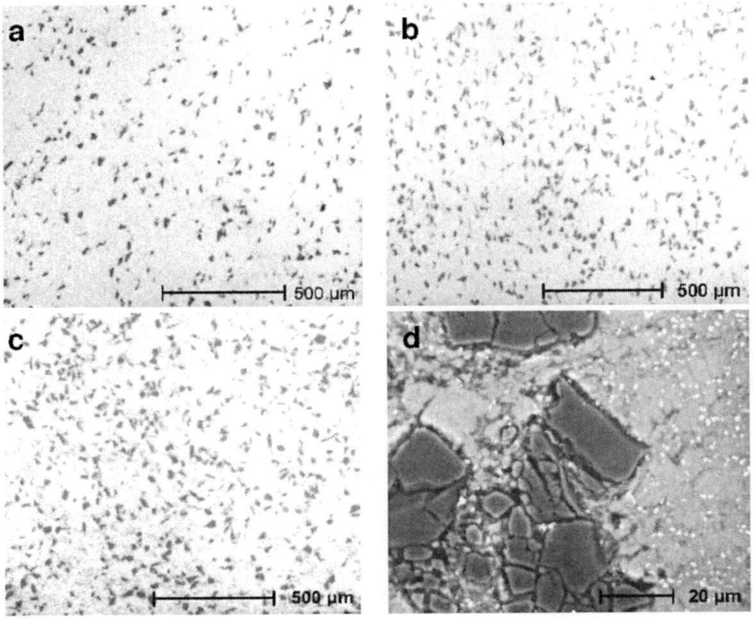

Figure 11. LM. Composite microstructure (d_{SiC}=15 µm). a) 5 vol. % SiC, b) 10 vol. % SiC, c) 15 vol. % SiC, and example of prolonged mixing time effect, d) SEM. 15 vol. % SiC, 150 min mixing time.

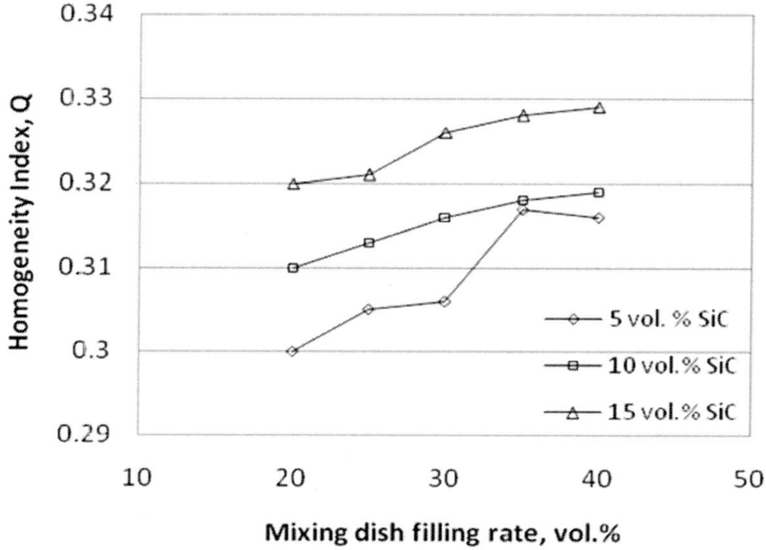

Figure 12. Plot of homogeneity index values *vs.* mixing dish filling rate for the alloys with 10 μm size reinforcing phase particles.

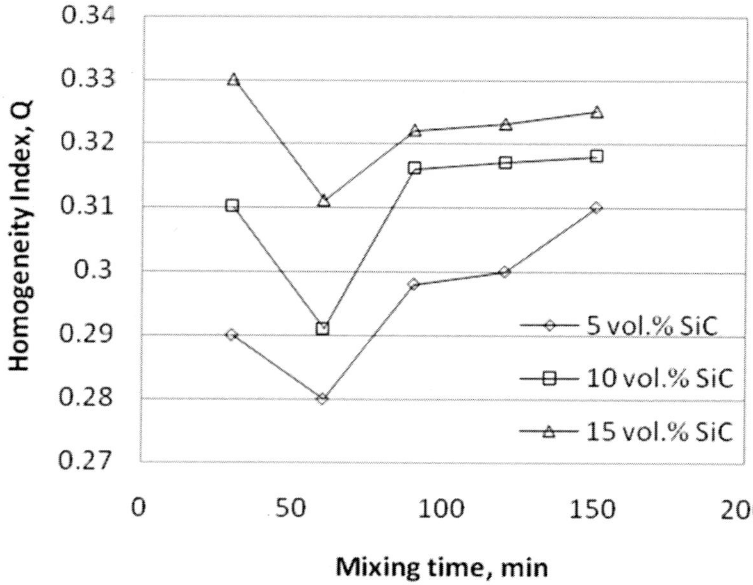

Figure 13. Plot of homogeneity index values *vs.* mixing time for the alloys with 10 μm size reinforcing phase particles.

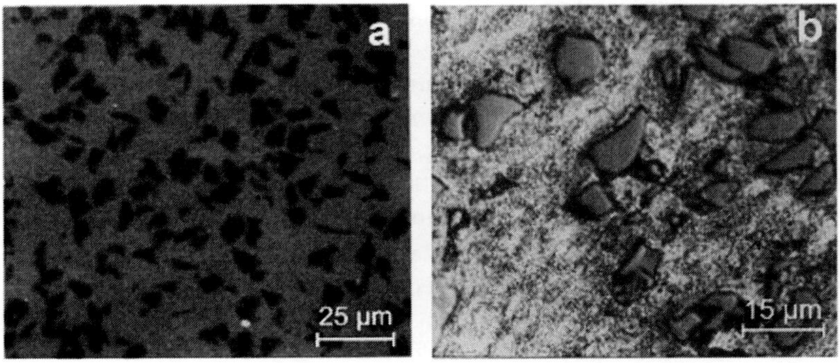

Figure 14. Composite microstructure (d_{SiC}=10 μm). a) LM. Optimal homogeneity reached-15 vol.% SiC and b) SEM. Effect of prolonged mixing time -10 vol.% SiC.

Mostly used consolidation techniques for production of DRA materials are hot pressing and hot vacuum pressing. Difference between these two is that in the first case mixture of powders or porous compact is encapsulated, degassed and then hot pressed in air, while in the second case degassing is done without encapsulation of the material just before the hot pressing under vacuum. It is accepted that the latter one offers better results.

To control a large number of parameters and to understand certain mechanisms of above mentioned processes is very important for the production of composites. By presenting the densification process with densification maps it is much easier to comprehend these issues (German, 1996). Practical value of these maps is that they link the influence of the crystal structure and atomic bonds on the plastic flow of the materials. This further enables more simple analysis of mechanisms which are very important for the experiment design.

Densification maps presented in Figure 15 (Bozic et al., 2009) are determined for the constant pressure of 35 MPa and for the constant size of matrix powder particles (2R=125 μm). A model that is used is universal and it allows the comparison of densification mechanism for different applied pressures and different particle sizes. Other densification process parameters can easily be changed by changing boundaries between zones of dominating densification mechanisms. Due to the modular model structure, complete modulus can be corrected or changed. This fact is particularly useful since it enables potential corrections of the model.

A model used in this study was made for analyzing different densification mechanisms during the hot pressing, such as plastic flow, power-law creep and diffusion. From the chosen examples it can be noted that the process of densification is slowing down as the fraction of the reinforcing phase increases (slope of the dashed lines) which is in agreement with the experimental studies. A change in densification mechanism during hot pressing induced by the change of fraction of reinforcing phase can also be observed from this maps (different size of the zones for certain densification mechanism).

Densification maps provide a possibility of predicting dominant deformation mechanism, densification rate and time of pressing for the given conditions of hot pressing process.

Densification parameters are slightly changed compared to the optimal hot pressing parameters for the DRA materials production. Poreless structure was obtained at higher temperature (470 °C) and during longer time (5h). Only under these conditions densification can be fully performed when low pressure (35 MPa) is applied.

In the presented case, complete densification was achieved for aluminum alloy and DRA composite with 15 μm SiC, but not for the composite containing the smallest (0.7 μm diameter) SiC particles.

Microstructures of the hot-pressed composite materials show that mixing parameters used in the present work enabled formation of relatively uniformly distributed SiC particles in the aluminum matrix (Fig. 16 a and b). Porosity existing in the composites reinforced with 0.7 μm SiC, which has been detected after density measurements (Table 2) is illustrated in Fig. 17. Formation of agglomerates during mixing was the primary reason for such behaviour.

6. INFLUENCE OF THE REINFORCING PHASE ON MECHANICAL AND FRACTURE CHARACTERISTICS OF DRA MATERIALS

Prior to compressive testing, all hot-pressed materials were subjected to two types of thermal treatments: under-ageing and peak-aging. Initially, all samples have been homogenized by solution thermal treatment (1h at 475 °C) followed by water quenching to room temperature. Under-ageing was performed at 120 °C for 1h, while the peak-ageing was done at the same temperature of 120 °C for 24h for the CW67 alloy and 22h for the composites. Presence of SiC particles significantly influences the heterogeneous nucleation of strengthening phases. Capability for heterogeneous nucleation can be partially attributed to high dislocation density in the composite matrix, which is present due to the large mismatch between the thermal expansion coefficients of the matrix and the reinforcing phase. It is well known that dislocations act as nucleating sites for precipitation during aging, but they also promote accelerated ageing in the matrix. For that reason, peak-aging conditions for composites are reached in shorter times than for the monolithic alloy (Fig. 18).

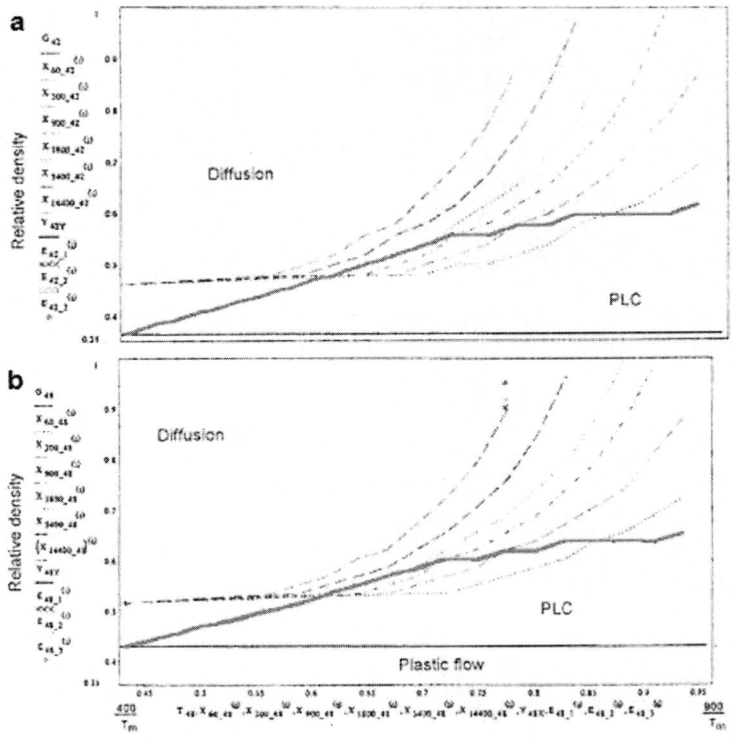

Figure 15. Deformation mechanism maps (*X t_D0* is a time curve, where *t* is the time of densification and *D 0* is the starting density of the powder, and *E D0_n* is the group of experimental values; thick line represents the boundary between power-law creep and diffusion zones. a) Pure Al powder; b) Al-10 vol.% SiC composite. Al particle radius in the composite was *2R=125 μm*, and SiC particle radius was *2R=33 μm*. Applied pressure was P=35 MPa (**Bozic et al., 2009**)

Table 2. Density values of the hot pressed aluminum alloy (CW67) and two selected composites.

Sample	Theoretical density, ρ_t (g/cm^3)	Density, ρ (g/cm^3)	Percentage of theoretical density
CW67	2.88	2.87	100
CW67-0.7 μm SiC	2.90	2.80	97.5
CW67-15 μm SiC	2.93	2.93	100

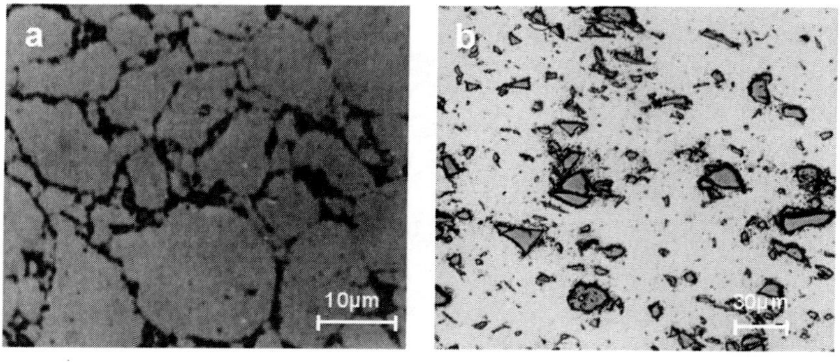

Figure 16. Microstructures of hot-pressed samples. a) LM. Composite containing 10 vol.% - 0.7 μm SiC, b) SEM. Composite containing 10 vol.% - 15 μm SiC.

Influence of different SiC particle sizes on the yield strength (σ_{yc}), ultimate strength (σ_{uc}) and ductility (strain to failure) (ε) under compression at room temperature for both under-aged (UA) and peak-aged (PA) samples is listed in Table 3.

Values of yield strength (measured at 0.2 % offset strain) for all composites, irrespective to heat treatment, are higher than those of the monolithic alloy. On the other hand, the ultimate strength and ductility values of the composite samples are lower than those obtained for the unreinforced alloy. It can also be noted that 0.2 % offset strain occurred at much higher fraction of the total strain-to-failure in the composite than it did in the case of the CW67 alloy. Main reason for such behavior could be the influence of the SiC particle sizes, their nature and agglomerate formation. These results also imply that smaller number of particles is damaged when the applied stress reaches yield stress in the composites. Stress will continue to accumulate after reaching the yield strength, leading to rapid failure and therefore low values of the ultimate strength and ductility.

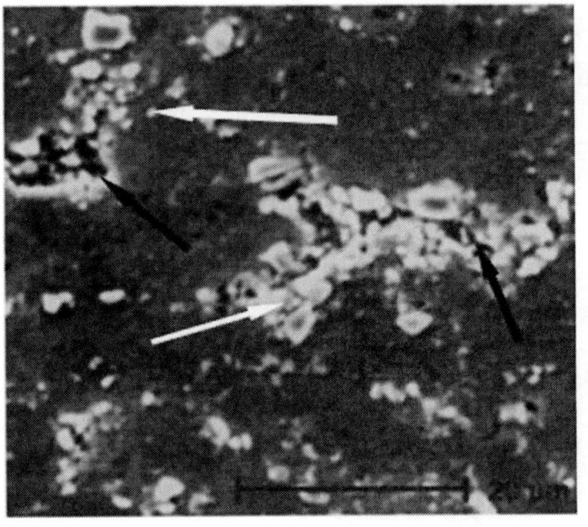

Figure 17. SEM. Microstructure of hot-pressed composite containing 10 vol.% - 0.7 μ SiC. White arrows -particle agglomeration; Black arrows-pores.

Analysis of results indicates that the samples submitted to peak-aging generally exhibited higher yield strength and ultimate strength values but also lower ductility when compared to under-aged samples.

The highest compressive property values are obtained for the composite reinforced with 10 μm SiC particles. It is worth noting that the lowest values are obtained for the composite material with 0.7 μm SiC particles. This result is unexpected since it is known that particle size decreasing is associated with improved mechanical characteristics of the particle reinforced composites.

Generally, mechanical behavior of composites is affected by the presence of reinforcing particles dispersed in the metal matrix. When analyzing the yield behaviour of these materials, it is important to recognize that the yielding strength of the composite is determined by the yield strength of the matrix. Most probably yield strength may not be the same as yield strength of the monolithic alloy processed and heat-treated in a similar way (Doel & Bowen, 1996). The local yield strength of the matrix may, in fact, be higher than that of the monolithic material due to the increased dislocation density and reduced grain size. Namely, matrix grain size is reduced by particles acting as nucleation sites during solidification. Also, grain size is smaller due to formation of subgrains after dislocation rearrangement into boundaries within the grain (Rees, 1998). These subgrains are formed in grains with high dislocation density surrounding a SiC particle. This rearrangement is a recovery process driven by the energy stored within the distorted matrix at the interface of two phases. Also, SiC particles act as obstacles for dislocation motion.

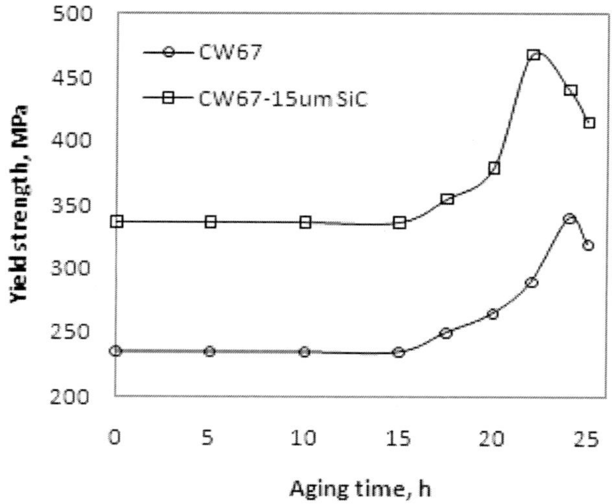

Figure 18. Yield strength of the CW67 alloy and its corresponding composite *vs.* aging time.

There are several factors affecting the local stress in the matrix. First of all, there is a tensile residual stress field in the matrix due to the difference in coefficient of thermal expansion (CTE) between aluminum alloy and SiC. Partial relief of this stress can result in an increase of dislocation density (Miller & Humphreys, 1991). Further, any applied load cannot be evenly distributed between the matrix and the SiC particles because of difference in stiffness. Several authors have previously reported that the reinforcing phase can support more applied load than the matrix (Doel & Bowen, 1996). Thus, local stress in the matrix may be lower than the nominal applied stress. In addition, SiC particles can withstand this stress until it becomes sufficiently high to cause particle fracture or failure either at or near the interface. It is believed that damage will first occur by particle fracture because the interface between SiC particles and aluminum is very strong. Hence, as soon as the particle breaks the stress in the undamaged material increases. In time, micro voids are beginning to form in the matrix as a result of dislocation pile-up, due to the applied load. These voids continue to grow causing failure of composite materials.

Table 3. Compressive properties of pure aluminum alloy and reinforced materials in under-aged (UA) and peak-aged (PA) conditions.

Material	σ_{yc} (MPa)		σ_{uc} (MPa)		ε (%)	
	UA	PA	UA	PA	UA	PA
CW67	235	340	860	950	38	35
CW67-0.7 μm SiC	312	400	720	740	22	19
CW67-10 μm SiC	400	490	850	930	27	23
CW67-15 μm SiC	331	437	743	830	24	20

In order to comprehend the exact influence of SiC particle size on the compressive properties of the composite, a comparison was made between the properties of three composites in both under-aged and peak-aged condition.

Composite sample with 0.7 μm SiC particle size had the smallest increase of yield strength, presumably due to the presence of the agglomerates in the structure (Fig. 17), originating from the starting powder (Fig. 6). In this case, matrix around the agglomerates is subjected to higher stress because they cannot support the same amount of stress as non agglomerated particles. Therefore, matrix is forced to yield at lower applied stress and, therefore, yield strength of the composites is also lower. Fracture surface of this composite show evidence of both trans-crystalline and inter-crystalline fracture mechanisms (Fig. 19). A factor that might have a detrimental effect on the compressive properties of this particular composite material is the homogeneity of the particle distribution.

Composites reinforced with 10 and 15 μm SiC particles are generally both stronger and more ductile than the material reinforced with 0.7 μm particles. Of the two analyzed composites, one containing 10 μm SiC particles has the highest increase in yield strength and the lowest decrease in ultimate compressive strength and ductility for both ageing conditions.

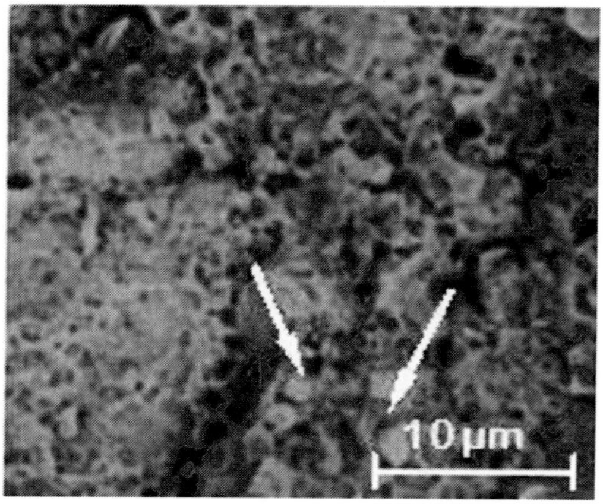

Figure 20. SEM. Fracture surface of composite with 0.7μm SiC particles in peak-aged condition. White arrows appointing agglomerates.

Main reason for the superior mechanical characteristics of the medium sized particles composites is uniformity of reinforcing phase particle dimensions (Fig. 20) and structure without agglomerates.

By making a mixture with narrower particle size distribution, the number of fracture initiation sites in DRA material is reduced making the matrix/SiC decohesion a dominant fracture mechanism (Fig. 20).

Formation of the crack and the crack path depend on the microstructure homogeneity of the alloys. Uniform path of the existing crack is typical for the less homogenous alloys, while the zig-zag path characterizes homogenous alloys (Fig. 21). The latter type of crack propagation is a consequence of homogenous distribution of secondary (SiC) particles which block the propagating crack and deflect it. By deflecting the crack, the energy level at the crack tip which is necessary for its propagation decreases disabling its further propagation. Therefore, more energy (larger applied force) has to be introduced to the system for the continuation of crack movement. This fact explains why these, more homogenous alloys, have better mechanical properties (Fig. 22).

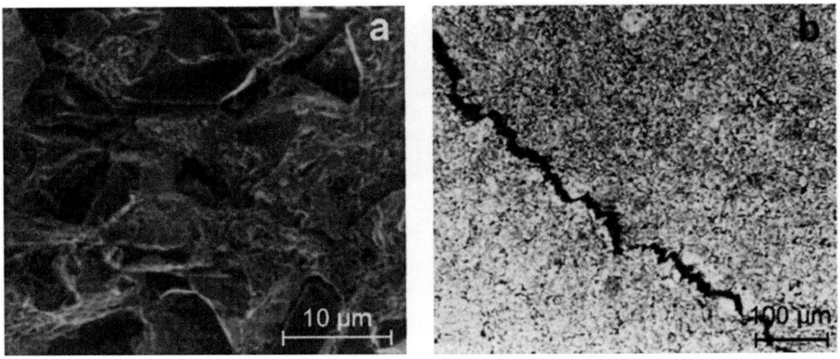

Figure 21. Fracture surface of DRA materials after 60 min mixing. a) SEM. CW67- 10 vol.% SiC, d_{SiC}=10μm; b) LM. CW67- 10 vol.% SiC, d_{SiC}=10μm.

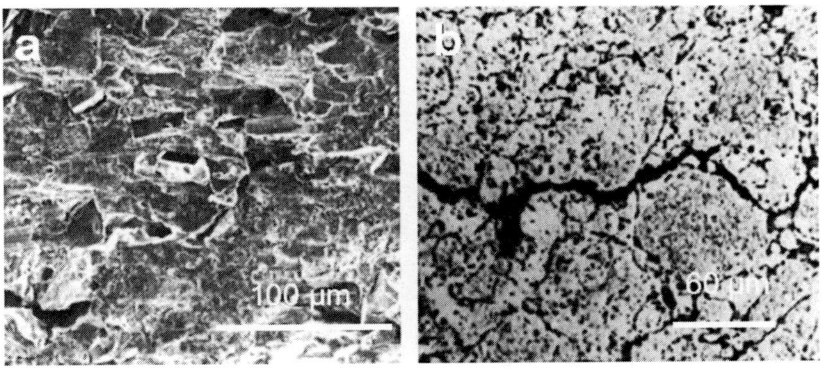

Figure 22. Fracture surface of DRA materials after 60 min mixing. a) SEM. CW67- 10 vol.% SiC, d_{SiC}=15μm; b) LM. CW67- 5 vol.% SiC, d_{SiC}=0.7 μm.

In the case when the reinforcing particles are not evenly distributed in the alloy matrix, crack propagates easily through the matrix (Fig. 23 a and b). This behaviour corresponds to poor mechanical properties of the DRA material with such microstructure. When mixing time is too long a formation of SiC agglomerates is noticeable. Due to the very weak bonding between the SiC particles in the agglomerates, a particle decohesion will easily occur under applied stress and the crack will appear (Fig.24).

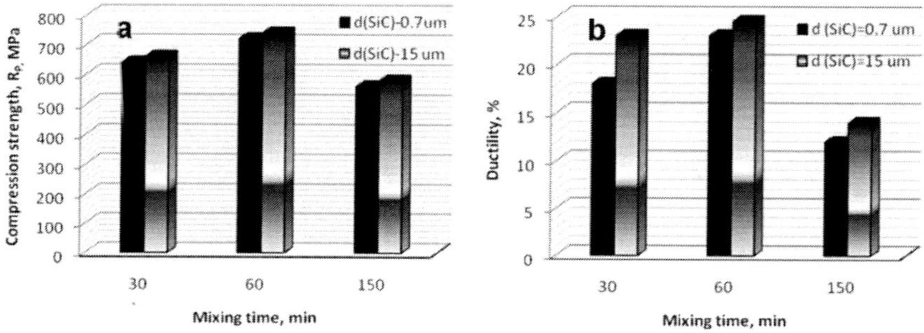

Figure 23. CW67- 5 vol. % SiC composite compression properties dependence on time of mixing. a) compression strength; b) ductility.

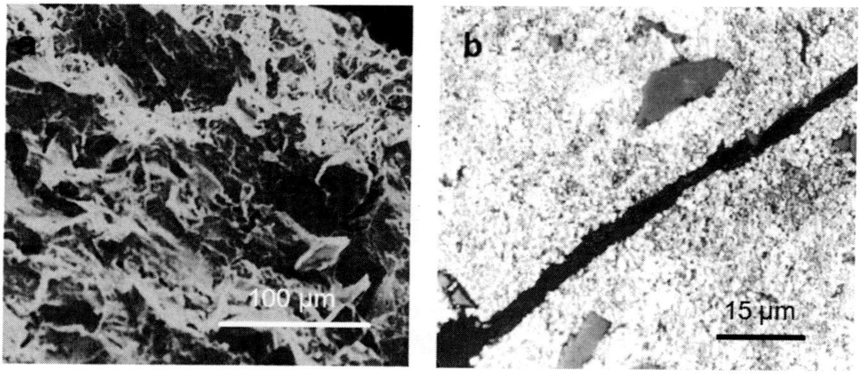

Figure 24. Fracture surface of DRA material after 30 min mixing. a) SEM. CW67-10 vol.% SiC, d_{SiC}=15 µm and b) LM. CW67-10 vol.% SiC, d_{SiC}=15 µm.

As observed by Humphreys (1988), an addition of a brittle reinforcement phase in high strength aluminum alloys may even decrease the ultimate tensile strength. Indeed, in the present composites this was found to be the case. Indeed, in the present composites this was found to be the case. The values of the ultimate strength were lower than the values for the monolithic material (Table 3). The presence of SiC particles can be detrimental to the ultimate compressive strength of the composite materials because of more additional cracking mechanisms compared to monolithic alloy. These mechanisms are: particle cracking (Fig. 25), particle matrix debonding (Fig. 20 a) and particle agglomerate decohesion (Fig. 19). The latter two mechanisms are of a secondary importance when particles are well distributed and strongly bonded. Thus, it is particle cracking that has a major influence on the ultimate compressive strength of SiC/Al composite materials.

Extensive studies of the room temperature mechanical behavior of these composites have been conducted, but the information about mechanical properties of DRA materials at elevated temperatures is still limited. Therefore, some results of the temperature effect on the compressive properties and fracture behaviour of a SiC particle reinforced CW67 aluminium alloy will be presented.

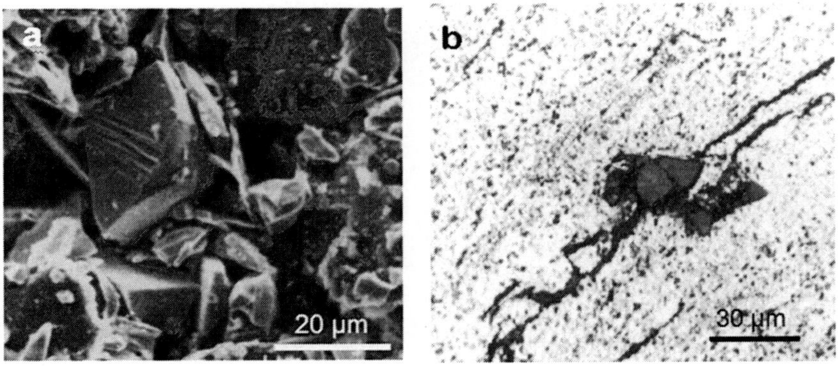

Figure 25. Fracture surface of DRA material after 150 min mixing. a) SEM. CW67-5 vol.% SiC, d_{SiC}=15 μm; b) LM. CW67-5 vol.% SiC, d_{SiC}=15 μm.

With increasing temperature matrix becomes softer and stress accommodation around the particles caused by difference in CTE becomes easier compared to the room temperature. The constraint around the particles in the matrix can be relaxed by a dynamic recovery process. It is therefore expected that any misfit strain gradient can be relaxed by the recovery process which could lead to a decrease in work hardening and strength, resulting in a small difference in strength between the composite and the monolithic alloy at higher temperatures (Fig. 26 a, b). Increase of composite ductility with temperature (Fig. 26 c) can be a consequence of the improved ductility of the matrix. However, ductility of the composite is lower than that of the monolithic alloy, which is caused by the particle cracking or particle matrix debonding effects.

Figure 26. SEM. Fracture surface of the composite with 15 vol.% SiC.

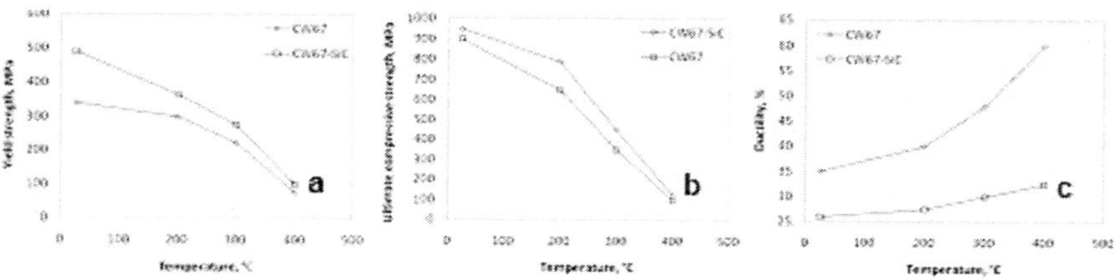

Figure 27. a) to c) Compressive properties of CW67 and CW67- 15 vol. % SiC composite as a function of temperature.

Microstructures of composites strained at room temperature show that SiC particles may fracture more frequently in the regions of agglomerates or by particle cracking of large particles, rather than by debonding between the matrix and SiC particles.

Presence of cavities is observed in microstructures of samples tested at elevated temperatures at the matrix/reinforcing phase interface (A in Figs. 27 and 28 b), particle boundaries (B in Fig. 27 b), and in the regions where the SiC particles are agglomerated (C in Fig. 27).

At 200°C, small voids were found at the particle boundaries with tiny linkage between voids. At 300°C and 400°C these voids become larger. It has been observed (Fig. 27 a-c) that debonding between particles and the matrix accelerates above 200°C.

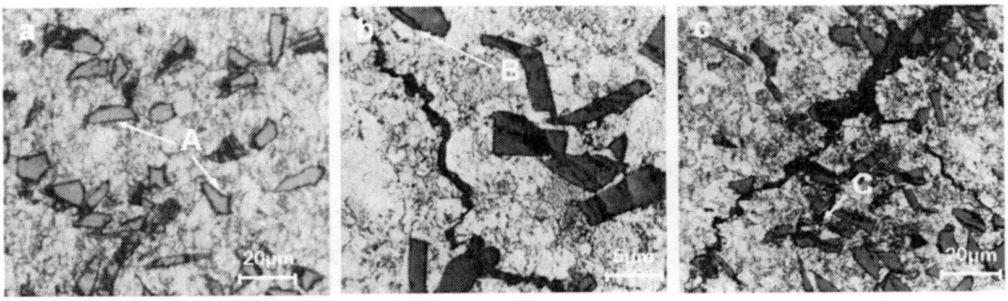

Figure 28. LM. Longitudal sections of the composites tested at: a) 200°C, b) 300°C and c) 400°C.

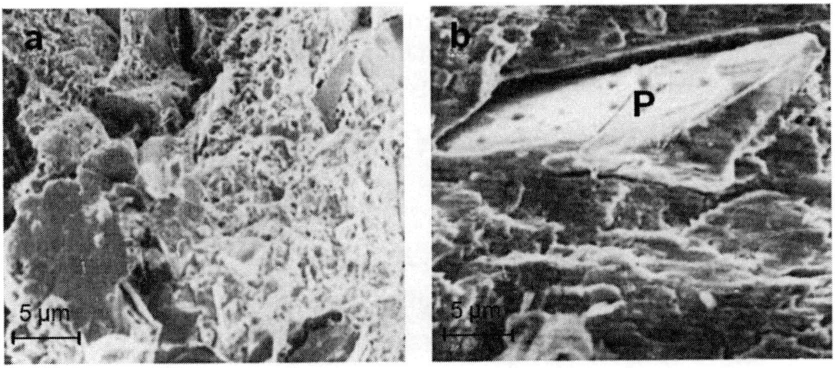

Figure 29. SEM. a) and b) Fracture surface of the composite CW67-SiC tested at 400°C. (Mg, Cu)Zn_2precipitates on SiC (indicated with P).

7. CONCLUSIONS

The basic conclusion is that much more research is needed for further enhancement of the DRA materials properties. Computer simulation and mathematical modeling of composite fabrication would enable faster achieving of results and therefore these methods need to be addressed.

In this area some scattered work has been done but it still needs further development. By finding and establishing the proper models actual experimentation and prediction of properties, shape, size, volume fraction of the participating phases and mixing parameters would largely be cut down.

As mentioned before, dispersed phase size, shape, volume fraction, wettability and distribution play the most important role in the properties attained in metal-matrix composites. A lot of research is focused on optimizing these parameters and based on the results, property designing of composites is possible. However, all the experimentation has stopped at decreasing the dispersed phase level below 10μm, possibly due to the difficulty in dispersing uniformly and without coagulation of finer particles in the matrix. This is especially pronounced when the liquid metallurgy route is adopted for making the composites. If the coagulation problem cannot be solved when the particle size is decreased below 10 μm, alternate methods of fabrication have to be attempted. Most promising method so far, seems to be in-situ production of composites, which is recently a very modern trend in science. In the present age of nanomaterials, methods of dispersion or making composites with nano sized dispersed phase holds a lot of potential for commercial exploitation in the future.

ACKNOWLEDGEMENTS

Authors are in debt to the Ministry of Science of the Republic of Serbia for the financial support realized through the Project No. 172005.

REFERENCES

1. Adler P.N., Delasi R., Geschwind G., Influence 7075 microstructure on mechanical properties and stress-corrosion susceptibility of 7075 aluminum-alloy, Metallurgical Transaction, 3 12 (1972), 3191 3200 .

2. D. Bozic, M. Pavlovic, M. Mitkov, M. Nikacevic, Mechanical and structural properties of CW67 alloy produced by powder metallurgy techniques (in Serbian), Metalurgija 3 (1997), 201

3. Bozic D., Dimcic O., Dimcic B., Vilotijevic M., Riznic-Dimitrijevic S., Modeling of densification process for particle reinforced composites, Journal of Alloys and Compounds, vol. 48712 1-2 (2009), 511 516 .

4. T. J. A. Doel, P. Bowen, Tensile properties of particulate-reinforced metal matrix composites, Composites Part A, vol. 278 (1996), 655 665 .

5. Flom Y.and Arsenault R.J., Effect of particle size on fracture toughness of SiC/Al composite material, Acta Metall 37 1989 2413 2423 .

6. German R.M., Powder Metallurgy Science, Second Edition, 1994 Metal Powder Industries Federation, Princeton, NJ.

7. German R. M., Sintering Theory and Practice, John Wiley and Sons Inc., New York,NY; 1996

8. Ž. Gnjidić, D. Božić, Tehnika, vol. 8, 6 (1999), 5

9. Gnjidić Ž., Božić D., and Mitkov M., The influence of SiC particles on the compressive properties of metal matrix composites, Mater Charact 47 2001 129 138 .

10. Gray W.A., The packing of solid particles, Champan and Hall Ltd., London (1973).

11. Humphreys F. J., In: Proceedings in Mechanical and Physical behavior of metallic and ceramic composites, ed. Anderson S et al. 51 (1988).

12. M. G. Mc Kimpson, A. N. Heimi, X. Hung, In: Proceedings of structural applications of mechanical alloying, vol. 203 Materials Park, Ohio: ASM; 1999.

13. W. S. Miller, F. J. Humphreys, Strengthening mechanism in metal matrix composites, Scripta Metall. Mater. 25 1991 pp.

14. Rees D.W.A., deformation and fracture of metal matrix particulate composites under combined loading, Composites, Part 1998, 29A: 171 82 .

15. Tietz T.E. and Palmour G., Advances in powder technology vol. 9 ASM, Metals Park, Ohio (1986) p. 189.

9

DEVELOPMENT OF Β TYPE TI23MO-45S5 BIOGLASS NANOCOMPOSITES FOR DENTAL APPLICATIONS

Karolina Jurczyk [1], Andrzej Miklaszewski [2], Mieczyslawa U. Jurczyk[3] and Mieczyslaw Jurczyk [2,*]

[1] Department of Conservative Dentistry and Periodontology, Poznan University of Medical Sciences, Bukowska 70, Poznan 60-812, Poland

[2] Institute of Materials Science and Engineering, Poznan University of Technology, Jana Pawla II 24, Poznan 61-138, Poland

[3] Division of Mother's and Child's Health, Poznan University of Medical Sciences, Polna 33, Poznan 60-535, Poland

ABSTRACT

Titanium β-type alloys attract attention as biomaterials for dental applications. The aim of this work was the synthesis of nanostructured β type Ti23Mo-*x* wt % 45S5 Bioglass (*x* = 0, 3 and 10) composites by mechanical alloying and powder metallurgy methods and their characterization. The crystallization of the amorphous material upon annealing led to the formation of a nanostructured β type Ti23Mo alloy with a grain size of approximately 40 nm. With the increase of the 45S5 Bioglass contents in Ti23Mo, nanocomposite increase of the α-phase is noticeable. The electrochemical treatment in phosphoric acid electrolyte resulted in a porous surface, followed by bioactive ceramic Ca-P deposition. Corrosion resistance potentiodynamic testing in Ringer solution at 37 °C showed a positive effect of porosity and Ca-P deposition on nanostructured Ti23Mo 3 wt % 45S5 Bioglass nanocomposite. The contact angles of glycerol on the nanostructured Ti23Mo alloy were determined and show visible decrease for bulk Ti23Mo 3 wt % 45S5 Bioglass and etched Ti23Mo 3 wt % 45S5 Bioglass nanocomposites. *In vitro* tests culture of normal human osteoblast cells showed very good cell proliferation, colonization, and multilayering. The present study demonstrated that porous Ti23Mo 3 wt % 45S5 Bioglass nanocomposite is a promising biomaterial for bone tissue engineering.

KEYWORDS

titanium; 45S5 bioglass; nanocomposite; powder metallurgy; surface properties; etching

1. INTRODUCTION

Since 1965, titanium has found an increasing application as an implant material in medicine. Commercial purity titanium has high corrosion resistance and outstanding biocompatibility [1,2,3]. One reason for these advantages may be a protective oxide layer, which forms spontaneously on the implant surface [4]. Titanium exists in two allotropic forms. At low temperatures, it has a closed packed hexagonal crystal structure (hcp), which is commonly known as α, whereas above 882 ± 2 °C, it has a body centered cubic structure (bcc) termed β. The alloying elements such as Al, O, N, *etc.* tend to stabilize the α phase while elements V, Mo, Nb, Fe, Cr, *etc.* stabilize the β phase [1,3].

Titanium and its alloys also attract a lot of attention in dental applications [5,6,7]. Pure titanium and Ti-6Al-4V alloy are the main materials in the dental field as well as in the surgical one. Implant sensitivity to titanium alloys is very seldom, despite components such as vanadium which are described to be cytotoxic [7]. By the elimination of toxic elements, it is possible to prepare Ti-type alloys with excellent biocompatibility. Ti-6Al-7Nb, which has been developed for surgical implants, is also attractive for dental applications [8]. Recently, Ti-40Zr, Ti-5Al-13Ta and Ti-43.1Zr-10.2Al-3.6V have been proposed [9,10]. Moreover the development of Ti-Mo-A (A = Ga, Ge, Al) and Ti-Ta, Ti-Ta-Zr, Ti-Sc-Mo alloys, as shape memory titanium alloys, for biomedical use is noticeable [11]. All above mentioned are β type titanium alloys.

New perspectives appear with nanostructure materials, which exhibit better mechanical and physico-chemical properties in comparison to their microcrystalline counterparts [12,13,14]. Current goals in the development of new Ti-based biomaterials are: (i) to avoid potentially toxic elements, such as vanadium, to further improve biocompatibility; (ii) to produce titanium alloys with a high fatigue strength. β-titanium alloys partially fulfill these requirements [1,5,8,9,10,11]. One of these improved β titanium alloys is Ti12Mo6Zr2Fe, which is currently used for implant materials.

Dental implants for load bearing applications require the use of materials that are both bioactive and have significant mechanical strength [8]. There are no existing materials that readily fit these two criteria. While titanium and titanium-based alloys have excellent mechanical properties and are generally well tolerated in a physiological environment, they have negligible capacity for osteointegration. Two approaches have been made to improve the osteointegration: a surface modification approach and a composite approach [15,16]. In the surface modification approach, the surface of the Ti surgical substrates is modified by chemical or electrochemical methods to make the surface bioactive, or a bioactive coating is applied using, for example, electrophoretic deposition technique (EPD) [15]. In the composite approach, the primary material is combined with bioactive materials, such as 45S5 Bioglass. Conventional powder metallurgy methods [14,16] or plasma-assisted processes [17] are commonly used for fabricating composites.

The biocomposites prepared from powder mixtures of titanium (α-Ti), hydroxyapatite (HA), and bioactive glass (BG) (SiO_2-CaO-P_2O_5-B_2O_3-MgO-TiO_2-CaF_2) were investigated [18]. The results showed that complex reactions among the starting materials mainly depended on the initial Ti/HA ratios as well as the sintering temperatures.

Recent studies in nanometer sized bioactive glasses (nBG) have indicated their possibility for biomedical applications. The potential of top-down processing of 45S5 BG by wet comminution in a stirred media mill was investigated [19]. The products were assessed by *in vitro* hydroxy-carbonate apatite (HCAp) formation in simulated body fluid, which is a marker for bioactive behavior. Bioactive submicron particles can be produced in organic solvents (*n*-butanol, *n*-pentanol, *n*-hexanol). However bioactivity is lost after processing in water. Wet comminution under controlled conditions is an appropriate technique in the production of submicrometer bioactive glass particles with enhanced bioactivity.

Moreover, bioceramic (HA, 45S5 Bioglass) coatings, applied in order to improve the surface bioactivity of titanium and its alloys, often flake off as a result of poor ceramic/metal interface bonding, which may lead to failure of the surgery [20]. The problems mentioned above can be solved by the fabrication of Ti-bioceramic composites. Recently the studies on the fabrication and characterization of Ti-based composites reinforced with bioceramic (HA or 45S5 Bioglass) particles have been reported [16,21,22].

The properties of composites depend very much upon structure, shape, volume fraction, and the interface among the constituents, as well as upon the chemical composition and their final microstructure. The *in vivo* studies indicate that Ti-20 vol % HA composite has good biocompatibility and can integrate with bone [22]. The osteointegration ability of the composite is better than that of pure titanium, especially in the early stage after the implantation, which may be due to the presence of HA or 45S5 Bioglass in the Ti-matrix composite [23].

In this work the mechanical alloying (MA) process followed by the pressing and sintering was applied for the preparation of the nanocrystalline β-type Ti23Mo alloy and Ti23Mo-*x* wt % 45S5 Bioglass nanocomposites (x = 3 and 10). Finally, the bulk samples were anodically etched. The bioactive ceramic Ca-P layer was cathodically deposited too. Structure, microstructure, composition, porosity, surface morphology, corrosion resistance, wettability, and *in vitro* cytocompatibility were investigated. Yet to the authors' knowledge, there have been no papers regarding the addition of 45S5 Bioglass into β Ti23Mo nanocomposite to have appeared until now.

2. EXPERIMENTAL DETAILS

2.1. Chemicals and Materials

The following commercial powders were used: titanium, CAS number 7440-32-6 (<45 μm, 99.9%, Alfa Aesar GmbH & Co KG, Karlsruhe, Germany), molybdenum, CAS number 7439-98-7 (44 μm, 99.6%, Sigma Aldrich, Saint Louis, MO, USA), 45S5 Bioglass (45% SiO_2, 24.5% Na_2O, 24.5% CaO, 6% P_2O_5, 53 μm; all in wt % from Mo-Sci Health Care L.L.C., Rolla, MO, USA).

The experiments were carried out on Ti23Mo alloy and two composite materials. For brevity, in this work, materials are denoted as follows:

- Ti23Mo nanocrystalline alloy is labeled as Ti23Mo.
- Ti23Mo-3 wt % 45S5 Bioglass nanocomposite is labeled as Ti23Mo 3BG.
- Ti23Mo-10 wt % 45S5 Bioglass nanocomposite is labeled as Ti23Mo 10BG.

2.2. Sample Preparation

Mechanical alloying (MA) was performed using SPEX 8000 Mixer Mill (SPEX® SamplePrep, Metuchen, NJ, USA). Argon was a protective atmosphere. Round bottom stainless vials were used. Elemental powders (titanium, CAS number 7440-32-6 (<45 μm, 99.9%, Alfa Aesar), molybdenum, CAS number 7439-98-7 (44 μm, 99.6%, Sigma Aldrich) and 45S5 Bioglass (45% SiO_2, 24.5% Na_2O, 24.5% CaO, 6% P_2O_5, 53 μm; all in wt % from Mo-Sci Health Care L.L.C.)) were weighted, blended and poured into vials in a glove box (Labmaster 130) filled with automatically controlled argon atmosphere (O_2 < 2 ppm and H_2O < 1 ppm). The weight ratio of hard steel balls (10 mm diameter) to powder weight ratio equaled 15:1. The MA process lasted 30 h in all cases. In order to prevent severe cold welding during high-energy milling, the ball milling was stopped every 2 h to dissipate the heat and to reduce an excessive rise in temperature. In the next step, the produced powders with size distribution of the particles from 40–150 μm were placed into the matrix and uniaxially pressed at a pressure of 500 MPa. Finally, the green compacts were heated over 1 h to 800 °C and kept at this temperature for 30 min for particle sintering. After that, the sinters were slowly cooled down to room temperature (RT) together with the furnace. The sintering was done at 10^{-2} Pa vacuum in an alumina tube (McDanel Adv. Ceramic Technologies, Beaver Falls, PA, USA). The final bulk sinters have diameter and height of 8 and 4 mm, respectively (Figure 1).

The Solartron 1285 potentiostat (Solatron analytical, Farnborough, UK) was applied in the electrochemical etching stage of the sinters, followed at 10 V *vs.* open-circuit-potential (OCP) for 60 min in 1 M H_3PO_4 + 2% HF electrolyte. In the following stage, calcium phosphate was deposited on the surface. We applied cathodic deposition at −5 V *vs.* OCP in the electrolyte containing 0.042 M $Ca(NO_3)_2$ + 0.025 M $(NH_4)_2HPO_4$ + 0.1 M HCl for 1 h.

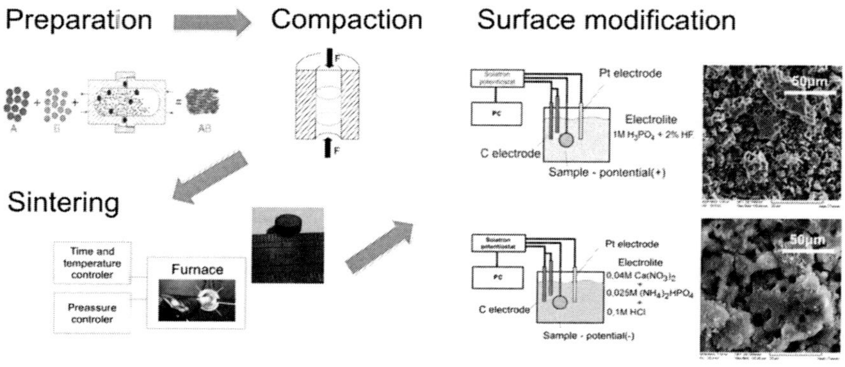

Figure 1. Experimental setup Ti23Mo-*x* wt % 45S5 Bioglass nanocomposite synthesis and electrochemical treatment procedure.

2.3. Material Characterization

The crystallographic structure of the samples during different processing stages was investigated at room temperature using a Panalytical Empyrean X-ray diffraction (XRD) with $CuK\alpha_1$ (λ = 1.54056 Å) radiation (Panalytical, Empyrean model, Almelo, The Netherlands). The conditions of XRD measurements were: voltage 45 kV, anode current 40 mA, 2θ range 20°–80°, time per step 12.54 s/step, step size 0.0167°. The average crystallite sizes d were estimated by Scherrer method: $\beta = 0.9\ \lambda/d\ \cos\theta$, where β is the full-width at half maximum intensity of a Bragg reflection excluding instrumental broadening, θ the Bragg angle, and λ the wavelength of the X-ray radiation. Scanning electron microscope (SEM, VEGA 5135 Tescan, Brno, Czech Republic) with energy dispersive spectrometer (EDS, PTG Prison Avalon, Princeton Gamma Tech., Princeton, NY, USA) was used to characterize the chemical composition of the prepared samples. EDS was calibrated using a typical Cu calibration procedure. Additionally, the Ti (99.9% from Alfa Aesar), Mo (99.6% from Sigma Aldrich) and SiO_2, Na_2O, CaO, P_2O_5 (all from Sigma Aldrich) specimens were used as reference.

The porosity of the porous materials was calculated by the formula $p = (1 - \rho/\rho_{th}) \times 100\%$, where ρ and ρ_{th} are the density of the porous material and its corresponding theoretical density calculated for the rule of the mixtures, respectively. The density of the sintered samples was determined by Archimedes method. The Vickers microhardness of the bulk samples was measured using a microhardness tester by applying a load of 300 g for 10 s on the polished surfaces of the samples. For each sample, 10 separate indents were created on the investigated surface.

The surface morphology of the samples was investigated using an Optical Profiler NT 1100 (Veeco WykoR®, Mannheim, Germany). Using the 3D optical profiler, the following standard roughness parameters were estimated: arithmetic mean roughness (μm)—R_a, maximum height of the profile (μm)—R_t, ten point mean roughness (μm)—R_z and hybrid 3D parameters: root mean square (°)—S_{dq}, surface area ratio gradient (%)—S_{dr}.

The Solartron 1285 potentiostat (Solatron analytical) was applied for corrosion measurements. The corrosion resistance of different samples was measured in Ringer's solution (simulated body fluid with composition: NaCl: 9 g/L, KCl: 0.42 g/L, $CaCl_2$: 0.48 g/L, $NaHCO_3$: 0.2 g/L) applying potentiodynamic mode with scan rate 0.5 mV/s at temperature of 37 ± 1 °C, controlled by thermostat. The corrosion test was run in EG & G K0047 corrosion cell. The counter electrode consisted of two graphite rods, and a platinum electrode (SCE, Hydromet, Gliwice, Poland) was used as the reference electrode. The surface area exposed to the electrolyte was 0.5 mm^2. Polarization curves were obtained for each specimen in the potential range for −2 to 3 V. The corrosion potentials (E_{corr}) and corrosion current densities (I_{corr}) were estimated from the Tafel extrapolations of the corrosion curves, using CorrView software (Scribner Associates Inc., Southern Pines, NC, USA).

For the surface wettability measurements, contact angles Θ were determined by computer software analysis Musial (Elektronika Jadrowa, Krakow, Poland) from geometrical shape of droplets recorded by the optical system with a digital camera. Droplets placement were realized by a special micropipette with a constant volume of test liquid (glycerol (99.5% purity); 2 μdm^3) on a clear, polished and rinsed with alcohol surface, by it raise until the bottom touches the specimen. If the drop is large enough, the adhesion to the surface pulls it off from the tip. All the data were obtained under equilibrium conditions, procedure was repeated three times for all samples. For each recorded picture (*i.e.*, for each liquid drop), the geometrical shape analysis was repeated 10 times: the extreme values were rejected, and the arithmetic mean value was calculated for the accepted findings.

2.4. In vitro Cytocompatibility

The *in vitro* cytocompatibility tests were performed under static conditions. The discs of Ti23Mo 3BG nanocomposite after electrochemical etching and additional Ca-P deposition were sterilized by autoclaving at 120 °C for 15 min and were separately located at the bottom of 24-well microplates. Normal Human Osteoblast (NHOst) cells from Cambrex (CC-2538, Walkersville, MD, USA) were cultured onto each disc at a concentration of 5000 cells/well in 1 mL of culture medium under static conditions. The cells were cultured at 37 °C in a 5% CO_2 incubator for 1 and 5 days. The medium was replaced every day. Then, the cells were fixed with a 25% glutaraldehyde solution for 10 min and stained with a 10% Giemsa's staining solution for 10 min. The specimens were sputter-coated with gold and examined using scanning electron microscope (SEM).

3. RESULTS

3.1. Structure Properties

Figure 2a,b shows the XRD patterns of the starting microcrystalline titanium (reference code 01-071-4632) and molybdenum (reference code 01-071-4645) powders. After 30 h of mechanical alloying the powders are amorphous (Figure 2d). The high plastic deformations of the powders results in high density of dislocation lines and subsequently subgrain formation, which finally leads to amorphisation [24,25]. The low temperature sintering results in crystallization and nanostructure formation.

XRD analysis of Ti23Mo (30 h MA and annealed at 800 °C for 0.5 h) showed the presence of almost single phase β-Ti type structure with cell parameter a = 3.234 Å (Figure 2e). When 45S5 Bioglass (Figure 2c) is added, the volume of the regular structure of Ti23Mo decreases, as manifested by a shift of the diffraction peaks of the (1,1,0), (2,0,0) and (2,1,1) crystal planes towards larger angles in comparison with pure Ti23Mo alloy (Table 1). According to the Scherrer method of XRD profiles, the mean crystallite size of heat treated Ti23Mo alloy was 8 nm. Ti23Mo 3BG nanocomposite is mostly single phase cubic β-type phase material (reference code 01-077-3482) with some impurities as α-Ti, MoTi (reference code 01-071-9821) and SiC (reference code 01-075-

834; Figure 2e). With the increase of the 45S5 Bioglass contents in Ti23Mo nanocomposite increase of α-phase is noticeable.

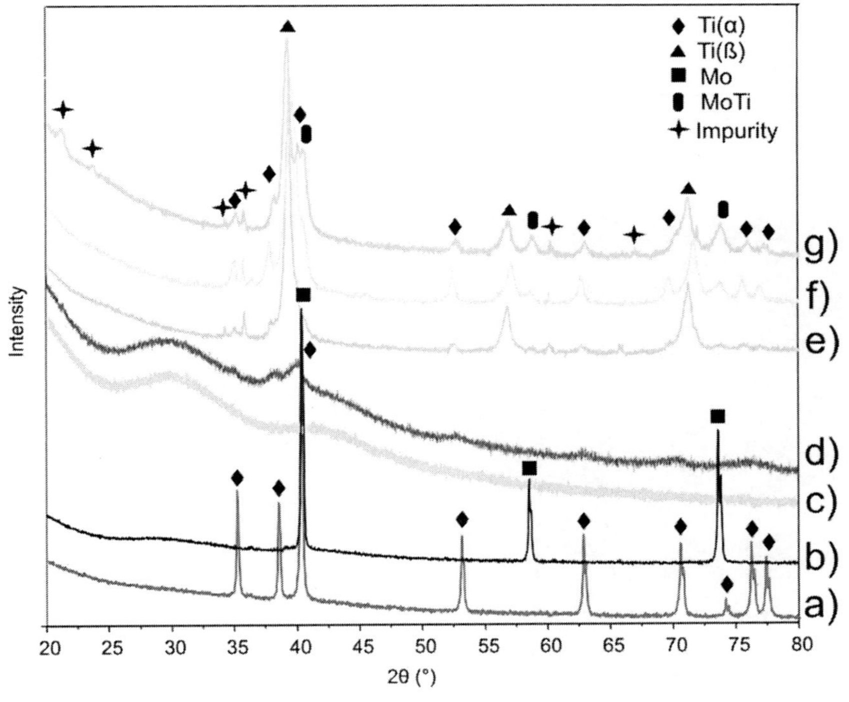

Figure 2. X-ray diffraction spectra of Ti (**a**); Mo (**b**) and 45S5 Bioglass (**c**) powders, their mixture after 30 h of mechanical alloying (**d**); and after annealing at 800 °C for 0.5 h: Ti23Mo (**e**); Ti23Mo 3BG (**f**); Ti23Mo 10BG (**g**).

The smooth and porous Ti23Mo alloy surface is presented in Figure 3a. The relative density of the bulk Ti23Mo 3BG nanocomposite was measured to be 91%. The results of EDS analysis of the surface of sintered Ti23Mo 3BG nanocomposite are shown in Figure 3d. It can be confirmed that synthesized nanocomposite mainly consists of Ti-Mo matrix with elements of O, Na, Si, Ca and P.

Table 1. Structural parameters (a, c, V), crystallite size (d), theoretical density (ρ_{th}) and microhardness ($HV_{0.3}$) for bulk Ti23Mo-x wt % Bioglass nanocomposites.

Sample	Phase	a (Å)	c (Å)	V (Å^3)	d (nm)	ρ_{th} (g/cm^3)	$HV_{0.3}/10$
Ti23Mo	-	3.248	-	34.265	19.1 ± 2.1	6.701	350
Ti23Mo 3BG	β	3.234	-	33.834	16.2 ± 1.5	6.581	365
Ti23Mo 10BG *	β	3.249	-	34.296	15.4 ± 0.7	-	-
	α	2.960	4.724	35.844	-	6.301	740

* 2-phase material.

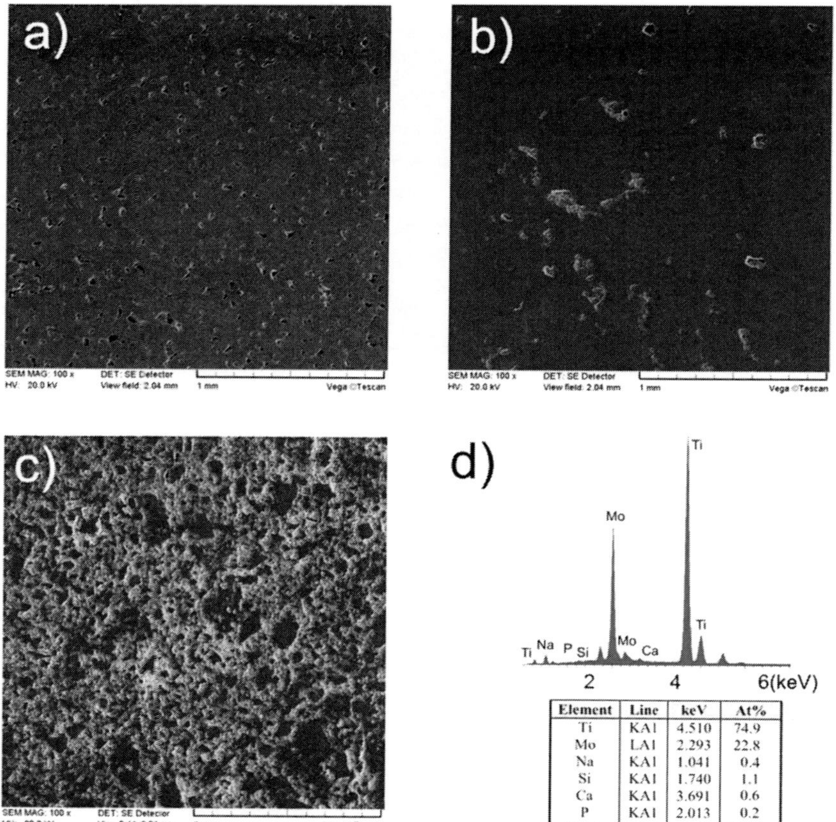

Element	Line	keV	At%
Ti	KA1	4.510	74.9
Mo	LA1	2.293	22.8
Na	KA1	1.041	0.4
Si	KA1	1.740	1.1
Ca	KA1	3.691	0.6
P	KA1	2.013	0.2
Total	-	-	100.0

Figure 3. Scanning electron microscopy (SEM) micrographs of Ti23Mo alloy (**a**); Ti23Mo 3BG (**b**) and Ti23Mo 10BG nanocomposites (**c**); Energy dispersive spectroscopy (EDS) analysis of the surface of sintered Ti23Mo 3BG nanocomposite is shown in (**d**).

Electrochemical anodic oxidation results in pore formation on the surface of Ti23Mo 3BG nanocomposite. The porosity of bulk Ti23Mo 3BG nanocomposite is about 9%. In the etched surface large macropores with size up to 20 μm (Figure 4a) and smaller mesopores with size smaller than 0.5 μm are formed (not shown). The macropores are created at the places, where previously remnant pores existed as a result of the sintering, while the mesopores on the surface are the result of the large volume grain boundary etching. The prepared rough surface states a good base for tissue growth and their strong fixing or for intermediate bioceramic (Ca-P) layer formation.

Next the Ca-P deposition (Figure 4b) leads to intermediate bioactive layer formation. The electrolyte with Ca/P ratio equal to 1.67 was used. The obtained layers are porous and rough, which result in increased surface area and enlarged area for contacts with tissue. The Ca-P is deposited in pits and pores, growing into them, therefore fixating the layer in the metallic background. EDS analysis of the deposited Ca-P layer (Figure 4c) shows that Ca content is 35.00 at % and P content is 21.69 at %, which means that the Ca/P atomic ratio is equal to 1.61, which corresponds almost to the value of hydroxyapatite (reference code 00-024-0033). On the spectrum, the calcium, phosphorus, and oxygen peaks dominate.

XRD results (Figure 5) show sinter before Figure 5a, after electrochemical etching Figure 5b and after electrochemical etching with additional Ca-P deposition Figure 5c. The sintered Ti23Mo 3BG nanocomposite has β-type structure with some trace of α-phase. On the XRD spectrum of Ti23Mo alloy the impurity is marked (Figure 5a). For etched Ti23Mo 3BG nanocomposite SiC phase is absent, which only confirms its surface character. In the anodized surface, besides pores, α-Ti phase is formed. XRD spectrum did not reveal strong peaks belonging to titanium oxides. It is likely during the electrochemical etching process, that thin amorphous titanium oxides were formed. After Ca-P deposition, the surface layer is composed of the apatite and small trace of MoTi phase.

The corrosion resistance of the different surfaces of the Ti23Mo 3BG composite in Ringer solution was investigated (Table 2, Figure 6). Using CorrView software (Solatron analytical), from the Tafel extrapolations of the recorded potentiodynamic corrosion curves, corrosion current density (I_{cor}) and corrosion potential (E_{cor}) were determined. The addition of 45S5 Bioglass to Ti23Mo alloy had also a positive effect on the corrosion resistance in Ringer's solution. Ti23Mo composite with 3 wt % of 45S5 Bioglass has better corrosion resistance (I_{cor} = 64.44

μA/cm², E_{cor} = −0.79 V) than nanostructured Ti23Mo alloy (I_{cor} = 70.98 μA/cm²). The best corrosion resistance is shown by Ti23Mo 3BG nanocomposite after electrochemical etching and Ca-P deposition (I_{cor} = 0.0168 μA/cm², E_{cor} = −0.44 V). After the electrochemical etching the surface is much rougher in comparison to the polished one. The oxides prepared in the anodic oxidation and Ca-P phase deposition from the electrolyte on the surface of Ti23Mo 3BG nanocomposite showed improved corrosion resistance. Additionally, the corrosion potential (E_{cor}) is shifted to the nobler direction (E_{cor} = −0.44 V).

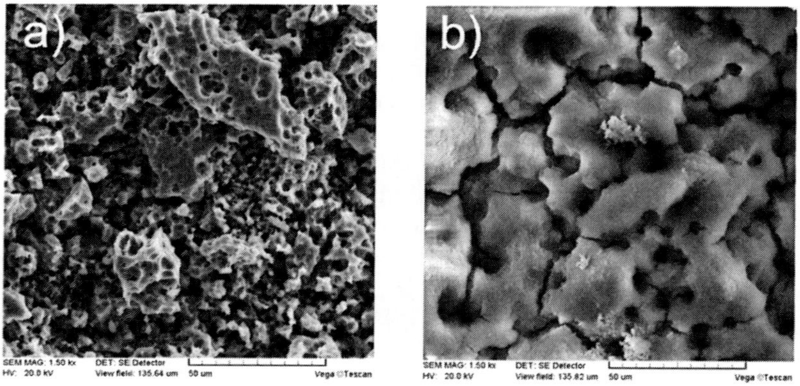

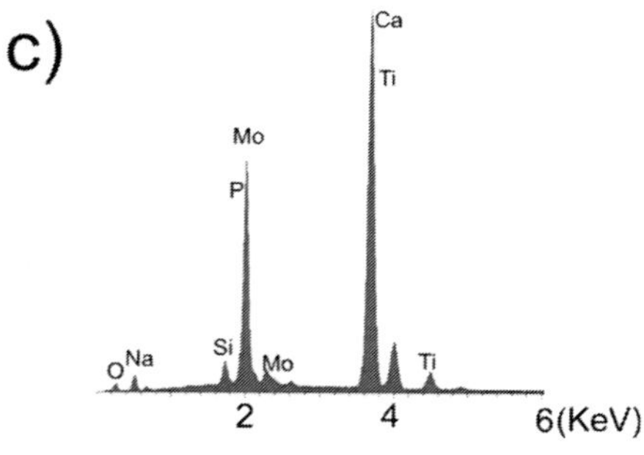

Element	Line	keV	At%
P	KA1	2.013	21.7
Ca	KA1	3.691	35.0
Ti	KA1	4.510	1.9
Mo	LA1	2.293	1.5
O	KA1	0.523	38.2
Si	KA1	1.740	1.6
Total	-	-	**100.0**

Figure 4. SEM micrographs of Ti23Mo 3BG after anodic oxidation (**a**) and additional Ca-P deposition (**b**) with EDS spectrum (**c**).

3.2. Surface Properties

Surface roughness is a considerably important property for the attachment of cells to the implant. It has been previously reported that not only micro- but also nano-topography can support the proliferation of different types of cells [12]. Optical profiler 3D topography and profiler surface scans and X-profiles of the bulk Ti23Mo 3BG nanocomposite, at the different processing stages, are presented in Figure 7 and Figure 8, respectively.

Table 2. Corrosion current density I_{corr}, corrosion potential E_{corr} and glycerol contact angles Θ of the nanocrystalline Ti23Mo and Ti23Mo 3BG nanocomposites before and after electrochemical etching as well as after additional Ca-P deposition.

Sample	I_{corr} (μA/cm^2)	E_{corr} (V)	Glycerol Contact Angle Θ (°)
Ti23Mo	70.93	−0.79	85.64 ± 0.72
Ti23Mo 3BG	64.44	−0.79	41.41 ± 0.94
Ti23Mo 3BG etched	5.43	−0.53	26.11 ± 1.62
Ti23Mo 3BG etched and Ca-P deposited	0.02	−0.44	72.41 ± 0.56

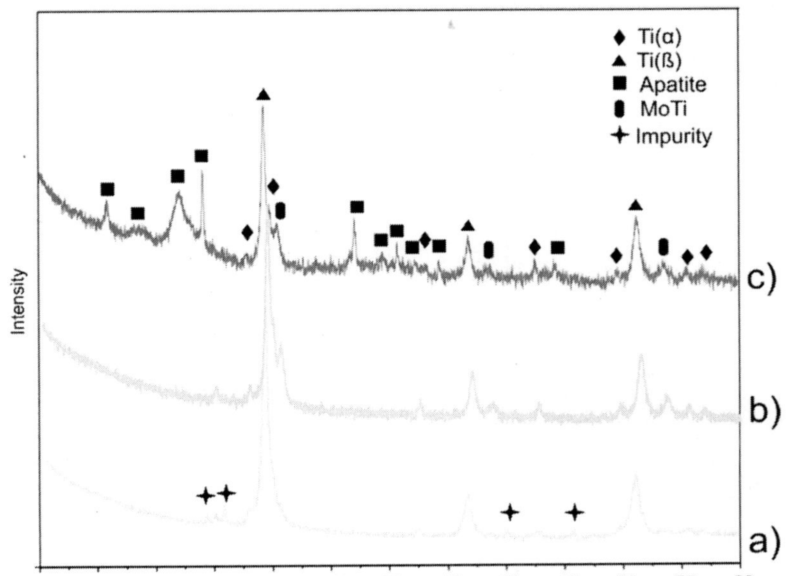

Figure 5. X-ray diffraction (XRD) spectra of Ti23Mo 3BG nanocomposite before (**a**) and after the surface biofunctionalization (**b**,**c**); etched (**b**) and Ca-P deposited (**c**).

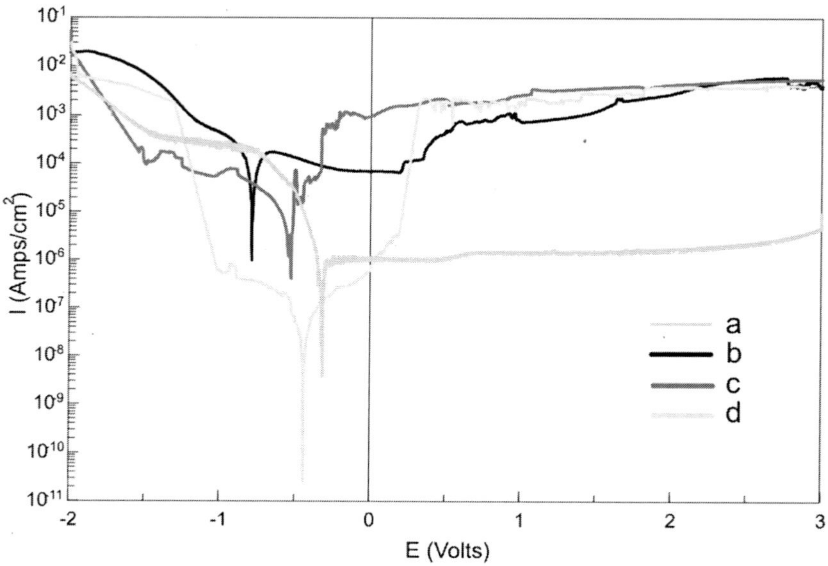

Figure 6. Potentiodynamic polarization curves of microcrystalline Ti (**a**); Ti23Mo 3BG nanocomposites before (**b**) and after electrochemical etching (**c**) as well as after additional Ca-P deposition (**d**) in Ringer's solution at 37 °C.

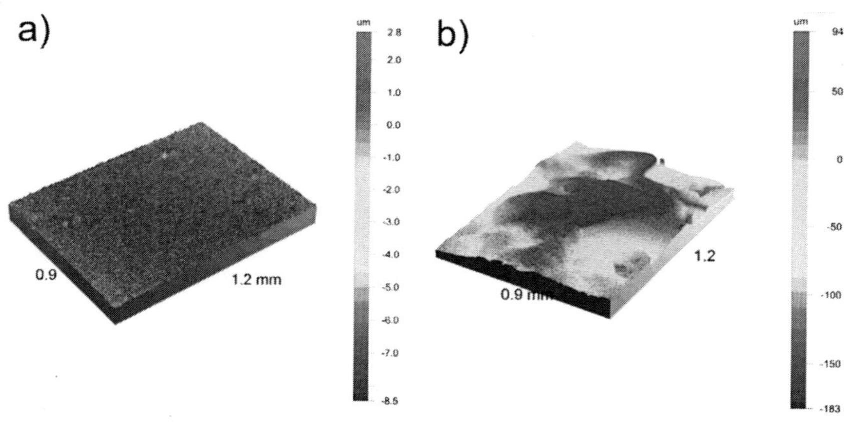

Figure 7. Optical profiler 3D topography of the Ti23Mo 3BG nanocomposites before (**a**) and after electrochemical etching with additional Ca-P deposition (**b**); 0.9 mm × 1.2 mm scan size.

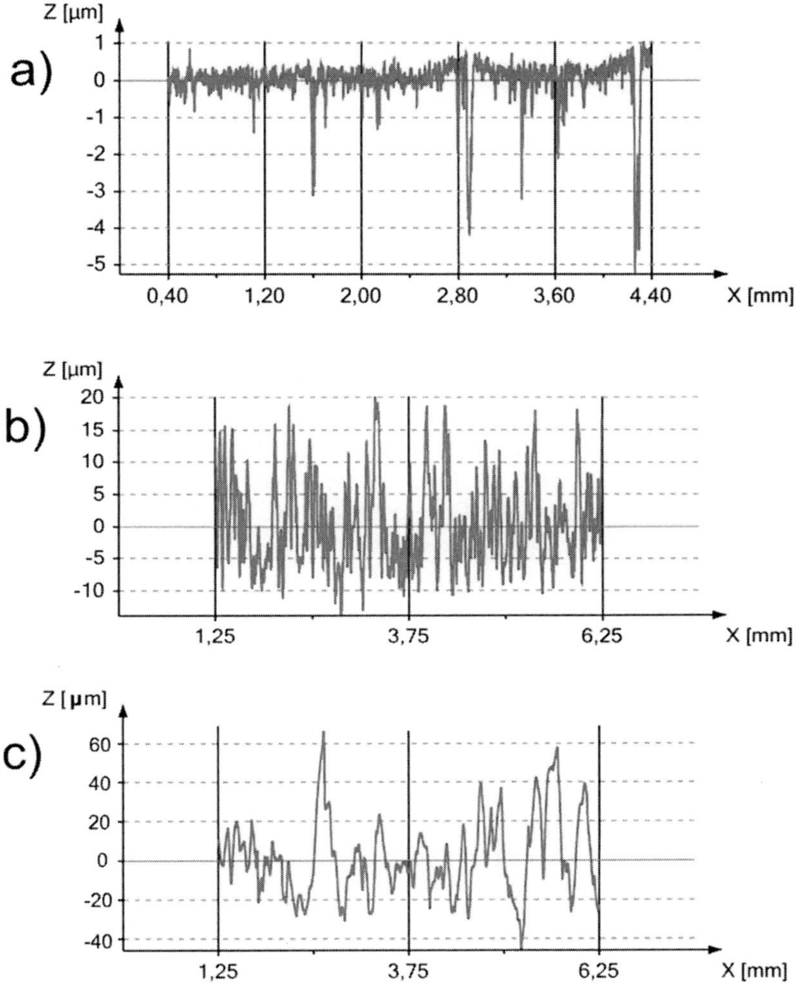

Figure 8. Optical profiler surface scans and X-profiles of the Ti23Mo 3BG nanocomposites before (**a**) and after electrochemical etching (**b**) as well as after additional Ca-P deposition (**c**).

At different processing stages changes occur in the surface porosity (Figure 7). Primarily the polished surface exhibited only few pores (Figure 7a), which resemble the remnants of the sintering process. Significant differences were observed after etching and additional Ca-P deposition (Figure 7b). The increase of surface together with surface composition modification plays a key role for the living cell attachment and proliferation.

The bulk Ti23Mo 3BG nanocomposite had R_a, R_t and R_z values of approximately 0.29, 11.29 and 8.24 μm, respectively (Table 3). In contrast, all of the roughness parameters increased for the etched and Ca-P deposited Ti23Mo 3BG nanocomposite surface (Figure 7b). This sample surface had an average surface roughness with R_a, R_t and R_z values in the range of 35.80–276.55 μm. The optimal pore size for the cell attachment, differentiation, and ingrowth of osteoblasts and vascularization has been reported to be approximately 200–500 μm [26].

In Figure 8, the respective X-line profiles of the studied materials are shown. Large pores are formed in the surface during the etching process. This surface is able to support cell growth and proliferation, which provides a surface with large hillocks and a very non-uniform surface profile after culturing of the osteoblasts. The profile is an effect of the multilayered nature of growing cells.

Table 3. 2D (R_a, R_t, R_z) and hybrid (S_{dq}, S_{dr}) parameters for the Ti23Mo 3BG nanocomposite on different processing routes; parameters taken from surface area of 1.08 mm^2.

Processing Route	R_a (μm)	R_t (μm)	R_z (μm)	S_{dq} (°)	S_{dr} (%)
before etching *	0.29	11.29	8.24	8.97	1.22
after etching	6.42	42.47	35.54	74.37	341.72
after etching with Ca-P deposition	35.80	276.55	247.74	69.84	255.66

* polished.

The hybrid parameters (S_{dq}, S_{dr},) reflect slope gradients and exhibit comparable behavior to the roughness parameters. The S_{dq} value that affects the wetting of the surface by fluids significantly increases for the etched Ti23Mo 3BG sample. An increase in the surface slope (S_{dq}) may improve the fixing force of the cells, which ultimately provides a more stable connection between the implant and bone. Finally, S_{dr}, which expresses the increase of the interfacial surface area relative to the area of the projected (flat) x, y plane, may improve the area for the attachment of cells. In our case, there is a considerable increase of S_{dr} for the etched and Ca-P deposited Ti23Mo 3BG surface (255.66%) compared to the polished surface (1.22%).

As shown in Figure 9 surface wettability assay recorded a lower glycerol contact angle of Ti23Mo 3BG bulk nanocomposite (41.41° ± 0.94°) than that of the nanocrystalline Ti23Mo alloy (85.64° ± 0.72°). Etched nanocomposite exhibited the enhanced surface hydrophilicity (26.11° ± 1.62°). Additional Ca-P deposition after electrochemical etching showed, that contact angle significantly increases (72.41° ± 0.56°), which means that wettability decreases (Table 2). The change of two or more surface characteristics at the same time, such as surface roughness and chemistry, whether deliberate or not, complicates the evaluation of the roles of the parameters on the wetting behavior and the biological performance [27].

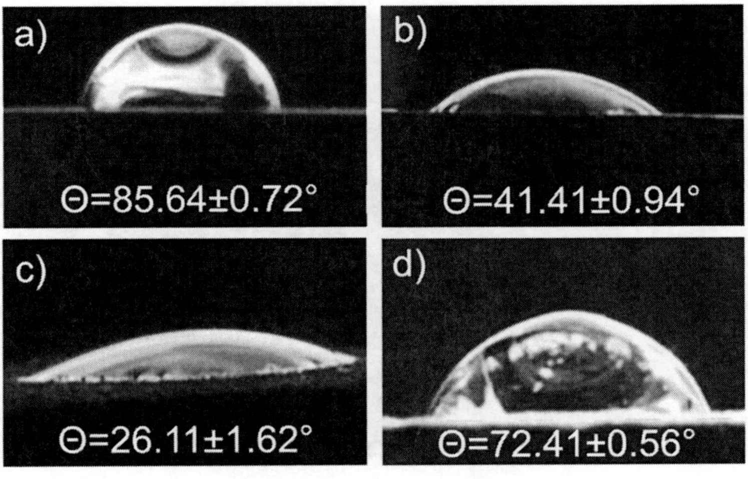

Figure 9. Contact angles of glycerol on the nanostructured bulk Ti23Mo alloy (**a**) and Ti23Mo 3BG nanocomposite before (**b**) and after electrochemical etching (1 M H_3PO_4 + 2% HF; 10 V/30 min) (**c**) as well as after additional Ca-P deposition (**d**).

3.3. In vitro Cytocompatibility

The culture was prepared on the Ti23Mo 3BG nanocomposite sample before electrochemical etching (Figure 10a–c), after electrochemical etching only (Figure 10d–f) and after etching with Ca-P deposition (Figure 10g–i). The collected data reveals a significant difference in the morphological characteristics of the cells on the porous and polished materials even after the first day of cell culturing (Figure 10). The surface irregularities (protrusions, hillocks and pores) improve osteoblast adhesion, thus cells attach on the micro- and nano-surface irregularities. It has been proven that cell attachment and proliferation depend on the surface topography and roughness [28]. The osteoblasts that grew on the Ti23Mo 3BG nanocomposite sample before electrochemical etching exhibited adhesion to the material surface after one day and covered most of the surface after five days (Figure 10a–c). In the case of the electrochemically etched sample, after 24 h of incubation, cells penetrated into the structure of the surface instead of forming filapodia (Figure 10d). After culturing the cells for five days, the cells on the etched Ti23Mo 3BG nanocomposite were observed to be well spread, both on the surface of the porous scaffold and inside the pores (Figure 10e,f).

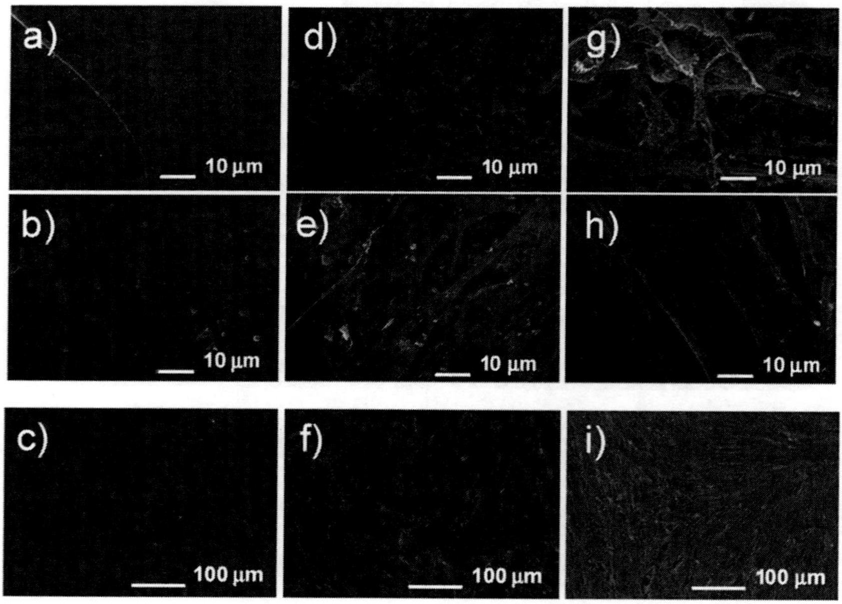

Figure 10. Osteoblast culture on the surface of the Ti23Mo 3BG nanocomposite before etching (**a**,**b**,**c**), after etching (**d**,**e**,**f**) and after etching with additional Ca-P deposition (**g**,**h**,**i**) after 1st (**a**,**d**,**g**) and 5th day ((**b**,**c**), (**e**,**f**) and (**h**,**i**)—different magnifications).

On the other hand, in the case of the etched and Ca-P deposited Ti23Mo 3BG sample, after the first day of incubation, cells showed good adhesion to the surface of the studied sample in the form of filapodia (Figure 10g). A monolayer was formed on the surface of the sample after five days of incubation (Figure 10h,i). The covering of the implant surface with Ca-P layer is the final stage of the surface treatment.

4. DISCUSSION

Nanotechnology is used for improving biomaterials or creating new biomaterials used for specific dental applications [29]. Nanostructured biomaterials can be personalized by engineering their structure, shape, size, and surface properties in order to be applied in precise anatomical sites. Based on the *in vitro* results, nanostructured metals undoubtedly have the potential to yield a faster and more stable integration of dental implants [29,30,31]. The *in vivo* effects of several nanoscale surface modification approaches and a potential applicability of these techniques to already commercialized dental implants are currently under investigation [28,29].

Current research focuses on improving the mechanical performance and biocompatibility of metal-based systems through changes in alloy composition, microstructure, and surface treatments [12,15,22,26,32,33]. In the case of titanium, a lot of attention is being paid to enhance the strength characteristics of commercial purity grades in order to avoid potential biotoxicity of alloying elements, especially in dental implants [34,35].

Improvement of the physicochemical and mechanical performance of Ti-based implant materials can be achieved through microstructure control, the top-down approach known as mechanical alloying technique [13,14,15,16,17,22,36,37]. Recent studies showed clearly that nanostructuring of titanium can considerably improve not only the mechanical properties, but also the biocompatibility [33]. Nanostructured materials can exhibit enhanced mechanical, biological, and chemical properties compared with their conventional counterparts [30,31,37].

While these biomaterials have been successful in encouraging bone ingrowth both *in vivo* and in clinical trials, the range of materials and microstructures available is still rather limited. To optimize Ti-based alloys for dental implant applications, several studies have focused on the design of biomaterials with optimized architecture to fulfill physico-chemical, mechanical as well as regeneration requirements. β-type titanium alloys are a class of interesting biomaterials that can exhibit a unique combination of physical, chemical and mechanical properties [38].

In this work, mechanical alloying was used to synthesize nanostructured Ti23Mo xBG nanocomposites. This technique enables alloying of elements that are difficult or impossible to combine by conventional melting methods (*i.e.*, Ti and Bioglass). The nanocomposite surface improvement was achieved by electrochemical treatment. This process was composed of two stages: anodic oxidation and Ca-P deposition. The applied etching electrolyte was composed of a mixture of 1M H_3PO_4 and 2% HF. Four stages of anodic oxidation were suggested [39]: (i) a compact oxide barrier layer formation; (ii) field-enhanced dissolution of the oxide layer; (iii) formation of porous structure in the slits and cracks (pore formation and dissolution of the oxide layer is possible as well) and (iv) the porous structure is consumed when the dissolution rate is larger than the pore formation rate.

It has been proven that the surface properties of the Ti-based implants determine successful osseointegration. The results published recently by Sul *et al.* showed that the morphology of different commercially available clinical titanium implants differed due to the surface modification techniques used during manufacture [40]. The bulk nanostructured Ti23Mo 3BG nanocomposite had R_a, R_t and R_z values of around 0.29, 11.29 and 8.24 μm, respectively. The etched Ti23Mo 3BG composite had average surface roughnesses, with R_a, R_t and R_z in the 6.42–42.47 μm range. After the Ca-P layer deposition a still relatively rough surface with Ca-P protrusions is well visible.

Implants, due to the corrosive environment of the tissue and body fluids, may undergo unexpected local corrosion attacks, leading to a release of the corrosion products to the tissue and its poisoning. The corrosion test results indicated that the nanostructured Ti23Mo alloy possesses a lower corrosion resistance and consequently a higher corrosion current density (I_{corr} = 70.93 μA/cm^2, E_{corr} = −0.7925) in the Ringer's solutions. The bulk Ti23Mo 3BG nanocomposite has a better corrosion resistance (I_{corr} = 0.0168 μA/cm^2, E_{corr} = −0.4423). For comparison, corrosion current density of microcrystalline Ti is 0.658 μA/cm^2 in Ringer's solution at 37 °C.

The surface wettability was confirmed to be a key parameter influencing the proliferation of osteoblasts [27,41]. Enhanced surface wettability would improve the cellular adhesion by optimizing the protein adsorption and providing a suitable surface presented to the filopodia. As a result, the electrochemically etched Ti23Mo 3BG nanocomposite, due to its good corrosion resistance and the surface wettability (contact angle Θ = 26.1°), shows a high extent of cell proliferation.

In order to test the toxicity of materials, as well as interaction between the tested materials and the cells, *in vitro*cytocompatibility tests are conducted. The first results of increased osteoblast adhesion on the nanostructured metals was published by Webster and co-workers [12,26,30,31]. In this work, the *in vitro* cytocompatibility test was performed on the samples with different surfaces. The residual porosity after powder compaction played an important role in the cell adhesion. A significant difference in the morphological characteristics of the cells on the etched and polished materials was visible after one day of cell culturing. The cells tend to adhere with their entire surface to the porous sample penetrating the pores, whereas more spherical cells were seen with a smaller contact surface but more filapodia on the polished surface. After the first day, the cells strongly attached to the surface of the etched nanocomposite Ti23Mo 3BG sample, not producing filapodia or only producing them in a smaller amount compared to the cell culture on the smooth surface. After five days of cell culture, a smaller portion of the observed area was covered by cells on the etched Ti23Mo 3BG composite in comparison with the analogous bulk sample.

Bioactive glasses are biocompatible and exhibit a strong interfacial bonding with bone. Their bioactivity is attributed to the formation on their surface of a hydroxycarbonated apatite (HCA) layer. The rate of tissue bonding appears to depend on the rate of HCA formation, which follows a sequence of reactions between the implanted material and the surrounding tissues and physiologic fluids [42,43,44,45]. Cao and Hench proposed a three-step mechanism of HCA formation when the bioactive glass comes into contact with physiologic fluids: ion exchange, dissolution, and precipitation [42]. In the last case, precipitation of the calcium and phosphate ions released from the glass together with those from the solution, form a calcium-phosphate-rich layer (CaP) on the surface [44].

Modification at the nanoscale of the surface properties of dental implants can be achieved by various techniques, in order to create a more efficient implant integration with the bone [46]. Alterations in the cell shape and cytoskeleton, thus influencing specific gene expression, have been noticed after modifying the surface texture or roughness of the implants [46,47]. Moreover it has been demonstrated that anodized titanium surfaces enhances osteogenic activity *in vitro* [48,49]. On the other hand, oxidative nanopatterning confers titanium-based metals the ability to selectively guide cell behavior, facilitating the growth of cells having the osteoblastic potential [50]. Therefore these techniques are crucial for tissue regeneration, leading in the future to higher predictability in tissue healing around dental implants.

5. CONCLUSIONS

In this work, new kinds of biomedical Ti23Mo xBG nanocomposite were developed by the introduction of 45S5 Bioglass powders into the Ti23Mo matrix. Additionally the formation of a porous nanocrystalline Ti23Mo 3BG nanocomposite with an electrochemically modified surface was demonstrated. On the etched surface a bioactive HA layer was deposited. The following conclusions can be withdrawn: (1) etching improves porosity of the surface; (2) Ti23Mo 3BG nanocomposite after etching and Ca-P deposition had an average surface roughness in the 35.80–276.55 μm range; (3) the corrosion resistance significantly changes after the electrochemical treatment and Ca-P deposition; (4) noticeable difference in the morphology of the NHOst cells on porous and polished materials is observed in SEM, even after one day of cell culturing; (5) the rough, electrochemically biofunctionalized surface (porous with Ca-P layer) supports osteoblast cell growth and proliferation.

The key factors for the success of implant integration with the surrounding hard tissues are surface topography and chemical composition. Therefore the biofunctionalized nanocrystalline Ti23Mo 3BG composite may be an important step forward in the development of such a structure, which will support the process of osseointegration of the implant.

AUTHOR CONTRIBUTIONS

Karolina Jurczyk, Andrzej Miklaszewski, Mieczyslawa U. Jurczyk and Mieczyslaw Jurczyk conducted the experimental and analytical works. Karolina Jurczyk designed, supervised the project, and wrote the manuscript. All the authors contributed to the critical reading, and editing of the final version of the manuscript.

CONFLICTS OF INTEREST

The authors declare no conflict of interest.

REFERENCES

1. Rack, H.J.; Qazi, J.J. Titanium alloys for biomedical applications. Mater. Sci. Eng. C 2006, 26, 1269–1277.

2. Geetha, M.; Singh, A.K.; Asokamani, R.; Gogia, A.K. Ti based biomaterials, the ultimate choice for orthopedic implants—A review. Prog. Mater. Sci. 2009, 54, 397–425.

3. Sidambe, A.T. Biocompatibility of advanced manufactured titanium implants—A review. Materials 2014, 7, 8168–8188.

4. Kim, S.E.; Lim, J.H.; Lee, S.C.; Nam, S.C.; Kang, H.G.; Choi, J. Anodically nanostructured titanium oxides for implant applications. Electrochim. Acta 2008, 53, 4846–4851.

5. Okuno, O. Titanium alloys for dental applications. J. Jpn. Soc. Biomater. 1996, 14, 267–273.

6. Kuroda, D.; Niinomi, M.; Morinaga, M.; Kato, Y.; Yashiro, T. Design and mechanical properties of new beta type titanium alloys for implant materials. Mater. Sci. Eng. A 1998, 243, 244–249.

7. Yang, I.H.; Kim, S.Y.; Rubash, H.E.; Shanbhag, A.S. Fabrication of submicron titanium-alloy particles for biological response studies. J. Biomed. Mater. Res. Part B Appl. Biomater. 1999, 48, 220–223.

8. Kobayashi, E.; Doi, H.; Takahashi, M.; Nakano, T.; Yoneyama, T.; Hamanaka, H. Castability and mechanical properties of Ti-6Al-7Nb alloy dental-cast. J. Jpn. Soc. Dent. Mater. Dev. 1995, 14, 406–413.

9. Kobayashi, E.; Doi, H.; Yoneyama, T.; Hamanaka, H.; Matsumoto, S.; Kudaka, K. Evaluation of mechanical properties of dental casting Ti-Zr based alloys. J. Jpn. Soc. Dent. Mater. Dev. 1995, 14, 321–328.

10. Doi, H.; Yoneyama, T.; Kobayashi, E.; Hamanaka, H. Mechanical properties and corrosion resistance of Ti-5Al-13Ta alloy castings. J. Jpn. Soc. Dent. Mater. Dev. 1998, 17, 247–252.

11. Niinomi, M. Recent research and development in titanium alloys for biomedical applications and healthcare foods.Sci. Technol. Adv. Mater. 2003, 4, 445–454.

12. Ward, B.C.; Webster, T.J. Increased functions of osteoblasts on nanophase metals. Mater. Sci. Eng. C 2007, 27, 575–578.

13. Jurczyk, K.; Jurczyk, M. Applications of nanomaterials in dentistry. In Handbook of Clinical Nanomedicine: Nanoparticles, Imaging, Therapy, and Clinical Applications; Bawa, R., Audette, G.F., Rubinstein, I., Eds.; Pan Stanford Publishing Pte. Ltd.: Singapore, 2015; Chapter 37.

14. Jurczyk, K.; Adamek, G.; Kubicka, M.M.; Jakubowicz, J.; Jurczyk, M. Nanostructured titanium-10 wt % 45S5 Bioglass-Ag composite foams for medical applications. Materials 2015, 8, 1398–1412.

15. Abdeltawab, A.A.; Shoeib, M.A.; Mohamed, S.G. Electrophoretic deposition of hydroxyapatite coatings on titanium from dimethylformamide suspensions. Surf. Coat. Technol. 2013, 206, 43–50.

16. Jurczyk, K.; Niespodziana, K.; Jurczyk, M.U.; Jakubowicz, J.; Jurczyk, M. Titanium-10 wt % 45S5 Bioglass nanocomposite for biomedical applications. Mater. Chem. Phys. 2011, 131, 540–546.

17. Miklaszewski, A.; Jurczyk, M.U.; Jurczyk, K.; Jurczyk, M. Plasma surface modification of titanium by TiB precipitation for biomedical applications. Surf. Coat. Technol. 2011, 206, 330–337.

18. Ning, C.Q.; Zhou, Y. On the microstructure of biocomposites sintered from Ti, HA and bioactive glass. Biomaterials2004, 5, 3379–3387.

19. Romeis, S.; Hoppe, A.; Detsch, R.; Boccaccini, A.R.; Schmidt, J.; Peukert, W. Top-down processing of submicron 45S5 Bioglass® for enhanced in vitro bioactivity and biocompatibility. Procedia Eng. 2015, 102, 534–541.

20. Yang, C.Y.; Wang, B.C.; Chang, E.; Wu, B.C. Bond degradation at the plasma-sprayed HA coating/Ti-6AI-4V alloy interface: An in vitro study. J. Mater. Sci. Mater. Med. 1995, 6, 258–265.

21. Zhu, S.; Wang, X.; Yoshimura, M.; Inoue, A. Synthesis of Ti-based glassy alloy/hydroxyapatite composite by spark plasma sintering. Mater. Trans. 2008, 49, 502–505.

22. Niespodziana, K.; Jurczyk, K.; Jakubowicz, J.; Jurczyk, M. Fabrication and properties of titanium-hydroxyapatite nanocomposites. Mater. Chem. Phys. 2010, 123, 160–165.

23. Kaczmarek, M.; Jurczyk, M.U.; Rubis, B.; Banaszak, A.; Kolecka, A.; Paszel, A.; Jurczyk, K.; Murias, M.; Sikora, J.; Jurczyk, M. In vitro biocompatibility of Ti-45S5 Bioglass nanocomposites and their scaffolds. J. Biomed. Mater. Res.2014, 102A, 1316–1324.

24. Benjamin, J.S. Mechanical alloying. Sci. Am. 1976, 234, 40–46.

25. Suryanarayna, C. Mechanical alloying and milling. Prog. Mater. Sci. 2001, 46, 1–184.

26. Webster, T.B.; Ejiofor, J.U. Increased osteoblast adhesion on nanophase metals: Ti, Ti6Al4V and CoCrMo. Biomaterials2004, 25, 4731–4739.

27. Gittens, R.A.; Scheideler, L.; Rupp, F.; Hyzy, S.L.; Geis-Gerstorfer, J.; Schwartz, Z.; Boyan, B.D. A review on the wettability of dental implant surfaces II: Biological and clinical aspects. Acta Biomater. 2014, 10, 2907–2918.

28. Guehennec, L.; Souedan, A.; Layrolle, P.; Amouriq, Y. Surface treatments of titanium dental implants for rapid osseointegration. Dent. Mater. 2007, 23, 844–854.

29. Roco, M.C. International strategy for nanotechnology research and development. J. Nanopart. Res. 2001, 3, 353–360.

30. Khang, D.; Lu, J.; Yao, C.; Haberstroh, K.M.; Webster, T.J. The role of nanometer and sub-micron surface features on vascular and bone cell adhesion on titanium. Biomaterials 2008, 29, 970–983.

31. Smith, L.J.; Swaim, J.S.; Yao, C.; Haberstroh, K.M.; Nauman, E.A.; Webster, T.J. Increased osteoblast cell density on nanostructured

PLGA-coated nanostructured titanium for orthopedic applications. Int. J. Nanomed. 2007, 2, 493–499.

32. Jakubowicz, J.; Adamek, G.; Jurczyk, M.U.; Jurczyk, M. 3D topography study of the biofunctionalized nanocrystalline Ti-6Zr-4Nb/Ca-P. Mater. Charact. 2012, 70, 55–62.

33. Estrin, Y.; Kasper, C.; Diederichs, S.; Lapovok, R. Accelerated growth of preosteoblastic cells on ultrafine grained titanium. J. Biomed. Mater. Res. A 2008, 90, 1239–1242.

34. Pye, A.D.; Lockhart, D.E.A.; Dawson, M.P.; Murray, A.A.; Smith, A.J. A review of dental implants and infection. J. Hosp. Infect. 2009, 72, 104–110.

35. Matusiewicz, H. Potential release of in vivo trace metals from metallic medical implants in the human body: From ions to nanoparticles—A systematic analytical review. Acta Biomater. 2014, 10, 2379–2403.

36. Tulinski, M.; Jurczyk, M. Nanostructured nickel-free austenitic stainless steel composites with different content of hydroxyapatite. Appl. Surf. Sci. 2012, 260, 80–83.

37. Jurczyk, M., Ed.; Bionanomaterials for Dental Applications; Pan Stanford Publishing: Singapore, 2012.

38. Niinomi, M.; Hattori, T.; Morikawa, K.; Kasuga, T.; Suzuki, A.; Fukui, H.; Niwa, S. Development of Low Rigidity β-type Titanium Alloy for Biomedical Applications. Mater. Trans. 2002, 43, 2970–2977.

39. Yu, X.; Li, Y.; Wlodarski, W.; Kandasamy, S.; Kalantar-Zadeh, K. Fabrication of nanostructured TiO2 by nnodization: A comparison between electrolytes and substrates. Sens. Actuators B 2008, 130, 25–31.

40. Sul, Y.T.; Johansson, C.B.; Petronis, S.; Krozer, A.; Jeong, Y.; Wennerberg, A.; Albrektsson, T. Characteristics of the surface oxides on turned and electrochemically oxidized pure titanium implants up to dielectric breakdown: The oxide thickness, micropore configurations, surface roughness, crystal structure and chemical composition. Biomaterials2002, 23, 491–501.

41. Liu, X.; Chu, P.K.; Ding, C. Surface modification of titanium, titanium alloys, and related materials for biomedical applications. Mater. Sci. Eng. R 2004, 47, 49–121.

42. Cao, W.P.; Hench, L. Bioactive materials. Ceram. Int. 1996, 22, 493–507.

43. Hench, L.L. Bioceramics: From concept to clinic. J. Am. Ceram. Soc. 1991, 74, 1487–1510.

44. Gerhardt, L.C.; Boccaccini, A.R. Bioactive glass and glass-ceramic scaffolds for bone tissue engineering—Review.Materials 2010, 3, 3867–3910.

45. Lefebvre, L.; Chevalier, J.; Gremillard, L.; Zenati, R.; Thollet, G.; Bernache-Assolant, D.; Govin, A. Structural transformations of bioactive glass 45S5 with thermal treatments. Acta Mater. 2007, 55, 3305–3313.

46. Masaki, C.; Schneider, G.B.; Zaharias, R.; Seabold, D.; Stanford, C. Effects of implant surface microtopography on osteoblast gene expression. Clin. Oral Implant. Res. 2005, 16, 650–656.

47. Ogawa, T.; Nishimura, I. Genes differentially expressed in titanium implant healing. J. Dent. Res. 2006, 85, 566–570.

48. Ercan, B.; Webster, T.J. Greater osteoblast proliferation on anodized nanotubular titanium upon electrical stimulation.Int. J. Nanomed. 2008, 3, 477–485.

49. Yao, C.; Slamovich, E.B.; Webster, T.J. Enhanced osteoblast functions on anodized titanium with nanotube-like structures. J. Biomed. Mater. Res. A 2008, 85, 157–166.

50. Variola, F.; Vetrone, F.; Richert, L.; Jedrzejowski, P.; Yi, J.H.; Zalzal, S.; Clair, S.; Sarkissian, A.; Perepichka, D.F.; Wuest, J.D.; et al. Improving biocompatibility of implantable metals by nanoscale modification of surfaces: An overview of strategies, fabrication methods, and challenges. Small 2009, 5, 996–1006.

INDEX

A

Air-Jet Erosion Tester, 46
applications, 1, 25, 27, 28, 32, 315, 316, 317, 333, 334, 337, 338, 340

B

bioglass, 150, 316

C

Cancer therapy, 32
carbon nanotubes, 39, 82, 85, 224, 232, 235, 244, 245, 246, 247, 248, 249, 250, 251, 252, 261, 264, 276, 277
Ceramic Nanoparticle, 268
Composite materials, 1, 80

E

electro-responsive, 137, 222, 223, 232, 233, 234, 241, 243, 252, 253
etching, 316, 319, 322, 325, 326, 327, 328, 329, 330, 331, 332, 334, 336

G

graphene, 75, 82, 85, 87, 133, 227, 228, 236, 237, 239, 253, 254, 255, 256, 257, 261, 262, 264, 265, 276, 277, 278

H

hybrid hydrogels, 70, 75, 81, 91, 97, 108, 121, 125, 126, 127, 234, 236, 263
hydrogels, 69, 70, 71, 72, 73, 74, 271, 272, 275, 277, 278, 279, 280, 281

M

medicine biology, 27

N

nanocomposite hydrogels, 69, 70, 75, 81, 82, 86, 91, 92, 95, 97, 99, 108, 113, 118, 125, 127, 128, 259, 260
nanomaterials, 27, 28, 29, 34, 261, 262, 263, 264, 265, 268, 272, 275, 276, 312, 338

nanoparticles, 27, 28, 29, 31, 32, 33, 34, 35, 36, 37, 38, 39, 40, 259, 260, 261, 265, 266, 267, 268, 269, 270, 271, 272, 275, 278, 280, 281, 282, 340
Nanotechnology, 27, 38, 40, 132, 144, 171, 249, 333

P

Polyester Composites, 43
powder metallurgy, 284, 285, 289, 290, 312, 315, 316, 317
Protein detection, 33

Q

quantum dots nanotubes, 27

S

surface properties, 72, 77, 81, 128, 134, 147, 227, 316, 333, 334, 335
Surfing films, 198

T

Tissue engineering, 29, 30, 37
titanium, 31, 40, 82, 147, 316, 317, 318, 319, 322, 325, 333, 334, 336, 337, 338, 339, 340, 341
Triboplasma Structural Elements, 189
Tribosystems, 177
tribounits, 181